Safety-Scale Laboratory Experiments for General, Organic, and Biochemistry

Third Edition

Spencer L. Seager
Weber State University

Michael R. Slabaugh
Weber State University

Brooks/Cole Publishing Company

I(T)P® An International Thomson Publishing Company

Pacific Grove • Albany • Belmont • Bonn • Boston • Cincinnati • Detroit
Johannesburg • London • Madrid • Melbourne • Mexico City
New York • Paris • Singapore • Tokyo • Toronto • Washington

COPYRIGHT © 1997 by Brooks/Cole Publishing Company
A division of International Thomson Publishing Inc.
I⍜P® The ITP logo is a registered trademark under license.

Printed in the United States of America

10 9 8 7 6 5 4 3 2

For more information contact:

BROOKS/COLE PUBLISHING
 COMPANY
511 Forest Lodge Road
Pacific Grove, CA 93950
USA

International Thomson Editores
Seneca 53
Col. Polanco
México, D. F., México
C.P. 11560

International Thomson Publishing Europe
Berkshire House 168-173
High Holborn
London WC1V 7AA
England

International Thomson Publishing GmbH
Königswinterer Strasse 418
53227 Bonn
Germany

Thomas Nelson Australia
102 Dodds Street
South Melbourne, 3205
Victoria, Australia

International Thomson Publishing Asia
221 Henderson Road
#05-10 Henderson Building
Singapore 0315

Nelson Canada
1120 Birchmount Road
Scarborough, Ontario
Canada M1K 5G4

International Thomson Publishing Japan
Hirakawacho Kyowa Building, 3F
2-2-1 Hirakawacho
Chiyoda-ku, Toyoko 102
Japan

ISBN: 0–314–20623–X

Contents

Introduction

To the Student: A significant amount of your training in chemistry will take place in the laboratory. The following instructions should be read carefully before you attend the first laboratory session. These instructions will help you make efficient use of your time while in the laboratory and also promote laboratory safety.

PRE-LAB PREPARATION

Carefully read the experiment to be performed before you come to the laboratory. Your instructor might require you to complete the Pre-Lab Review Sheet found before the Data Sheet of each experiment and turn it in before you begin your lab work. Even if this is not a requirement, you are still advised to complete the Review Sheet. The questions have been chosen to draw your attention to specific techniques and precautions that you should be aware of before you start the experiment. The experiments are designed to allow you to collect the data in 3 hours or less. Students unable to do this have usually failed to prepare properly for the laboratory. Make sure you arrive on time, since your instructor will provide additional instructions for experiments as needed.

SAFETY PRECAUTIONS

In the laboratory, you will work with a variety of substances and equipment. The following precautions should be followed to minimize the chances for accidents.

1. Carefully note any special safety precautions stated in the experiment. They are identified as **SAFETY ALERTS** and are enclosed in boxes for easy recognition.
2. Approved safety glasses or goggles must be worn *at all times* in the lab.
3. Some chemicals used in experiments are toxic. Therefore, no eating, smoking, or drinking are allowed in the laboratory. After leaving the lab, you should wash your hands before eating, smoking, or drinking.
4. Never taste or smell anything in the lab unless you are specifically instructed to do so.
5. Use the hood when instructed to do so. Do *not* release noxious gases into the open lab.
6. No unauthorized experiments are to be performed. Follow directions carefully, using only the amounts of chemicals specified.
7. Always pour concentrated acid into water; *never* pour water into the acid.
8. No visitors are allowed in the lab unless specific permission is given by the instructor.

9. Clean up all chemical or solution spills immediately. Your work area should always be left clean at the end of the lab period.
10. Protect both hands when you insert glass tubing or thermometers into stoppers. Make sure the glass is well lubricated before such insertions are attempted. Your instructor will show you the correct procedure.

EMERGENCY FACILITIES AND PROCEDURES

Despite safety precautions, accidents sometimes occur. Even though the chance is small, it is important to know what to do in the event you are involved. Therefore, you should acquaint yourself with the location and use of the following emergency facilities. Your instructor will describe or demonstrate their operation.

1. Safety shower   3. Safety eye-wash facility
2. Fire blanket   4. Fire extinguisher

The following procedures should be followed if you are involved in an accident. Also be prepared to assist other students in the laboratory. In some instances, people involved in accidents become disoriented and frightened and forget what to do. Be ready to help.

1. Report all injuries—no matter how slight—to your instructor.
2. Splashes of corrosive or toxic substances should be washed immediately from skin and clothing, using copious amounts of cold water. Speed is especially important if the material has splashed into the eyes.
3. The best immediate treatment for a burn is to hold the injured area under cold water or to cover it with ice.
4. If your clothing catches fire, use the safety shower or blanket.
5. If a fire occurs on a bench area, use the fire extinguisher.
6. If you cut or burn yourself, call your instructor immediately for first aid.

GENERAL LABORATORY PROCEDURES

The following procedures will help you use your time efficiently and will help minimize the waste of chemicals and other supplies. Other techniques will be described to you as needed in later experiments.

1. Cleaning glassware: Scrub inside and out with a brush, detergent, and tap water. Rinse away all suds with tap water. Rinse the inside of the glassware two or three times with *minimal* amounts of distilled water. (Distilled water is expensive and should be used sparingly.) Shake out as much rinse water as possible and dry the outside with a towel. If dry glassware is needed immediately, rinse the equipment twice with small amounts of acetone, then return the used acetone to the original container. The residual acetone in the equipment will vaporize quickly and leave no residue.
2. Disposal of used materials: Used chemicals and other materials must be disposed of appropriately. We will use three disposal methods in

the experiments of this manual. Some used chemicals can be flushed down the sink drain with water. When this method is to be used, you will be notified as follows:

DISPOSAL  [Identity] in Sink, where [Identity] is the material you are disposing.

Most used chemicals will be collected in labeled containers located in the lab. Your instructor or other qualified individuals will then properly dispose of the collected materials. This method will be indicated to you by a notification in bold type similar to the following:

DISPOSAL  Iron Chloride in container labeled "Exp. 4, Used Chemicals."

You would then dispose of your used iron chloride by putting it into the container with the label matching the one given in the notification. A few materials used in the lab can be put into an ordinary wastebasket or other similar solid-waste receptacle. This will be indicated by the following in bold type:

DISPOSAL  [Identity] in wastebasket.

If you are not sure about the proper way to dispose of a used chemical or other material, ask your lab instructor for directions.

3. Smelling vapors and gases: *Never* smell a chemical by holding the container to your nose. Hold it about 6 to 12 inches away and carry the gas or vapor to your nose in your cupped hand.
4. Chemical use: To avoid waste, obtain only the amount of chemical called for in the experiment. Do not remove reagent containers (solids or liquids) from the designated dispensing area.
5. Dispensing solids: Solid chemicals can be poured easily from containers by gently rotating the container back and forth during the pouring process. Do not return excess reagent to the original container.
6. Dispensing liquids: When moderate to large amounts of liquid reagents are needed, pour the amount needed from the reagent bottle into an appropriate container (beaker, test tube, and so on). Do not lay the stopper on the bench but hold it between your fingers as demonstrated by your lab instructor. When small amounts (drops) are needed, do not put your dropper into the container. Either use the dropper supplied with the container or pour a small amount into a small test tube or other container and fill your dropper from that supply. Never return a used liquid reagent to the original container.

DATA AND CALCULATIONS

Each experiment in this manual consists of three parts that provide information you will need to successfully complete the experiment. The *Introduction* contains background material and develops any theory, equation, etc used in the experiment. The *Experimental Procedure* provides the actual steps you will follow to collect data (weights, volumes, observations and the like). The *Calculation and Report* part of the experiment contains directions for doing calculations or treating your data in other ways to prepare the final report of the experiment.

Each experiment includes a Data and Report Sheet. Data collected by following the *Experimental Procedure* should be recorded during the laboratory session directly on the sheet in the tables labeled "data." Calculations and other treatment of the data should be done outside the laboratory according to the directions given in the *Calculations and Report* part of the experiment, and recorded in the tables labeled "report." Unless you are directed to do otherwise, submit the completed Data and Report Sheets, including the completed questions, to your instructor at the next laboratory session. Note that the data tables and the report tables are located near each other for convenience when you do your calculations. However, complete only the data tables while you are in the laboratory. You may find it useful to remove the Data and Report Sheet from its location near the end of the experiment and keep it available for easy access when you have data to record. This will avoid the need to flip to the end of the experiment each time you record data.

Scale of Experiments

The experiments included in this manual represent a compromise between microscale approaches that minimize the amounts of chemicals used and macroscale approaches that require the use of relatively large amounts of chemicals. Microscale approaches have economic and safety advantages because of the small quantities of both chemicals required and wastes generated. However, such approaches often require specialized glassware and other equipment that must be purchased, a requirement that eliminates some of the economic advantage. Also, a completely microscale approach often fails to give students experience in the use of certain basic types of equipment such as burets and pipets.

The safety-scale experiments in this manual have been generally scaled down from macroscale in terms of the amounts of materials used to provide some of the economic and safety advantages of microscale. However, the amounts are large enough to allow the use of regular, small-sized laboratory glassware. Thus, most reactions are done using small (10-cm) test tubes, but pipets, burets, and other basic pieces of glassware are also used where it is appropriate. The small quantities of liquids required for most experiments are measured in drops and dispensed from dropper bottles.

We have found this compromise in scale to be an effective approach to teaching the chemistry laboratory. The students adapt to it very well, and the stockroom personnel who prepare materials for the experiments appreciate the small quantities involved.

Common Laboratory Equipment

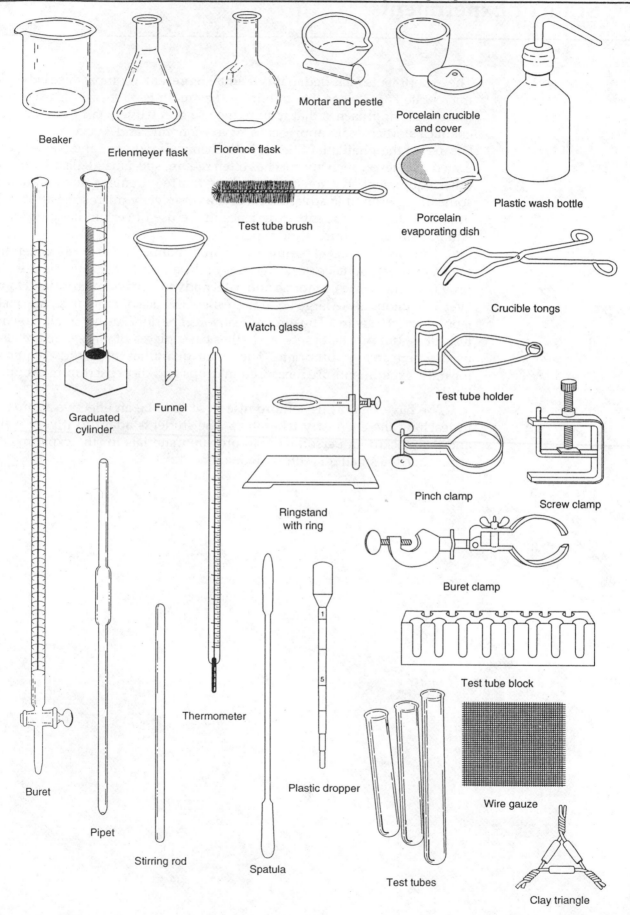

Beaker

Erlenmeyer flask

Florence flask

Mortar and pestle

Porcelain crucible and cover

Plastic wash bottle

Test tube brush

Porcelain evaporating dish

Buret

Graduated cylinder

Funnel

Watch glass

Ringstand with ring

Crucible tongs

Test tube holder

Pinch clamp

Screw clamp

Buret clamp

Test tube block

Thermometer

Plastic dropper

Wire gauze

Pipet

Stirring rod

Spatula

Test tubes

Clay triangle

Measurements and Significant Figures

In this experiment, you will

- Make measurements using devices having different uncertainties.
- Express measured quantities in a way that correctly shows the uncertainties of the measurements.
- Use significant figures to properly represent measured and calculated quantities.

- Investigate how to increase the number of significant figures in measured and calculated quantities by properly using measuring devices.

INTRODUCTION

Measurement is an important activity in most scientific studies. Every measurement contains an uncertainty that comes from the device or technique used to make the measurement. The numbers used to record a scientific measurement normally indicate the uncertainty in the measurement. For example, a mass recorded as 2.87 g indicates that the measurement has an uncertainty in the hundredths (.01) of a gram. This fact could be represented by recording the mass as 2.87 ± .01 g, but usually this is not done. The value is simply recognized as having an uncertainty of +1 or −1 in the last recorded number.

The numbers used to represent the certain part of a measurement (the 2 and 8 in the example), plus one number representing the uncertain part (the 7 in the example), are called **significant figures** or digits. Thus, the quantity 2.87 g contains three significant figures.

The necessity of using zeros to express measurements raises the question of when are zeros considered to be significant figures. The measured mass expressed as 2.87 g could also be expressed as .00287 kg. The significance of a measurement cannot be changed simply by changing the units used to express the measurement. Thus, .00287 kg must contain three significant figures just as 2.87 g does. This is an example of one rule concerning zeros. Zeros not preceded on the left by nonzeros do not count as significant figures. Other zeros, those located between nonzeros and those to the right of nonzeros, are counted as significant figures. Thus, 2.807 g and 2.870 g both contain four significant figures, and both indicate that the measurement uncertainty is +.001 or −.001 g.

In this experiment, you will make some measurements using different devices. You will express these measurements and results calculated from them using the correct number of significant figures.

A. Measurement with Ruler A

The area of a rectangle is equal to the product of the width (w) and length (l) (area = $w \times l$). The perimeter of a rectangle is equal to the sum of the four sides (perimeter = $w + w + l + l$). In this procedure, you will measure the length and width of four different rectangles. These quantities will be measured with a ruler that has divisions to the nearest centimeter. When measuring devices like rulers are used, the measurement uncertainty is expressed by estimating the value of the measured quantity to one decimal place more than the smallest scale division of the measuring device.

To calculate rectangular areas, you will have to multiply together two measured quantities. The area that results should be expressed using the correct number of significant figures. In the case of multiplication or division, the results of the calculation must have the same number of significant figures as the least significant measured number used in the calculation. For example, the product 1.1186×0.064 is equal to 0.07159. However, only two significant figures are justified in the answer to match the two significant figures in 0.064. Thus, the answer is 0.072, where the last significant figure (1) was rounded up to 2 because the first number being dropped (5) was equal to 5. In general, the last significant figure retained during rounding will be increased by 1 in the rounded answer when the first number being dropped is equal to or greater than 5. When the first number being dropped is less than 5, the last significant figure in the rounded answer is not changed.

Rectangle perimeters are obtained by adding a series of numbers. When numbers are added or subtracted, significant figure rules require that the answer be rounded so that it contains the same number of places to the right of the decimal as the smallest number of places in the quantities added or subtracted. For example, the sum $3.527 + 0.041 + 7.12$ is equal to 10.688. However, in this answer two places to the right of the decimal are justified to match the two places in 7.12, where the same rounding rules given earlier were used. Note that this gives an answer with four significant figures, even though the numbers added had four, two, and three significant figures, respectively.

PROCEDURE

1. Use a pair of scissors and carefully cut out ruler A from page 5.
2. Use ruler A to measure the length and width of rectangles W, X, Y, and Z that are drawn on page 5. Note the smallest division on ruler A and make appropriate estimates of the measured values. Record your measured values in Table 1.1 of the Data and Report Sheet, with the longest side designated as the length.

B. Measurement with Ruler B

PROCEDURE

1. Use a pair of scissors and carefully cut out ruler B from page 5.
2. Use ruler B to measure the length and width of rectangles W, X, Y, and Z that are drawn on page 5. Note the smallest division on

ruler B and make appropriate estimates of the measured values. Record your measured values in Table 1.5 of the Data and Report Sheet.

C. Improving the Significance of Measurements

The number of significant figures in a measured quantity and in quantities calculated from measured quantities depends on the way the measuring device is used.

PROCEDURE

1. Obtain 10 pennies from the stockroom.
2. Use ruler B to measure the diameter of a single penny; be sure to make an appropriate estimate and include it in your value. Record the measured value in Table 1.9 of the Data and Report Sheet.
3. Use ruler B to measure the thickness of a single penny and the thickness (height) of stacks of pennies containing 3, 5, 7, and 10 pennies. Include appropriate estimates in your measurements and record the values in Table 1.9 of the Data and Report Sheet.

CALCULATIONS AND REPORT

A. Measurement with Ruler A

1. Transfer the measured length and width values from Table 1.1 to Table 1.2 of the Data and Report Sheet.
2. Complete Table 1.2 by writing the number of significant figures found in each measured quantity.
3. Refer to the rules given earlier for multiplication calculations, and determine the number of significant figures that are justified for the calculated area of each rectangle. Record that number in Table 1.3.
4. Calculate the area of each rectangle and record the unrounded value and the value rounded to the correct number of significant figures in Table 1.3.
5. Refer to the rules given earlier for addition calculations, and determine the number of places to the right of the decimal that are justified for the calculated perimeter of each rectangle and record the unrounded value and the properly rounded value in Table 1.4.

B. Measurement with Ruler B

1. Transfer the measured length and width values from Table 1.5 to Table 1.6 of the Data and Report Sheet.
2. Complete Table 1.6 by writing the number of significant figures found in each measured quantity.
3. Refer to the rules given earlier for multiplication calculations, and determine the number of significant figures that are justified for the calculated area of each rectangle. Record that number in Table 1.7.
4. Calculate the area of each rectangle and record the unrounded and properly rounded values in Table 1.7.

5. Refer to the rules given earlier for addition calculations, and determine the number of places to the right of the decimal that are justified for the calculated perimeter of each rectangle. Record that number in Table 1.8.

6. Calculate the perimeter of each rectangle and record the unrounded and properly rounded values in Table 1.8.

C. Improving the Significance of Measurements

1. Transfer the measured thickness values for each stack from Table 1.9 to Table 1.10 of the Data and Report Sheet.

2. Determine the number of significant figures in each thickness value and record that number in Table 1.10.

3. Refer to the rules given earlier for division calculations, and determine the number of significant figures that are justified in a calculated value of the average thickness of a penny. This average is obtained by dividing the measured thickness of a stack by the number of pennies in the stack. The number of pennies in a stack is a counting number that is known exactly and does not influence the number of significant figures justified in the calculated value. Record the justified number of significant figures in Table 1.10.

4. Calculate the average thickness of a penny from the data for each stack and record the unrounded and properly rounded values in Table 1.10.

5. The volume of a penny is given by

$$V = \frac{1}{4}\pi d^2 t$$

where $\pi = 3.1416$, d = the measured diameter, t = the calculated thickness, and $\frac{1}{4}$ is an exact fraction that does not influence the number of significant figures in the calculated volume. Use the most significant calculated value for the average penny thickness t, recorded in Table 1.10, the measured diameter d, from Table 1.9, and determine the number of significant figures these will justify in a calculated volume V. Record this number in Table 1.11.

6. Calculate the volume of a penny and record the unrounded and properly rounded values in Table 1.11.

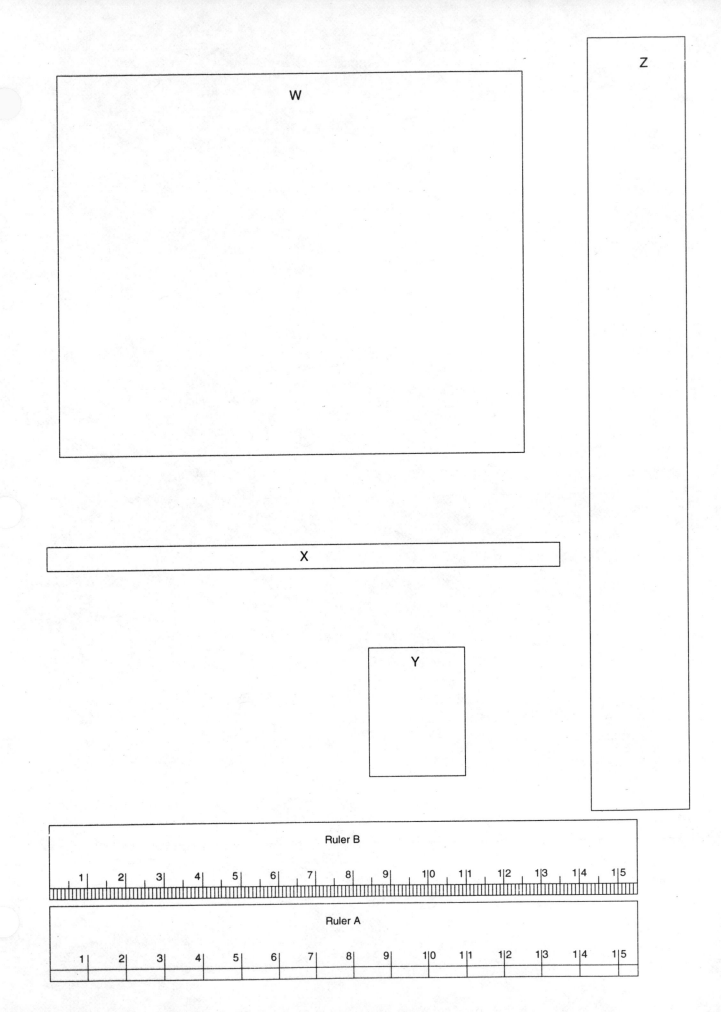

Experiment 1 ▪ Pre-Lab Review

Measurements and Significant Figures

1. Are any specific safety alerts given in the experiment? List any that are given.

    _____

2. Are any specific disposal directions given in the experiment? List any that are given.

    _____

3. A quantity has a measured value of 8.4126. Which number in the measured value has an uncertainty?

    _____

4. How many significant figures are contained in each of the following measurements?

    2.46 g _____   10.00 mL _____   0.0109 cm _____

5. What uncertainty (± an amount) is represented by the following measurements?

    1.0569 g _____   7.56 mL _____   1.815 cm _____

6. You are measuring a quantity with a measuring device on which the smallest scale division is 0.1 unit. A measurement appears to have a value of exactly 3.2 units. How should you record the measurement in order to properly indicate where the uncertainty is located?

    _____

7. Round the following numbers so they contain the number of significant figures indicated in parentheses.

    1.513 (3) _____   0.0155 (2) _____

    0.9866 (2) _____   12.689 (2) _____

    1.494 (1) _____   0.04020 (3) _____

8. Carry out the following calculations and write the result using the justified number of significant figures. Assume all numbers represent measured quantities.

    $0.521 \times 2.1 =$ _____

    $0.713 + 6.12 + 11.2 =$ _____

    $\dfrac{4.400}{3.92} =$ _____

    $5.472 - 4.001 + 0.0119 =$ _____

Experiment 1 ▪ Data & Report Sheet

Measurements and Significant Figures

A. Measurement with Ruler A

TABLE 1.1 (data)

	Rectangle W	Rectangle X	Rectangle Y	Rectangle Z
Measured length (cm)				
Measured width (cm)				

TABLE 1.2 (report)

Rectangle	Measured Length (cm)	Number of Sig. Figures in Length	Measured Width (cm)	Number of Sig. Figures in Width
W				
X				
Y				
Z				

TABLE 1.3 (report)

Rectangle	Justified Number of Sig. Figures in Calculated Area	Calculated Area Unrounded (cm^2)	Calculated Area Rounded (cm^2)
W			
X			
Y			
Z			

TABLE 1.4 (report)

Rectangle	Justified Number of Decimal Places for Calculated Perimeter	Calculated Perimeter Unrounded (cm)	Calculated Perimeter Rounded (cm)
W			
X			
Y			
Z			

B. Measurement with Ruler B

TABLE 1.5 (data)

	Rectangle W	Rectangle X	Rectangle Y	Rectangle Z
Measured length (cm)				
Measured width (cm)				

TABLE 1.6 (report)

Rectangle	Measured Length (cm)	Number of Sig. Figures in Length	Measured Width (cm)	Number of Sig. Figures in Width
W				
X				
Y				
Z				

TABLE 1.7 (report)

Rectangle	Justified Number of Sig. Figures in Calculated Area	Calculated Area Unrounded (cm^2)	Calculated Area Rounded (cm^2)
W			
X			
Y			
Z			

TABLE 1.8 (report)

Rectangle	Justified Number of Decimal Places for Calculated Perimeter	Calculated Perimeter Unrounded (cm)	Calculated Perimeter Rounded (cm)
W			
X			
Y			
Z			

C. Improving the Significance of Measurements

TABLE 1.9 (data)

Measured diameter (cm) _____

Number of pennies in stack	1	3	5	7	10
Measured thickness (cm)					

TABLE 1.10 (report)

Number of Pennies in Stack	Measured Stack Thickness (cm)	Number of Sig. Figures in Measured Thickness	Justified Number of Sig. Figures in Calculated Avg. Thickness	Calculated Avg. Thickness Unrounded (cm)	Calculated Avg. Thickness Rounded (cm)
1					
3					
5					
7					
10					

TABLE 1.11 (report)

Number of significant figures justified in a calculated penny volume _____

Calculated penny volume—unrounded (cm³) _____

Calculated penny volume—rounded (cm³) _____

QUESTIONS

1. Suppose the area of rectangle Y was calculated using the length from Table 1.1 and the width from Table 1.5. How many significant figures would be justified in the calculated area?

   a. 1   **b.** 2   **c.** 3   **d.** 4

   Explain your answer: _____

   _____

2. Suppose the perimeter of rectangle X was calculated using the length from Table 1.1 and the width from Table 1.5. How many significant figures would be found in the calculated perimeter after proper rounding?

   a. 1   **b.** 2   **c.** 3   **d.** 4

   Explain your answer: _____

   _____

3. A buret is a tubular device used to deliver measured volumes of liquids. The smallest division on the scale of a buret is 0.1 mL. Suppose a buret reading was exactly on the 10-mL mark. How should this reading be recorded?

   a. 10 mL   **b.** 10.0 mL   **c.** 10.00 mL   **d.** 10.000 mL

   Explain your answer: _____

   _____

4. Refer to Table 1.10. How can the number of significant figures in a measurement be increased without changing the measuring device?

   a. Estimate two decimals beyond the smallest scale division of the measuring device.
   b. Increase the size of the quantity being measured.
   c. Decrease the size of the quantity being measured.

   Explain your answer: _____

   _____

5. What is the shortest length that could be measured with ruler B that would contain four significant figures?

   a. Exactly 1 cm   **b.** Exactly 5 cm   **c.** Exactly 10 cm   **d.** Exactly 15 cm

   Explain your answer: _____

   _____

Experiment 2

The Use of Chemical Balances

In this experiment, you will

- Learn to use centigram balances, and electronic balances.
- Practice using the balances by measuring the mass of a coin, the mass of a group of several coins, and the mass of an unknown.

- Practice your graphing skills.
- Demonstrate the two techniques of weighing: direct weighing and weighing by difference.
- Practice the procedure used to weigh chemical samples.

≡ INTRODUCTION

In experimental work, it is often necessary to measure mass. In some instances, only approximate values are needed, while in others the mass must be determined quite accurately. The maximum accuracy possible in a mass determination is limited by the sensitivity of the balance that is used.

FIGURE 2.1
Chemical balances

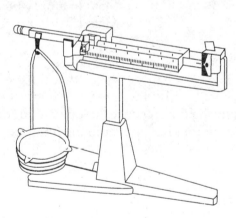

(a) Centigram balance (0.01 g)

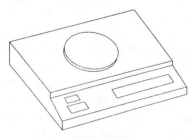

(b) Electronic balance, intermediate sensitivity (0.001 g)

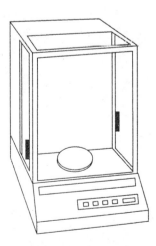

(c) Electronic balance, high sensitivity (0.0001 g)

A. The Centigram Balance

The centigram balance shown in Figure 2.1(a) has a sensitivity of 0.01 g. This means that the balance can be used to detect masses as small as 0.01 g, but no smaller. Do not attempt to use the balance until your lab instructor has demonstrated the proper techniques to utilize.

Objects may be weighed two different ways. A **direct weighing** is done by placing the object directly on the balance and obtaining the mass from the balance readings. Certain balance errors are eliminated by using a second technique called **weighing by difference.** When an object is weighed by difference, a container such as a beaker, a plastic weighing dish or piece of paper is placed on the balance and weighed. The object is then placed in the container, and the two are weighed together. The mass of the object is obtained by subtracting the mass of the empty container from the combined mass of object and container. Example 2.1 illustrates the advantage of weighing by difference.

EXAMPLE 2.1

Unknown to you, the centigram balance you are using has been zeroed incorrectly, and every reading is 0.09 g too high. You weigh a coin by direct weighing and by difference. The resulting data are given below. Determine the coin mass in each case. Which mass is more accurate?

Direct Weighing	Weighing by Difference
Coin mass = 5.76 g	Mass of beaker + coin = 32.31 g
	Mass of empty beaker = 26.64 g
	Mass by difference = 5.67 g

Solution

The coin mass by direct weighing is 5.76 g. The mass by difference is 32.31 g − 26.64 g = 5.67 g. We see that the mass by difference is 0.09 g less than that obtained by the direct method. The reason is that both masses measured in the difference technique are high by 0.09 g, but this error subtracts out and does not appear in the final result. The mass of 5.67 g is more accurate because it does not contain the 0.09 g error.

PROCEDURE

1. Weigh a coin on a centigram balance, using the direct-weighing technique. Record the coin mass in Table 2.1 of the Data and Report Sheet.
2. Weigh the same coin directly on another centigram balance. Record the mass in Table 2.1.
3. Weigh the same coin by difference. First, determine the mass of an empty 50-mL beaker. Then place the coin in the beaker and determine the combined mass. Record the data in Table 2.2.
4. Use a different centigram balance and again determine the coin mass by difference. Record the data in Table 2.2.
5. Obtain an unknown mass from the stockroom and record its identification number in Table 2.3.
6. Determine the mass of the unknown by direct weighing and by difference. Record the data in Table 2.3.

B. The Electronic Balance

Electronic balances such as those shown in Figure 2.1 have the greatest sensitivity of the balances you will use, and are the most delicate balances in the laboratory. They must be handled carefully, used correctly, and not abused. Your instructor will demonstrate the proper technique to use. The same balance should be used for any series of related weighings you might make.

PROCEDURE

1. Determine the mass of the same coin used in Part A by direct weighing and by difference, using an electronic balance. Record the data in Table 2.6, using the number of figures justified by the sensitivity of the balance you used.
2. Determine the mass of your unknown by direct weighing and by difference, using an electronic balance. Record the data in Table 2.6, using the number of figures justified by the sensitivity of the balance you used.

C. The Average Mass of a Coin

The speed of electronic balances makes them very useful in studies that involve the measurement of numerous masses. In this part of the experiment you will measure the mass of various numbers of one-cent coins, determine the value of the average mass mathematically, then graph the collected data and determine the value of the average mass graphically.

PROCEDURE

1. Obtain 10 one-cent coins from the stockroom.
2. Use an electronic balance to determine the mass of a single coin, then the mass of two coins together, then three coins together, etc., until you finally measure the mass of all ten coins together. Record the data in Table 2.8, using the number of figures justified by the sensitivity of the balance.

D. Weighing Solid Chemicals

In numerous future experiments, you will weigh out small samples of solid chemicals. In some cases, it will be necessary to know the sample masses quite accurately. In others, only approximate samples will be needed. Accurate masses will usually be determined with either electronic or centigram balances. Samples with approximate masses will be weighed on centigram balances. The procedures to follow are illustrated in Example 2.2. Approximate masses will not be recorded on the data sheet. Therefore, when a mass is to be recorded, you should recognize that it must be known accurately. In such cases, you must determine and record the mass as accurately as possible, consistent with the balance used. Be sure you read through the following example carefully before you attempt to do the weighings.

EXAMPLE 2.2

Explain how you would use chemical balances to make weighings consistent with the following directions. (NOTE: This is only an example. Do *not* do the weighings, just read through the example so you understand the procedures.)

a. Use a centigram balance and accurately weigh out a sample of oxalic acid in the range of 0.20 to 0.25 g. Record the data on the Data and Report Sheet.
b. Weigh out three samples of table salt weighing approximately 0.1 g.
c. Use an electronic balance and weigh a sample of sugar in the range of 0.5 to 0.6 g. Record the data on the Data and Report Sheet.

Solution

a. The directions indicate that this is an accurate weighing. Therefore, place a piece of paper or an empty container on the balance and weigh to 0.01 g. Record this mass, then adjust the weights to add about 0.2 g to the balance reading. Carefully add solid oxalic acid to the container until the balance just trips. Then adjust the weights to get the accurate mass of the container plus sample. Record this mass. Subtract the accurate container mass from the accurate container-plus-sample mass to get an accurate sample mass. Notice that no attempt was made to get a sample of some predetermined exact mass. We simply want the mass to be in the range given (0.20 to 0.25 g) and we want to know accurately what it is.
b. The directions indicate that these sample masses are approximate and will not be recorded. Place a piece of paper or a container on the chemical balance and adjust the weights until approximate balance is indicated. Then adjust the weights to increase the balance reading by 0.1 g. Carefully add table salt to the paper or container until the balance just trips. None of the balance readings are recorded.
c. This is an accurate weighing and should be done by difference. The procedure is similar to that used with centigram balances. Obtain slightly more than 0.5 g of sugar in a small test tube (you might do this by using a centigram balance to weigh it approximately). Put a piece of paper or empty container on a previously zeroed electronic balance. Record the accurate container mass, then add sugar to the container until the reading indicates that between 0.5 and 0.6 g have been added. Record the accurate combined mass of container and sugar.

PROCEDURE

1. Use a centigram balance to accurately weigh a sample of sodium chloride (table salt) in the range of 1.0 to 1.2 g. Record the appropriate data in Table 2.10 of the Data and Report Sheet.
2. Use an electronic balance to accurately weigh a sample of sodium chloride in the 1.0 to 1.2 g range. Begin with a test tube containing about 1.1 g of sodium chloride weighed approximately on a centigram balance. Record the data in Table 2.11.
3. Use a centigram balance and prepare a sample of sodium chloride with a mass of approximately 0.20 g. Show it to your instructor, who will then initial your Data and Report Sheet (Table 2.12).

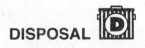

DISPOSAL   Sodium chloride in sink

CALCULATIONS AND REPORT

A. The Centigram Balance

1. Record in Table 2.4 all direct coin masses obtained in Table 2.1 of the Data and Report Sheet.
2. Use data from Table 2.2 and calculate the coin mass by difference measured on two balances. Record these masses in Table 2.4.
3. Fill in the remaining blanks of Table 2.4.
4. Record the unknown mass identification number in Table 2.5.
5. Use the data of Table 2.3 and calculate the mass of the unknown determined by direct weighing and by difference. Record the results in Table 2.5.
6. Fill in the remaining blanks of Table 2.5.

B. The Electronic Balance

1. Use the data of Table 2.6 and calculate the coin and unknown masses as determined by direct weighing and by difference. Record these results and the unknown identification number in Table 2.7.
2. Fill in the remaining blank of Table 2.7.

C. The Average Mass of a Coin

1. Use the data of Table 2.8 to calculate the average mass of a one-cent coin for each group of coins you weighed. This is done by dividing the total mass of the group by the number of coins in the group. Use the correct number of significant figures for each average mass calculated, and record the results in Table 2.9.
2. Refer once again to the data in Table 2.8. Plot the data of Table 2.8 on the graph paper provided on the Data and Report Sheet. If you need a review of the techniques used to draw graphs from data, refer to Appendix A of this manual. Plot the number of coins weighed in each group along the x (horizontal) axis of the graph paper, and the corresponding mass of each group along the y (vertical) axis of the graph paper. Remember, each value for the number of coins in a group and the corresponding mass of the group will form one point on your graph.
3. The resulting graph should be linear. Use a straight-edged ruler to draw the best straight line you can through the points. Determine the slope of the resulting straight line by dividing the rise (vertical distance) by the run (horizontal distance). See Figure A.4 of Appendix A for details of this procedure. The slope of the line will have the units g/coin, which is the units of the mass of a single coin. Because all of the coin groups have been used in graphing the line, this value is a form of average value per coin. Record this average value in Table 2.9.

D. Weighing Solid Chemicals

1. Use data from Table 2.10 and calculate the accurate mass of the sodium chloride sample weighed on the centigram balance. Record this result in Table 2.12, using the correct number of significant figures.
2. Use data from Table 2.11 and calculate the accurate mass of the sodium chloride sample weighed on the electronic balance. Record this result in Table 2.12, using the correct number of significant figures.

Experiment 2 ▪ Pre-Lab Review

The Use of Chemical Balances

1. Are any specific safety alerts given in the experiment? List any that are given.

   _____

2. Are any specific disposal directions given in the experiment? List any that are given.

   _____

3. An object has a mass of 2.62114 g. The object is weighed accurately on the three types of balances described in this experiment. What mass should be recorded for the object in each case?

   Centigram balance _____   Electronic balance (intermediate sensitivity) _____

   Electronic balance (high sensitivity) _____

4. A group of four one-cent coins is weighed on an electronic balance of intermediate sensitivity and has a mass of 10.147 g. What mass should be reported for the average mass of a single coin of the group?

   _____

   _____

   _____

5. The mass of the group of four coins described in question 4 is to be used as one point in a graph constructed as described in the Calculation and Report directions of Part C. What are the x and y values that would be used to represent this point on the graph?

   _____

   _____

   _____

6. Compare the procedures used in making direct weighings and indirect weighings (weighing by difference).

   _____

   _____

   _____

   _____

   _____

   _____

7. Which method, direct weighing or weighing by difference, is used when an accurate mass is to be obtained? Explain.

_____

_____

_____

_____

_____

8. How can you tell from the directions in the experiment whether a mass is to be determined accurately or approximately?

_____

_____

_____

_____

9. Compare the procedures used to weigh a chemical sample approximately and accurately.

_____

_____

_____

_____

_____

Experiment 2 ▪ Data & Report Sheet

The Use of Chemical Balances

A. The Centigram Balance

TABLE 2.1 (data)

Coin mass (direct) 1st balance _____

Coin mass (direct) 2nd balance _____

TABLE 2.3 (data)

Unknown ID number _____

Unknown mass (direct) _____

Mass of unknown + beaker _____

Mass of empty beaker _____

TABLE 2.5 (report)

Unknown ID number _____

Unknown mass (direct) _____

Unknown mass (by difference) _____

Difference in unknown mass
   by two methods _____

TABLE 2.2 (data)

	1st Balance	2nd Balance
Mass of coin + beaker	_____	_____
Mass of empty beaker	_____	_____

TABLE 2.4 (report)

	1st Balance	2nd Balance
Coin mass (direct)	_____	_____
Coin mass (by difference)	_____	_____
Difference in mass (direct) between two balances	_____	
Difference in mass (by difference) between two balances	_____	

B. The Electronic Balance

TABLE 2.6 (data)

Coin mass (direct) _____

Mass of coin + beaker _____

Mass of empty beaker _____

Unknown mass (direct) _____

Mass of unknown + beaker _____

Mass of empty beaker _____

TABLE 2.7 (report)

Coin mass (direct) _____

Coin mass (by difference) _____

Difference in coin mass
by two methods _____

Unknown ID number _____

Unknown mass (direct) _____

Unknown mass (by difference) _____

Difference in unknown mass
by two methods _____

C. The Average Mass of a Coin

TABLE 2.8 (data)

Number of coins in group	1	2	3	4	5	6	7	8	9	10
Mass of group	___	___	___	___	___	___	___	___	___	___

TABLE 2.9 (report)

Number of coins in group	1	2	3	4	5	6	7	8	9	10
Average coin mass	___	___	___	___	___	___	___	___	___	___
Average coin mass from slope of graph	___	___	___	___	___	___	___	___	___	___

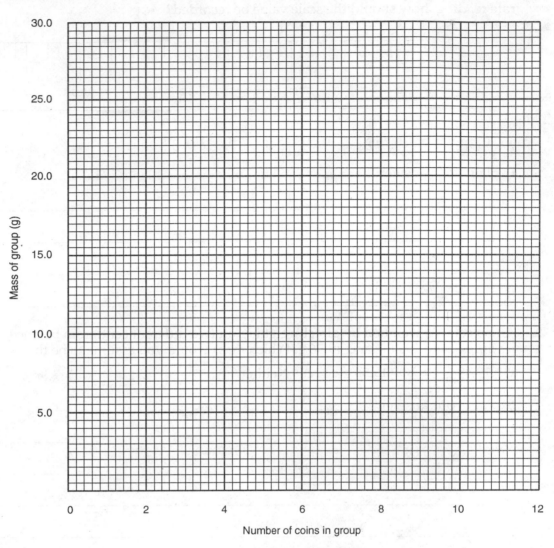

Mass of group (g)

Number of coins in group

D. Weighing Solid Chemicals

TABLE 2.10 (data)

Mass of sample + container _____

Mass of empty container _____

TABLE 2.11 (data)

Mass of sample + container _____

Mass of empty container _____

TABLE 2.12 (report)

Accurate sample mass
(centigram) _____

Accurate sample mass
(electronic) _____

Instructor approval of 0.20 g
approximate sample _____

QUESTIONS

1. The following represents a portion of the smallest scale of a centigram balance. For an accurate reading, how should the scale value be recorded?

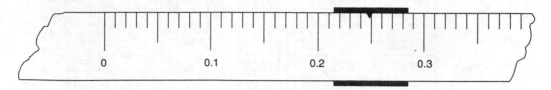

    a. 0.2 g   **b.** 0.25 g   **c.** 0.248 g   **d.** 0.24 g

Explain your answer: _____

_____

_____

_____

_____

2. An object was weighed by direct weighing on two different centigram balances. The balance readings were 19.48 g and 19.56 g. Which mass would you assume to be the more correct?

    a. 19.48 g   **b.** 19.56 g   **c.** Neither

Explain your answer: _____

_____

_____

_____

_____

3. A student weighed an unknown mass directly on an electronic balance of intermediate sensitivity, and got a result of 28.774 g. The student then realized a mistake had been made, and weighed the same unknown by difference on the same balance with the following results:

    Mass of beaker plus unknown = 59.121 g
    Mass of empty beaker = 30.346 g

When the sensitivity of the balance is taken into account, the two values obtained for the unknown mass are:

    a. Significantly different
    b. Different, but not significantly
    c. Identical

Explain your answer: _____

_____

_____

4. Two one-cent coins were weighed on an electronic balance of high sensitivity, and an average mass per coin of 2.4918 g was obtained from the data. How could you use an electronic balance of intermediate sensitivity to obtain an average mass per coin that contained the same number of significant figures as the value obtained using the high sensitivity balance?

    a. Weigh a group containing more than two coins.
    b. Weigh a two-coin group but estimate the balance reading to 0.0001 g.
    c. Weigh the two-coin group on two different intermediate sensitivity balances, and average the results.
    d. None of the three techniques described above will work

    Explain your answer: _____

    _____

    _____

5. Using the difference technique, you want to accurately weigh a sample of a solid with a mass of about 0.50 g. A container is placed on a centigram balance and weighs 0.71 g. What would you set the balance to read before adding the solid to the container?

    a. 0.50 g   b. 1.21 g   c. 0.21 g   d. 0.71 g

    Explain your answer: _____

    _____

    _____

    _____

The Use of Volumetric Ware and the Determination of Density

In this experiment, you will

- Learn how to make volume measurements using a graduated cylinder, pipet, and buret.
- Practice using the equipment listed above.
- Use volumetric ware to determine the average volume of a drop.

- Practice your graphing skills.
- Use the results of volume and mass measurements to determine the density of a liquid and a solid.

INTRODUCTION

The measurement of liquid volumes is an important part of many experiments. In some experiments, volumes must be measured quite accurately, while in others less accuracy is required. The maximum accuracy possible is determined by the type of volumetric equipment used (Figure 3.1).

FIGURE 3.1
Commonly used
volumetric glassware

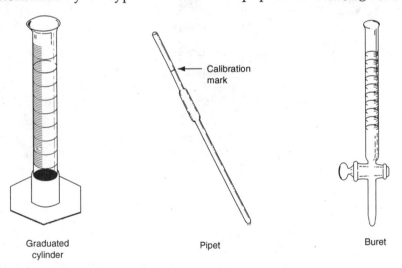

Graduated
cylinder

Calibration
mark

Pipet

Buret

EXPERIMENTAL PROCEDURE

A. The Graduated Cylinder

To properly use graduated cylinders and other volumetric ware, it is necessary to understand a little of the nature of the liquids involved. Water and most other liquids wet the surface of clean glass and, as a result, form a curved surface in glass containers. This curved surface, called a **meniscus,** becomes more apparent in narrow containers as shown in Figure 3.2. Volumetric readings are made at the bottom of the meniscus as represented in Figure 3.2(c).

FIGURE 3.2
The appearance of a
meniscus in volumetric
glassware

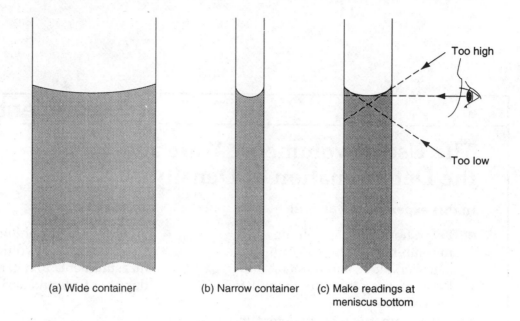

(a) Wide container   (b) Narrow container   (c) Make readings at
meniscus bottom

Graduated cylinders are designed to measure any liquid volume up to the cylinder capacity. The volume contained in a graduated cylinder is estimated to one decimal place more than the smallest division on the cylinder. For example, a 50-mL cylinder has divisions corresponding to 1 mL. Volumes can be estimated to the nearest 0.1 mL by estimating between the marks as shown in Figure 3.3.

FIGURE 3.3
Volume measurements
using a graduated cylinder

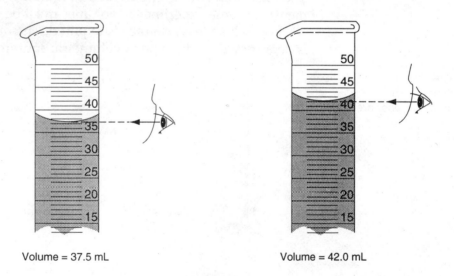

Volume = 37.5 mL   Volume = 42.0 mL

PROCEDURE

1. Obtain a 25- or 50-mL graduated cylinder from your desk equipment or the stockroom.
2. Weigh a clean, dry 50-mL beaker on a centigram or electronic balance and record the mass in Table 3.1 of the Data and Report Sheet.
3. Use your wash bottle to add distilled water carefully to the graduated cylinder until, in your judgment, it contains exactly 10.0 mL. Remember to read the volume at the bottom of the meniscus. Record this volume in Table 3.1.
4. Pour the water sample from the cylinder into the weighed beaker.
5. Weigh the beaker and contained water on the same balance used in Step 2, and record the mass in Table 3.1.

28   Experiment 3

6. Without emptying the beaker, add a second carefully measured 10.0-mL sample of water from your graduated cylinder.
7. Weigh the beaker and contained water on the same centigram balance used before and record the mass in Table 3.1.

B. The Pipet

Pipets are designed to deliver a specific volume of liquid and therefore have only a single calibration mark that is located on the narrow neck above the bulb (see Figure 3.1).

SAFETY ALERT

Pipets should never be filled by applying suction with your mouth. A suction bulb should always be used. The proper procedure will be demonstrated by your instructor.

PROCEDURE

1. Obtain a 10-mL pipet from your desk equipment or the stockroom. Record the pipet volume in Table 3.3, using four significant figures (record to the second decimal), in the Data and Report Sheet.
2. Weigh a clean, dry 50-mL beaker on a centigram or electronic balance and record the mass in Table 3.3.
3. Place about 50 mL of distilled water into a clean 100-mL beaker.
4. Use a suction bulb and draw the distilled water from the 100-mL beaker into the pipet to a level above the calibration mark.
5. Quickly remove the bulb and place your forefinger tightly over the top end of the pipet to keep the water from flowing out.
6. Control the flow of water out of the pipet by adjusting the pressure exerted by your finger. Carefully allow the water meniscus to drop to the level of the calibration mark (but no lower).
7. Touch the pipet tip to the side of the container of distilled water to remove the attached drop.
8. Allow the pipet to drain into the weighed beaker. When the pipet stops draining, touch the tip to the side of the beaker to remove some of the water in the tip. Any water that remains in the pipet after this procedure is not to be blown out. The pipet is designed to deliver the stated volume even though a small amount remains in the tip.
9. Weigh the beaker and water on the same balance used in Step 2. Record the combined mass in Table 3.3.
10. Without emptying the beaker, repeat Steps 4 to 8. Record the mass of the beaker containing the two water samples in Table 3.3.

C. The Buret

Burets are designed to deliver any precisely measured volume of liquid up to a maximum of the buret capacity. Once again, estimation between scale divisions is necessary. Both 25- and 50-mL burets have scale divisions corresponding to 0.1 mL. Therefore, estimates between scale divisions will be written in terms of the next smaller decimal, 0.01 mL. Buret scales increase downward, so estimations between divisions must also increase downward. Refer to Figure 3.4 for examples of buret readings and ask your instructor for assistance as needed.

FIGURE 3.4
Examples of buret
readings

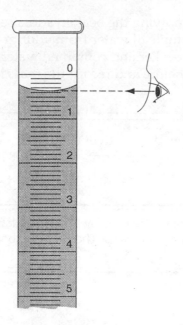

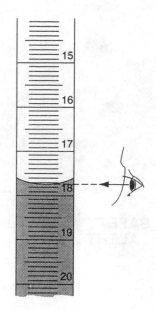

Reading = 0.36 mL

Reading = 17.73 mL

PROCEDURE

1. Obtain a 25- or 50-mL buret from the stockroom.
2. Mount the buret on a ringstand by means of a buret clamp as shown in Figure 3.5.

FIGURE 3.5
A buret mounted on a
ringstand

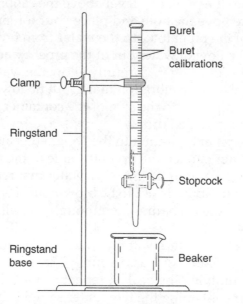

3. Fill the buret nearly to the top with distilled water.
4. Carefully open and close the stopcock a few times to acquaint yourself with its operation and also to remove any air bubbles from the tip of the buret. If any difficulties are encountered, ask your instructor for assistance.
5. When you are ready to proceed, refill the buret and adjust the level of the meniscus to be somewhere between 0.00 and 1.00 mL.
6. Read the water level to the nearest 0.01 mL and record this initial reading in Table 3.5 in the blank corresponding to Sample 1.
7. Weigh a dry 50-mL beaker on a centigram or electronic balance and record the mass in Table 3.5.
8. Carefully add 10 to 12 mL of water from the buret into the beaker.
9. Read the water level in the buret to the nearest 0.01 mL and record this final reading in Table 3.5 in the blank corresponding to Sample 1.
10. Weigh the beaker and contained water on the same balance used in Step 7 and record the mass in Table 3.5.
11. Do not refill the buret or empty the first water sample from the beaker.

12. Record the final buret reading from the first sample as the initial reading for the second sample in Table 3.5.
13. Add another 10 to 12 mL of water from the buret to the beaker.
14. Read the new water level in the buret to the nearest 0.01 mL and record it in Table 3.5 as the final reading for Sample 2.
15. Weigh the beaker and contained water on the same balance used before and record the mass in Table 3.5.

D. The Volume of a Drop

It is often convenient to measure the quantity of liquids in terms of a specific number of drops. However, the volume contained in a single drop depends on the device used to deliver the drop and the way the device is used.

PROCEDURE

1. Put about 20 mL of distilled water into a 50-mL beaker.
2. Obtain a plastic dropper and a 10-mL graduated cylinder from your desk equipment or the stockroom. Make certain the graduated cylinder is dry.
3. Fill your plastic dropper with distilled water from the 50-mL beaker and, while holding the dropper vertical, add 25 counted drops of distilled water to the 10-mL graduated cylinder.
4. Read the volume of water contained in the graduated cylinder and record the volume and the number of drops in Table 3.7. (NOTE: If necessary, review Figure 3.3 and the discussion that explains how to read graduated cylinders. Also, look carefully at your 10-mL cylinder. Some have smallest divisions of 0.1 mL, and others have 0.2 mL divisions. In either case, the volume readings should be estimated to the second decimal. Just be aware that in the case of the 0.2 mL divisions, a reading half way between two divisions corresponds to 0.1 mL, and not 0.05 mL. Ask your instructor for assistance if necessary.)
5. Without emptying the graduated cylinder, add an additional 25 drops of distilled water, being careful to hold the dropper in a vertical position above the graduated cylinder.
6. Read the total volume of water now contained in the graduated cylinder. Record the total volume and the total number of drops added (50) in Table 3.7.
7. Empty and dry your 10-mL graduated cylinder, then repeat Steps 3–6 but with the dropper held in a horizontal position when the drops of water are added to the graduated cylinder. Record the collected data in Table 3.7.
8. If you have not yet done Part C of the experiment, review Figures 3.4 and 3.5 and the directions for using a buret. Obtain a 25- or 50-mL buret from your desk equipment or the stockroom and set it up as shown in Figure 3.5.
9. Fill your buret with distilled water, open the stopcock and remove any air bubbles from the tip, then adjust the level of the meniscus in the buret to a value between 0.00 and 1.00 mL.
10. Read the level of water in the buret and record the reading in the blank labeled "Initial buret reading" in Table 3.7.

11. Carefully open the stopcock of the buret until individual drops are formed and fall from the tip. Allow 25 counted drops to fall from the tip, then close the stopcock and read the level of the meniscus. Record the reading in Table 3.7, along with the number of drops delivered from the buret.
12. Carefully open the stopcock of the buret again and allow another 25 drops to fall. Close the stopcock, read the buret, and record in Table 3.7 the reading and total number of drops delivered.
13. Repeat Step 12 twice and record in Table 3.7 the volume readings and total number of drops delivered.

E. Density of an Unknown Liquid

To evaluate a density, both the volume (v) and mass (m) of a sample of substance must be determined experimentally. The density (d) is the ratio of these two experimental quantities:

$$d = \frac{m}{v}$$

Eq. 3.1

PROCEDURE

1. Obtain a 30-mL sample of unknown liquid from the stockroom and record the identification number in Table 3.10
2. Clean and dry a 50-mL beaker.
3. Weigh the dry beaker on a centigram or electronic balance and record the mass in Table 3.10.
4. Use the suction bulb to draw 3 to 4 mL of the unknown liquid into your 10-mL pipet. Put your finger over the pipet top, then invert the pipet and allow the liquid to flow out the top into the sink. This procedure will rinse the entire pipet with the unknown liquid.
5. Use the suction bulb and the rinsed pipet to deliver a 10.00-mL sample of unknown liquid into the weighed beaker. Record the sample volume (pipet volume) in Table 3.10.
6. Weigh the beaker and contents on the same balance used in Step 3. Record the mass in Table 3.10.
7. Add another 10.00 mL of unknown liquid to the beaker without emptying it first.
8. Weigh the beaker and contents on the same balance used in Step 3 and record the mass in Table 3.10.
9. Rinse your pipet thoroughly with tap water and then distilled water before you store it with your equipment or return it to the stockroom.

DISPOSAL 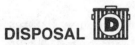 Unknown liquid in sink

F. Density of a Solid

Solids can be weighed quite easily, but volume determinations are somewhat more difficult. If the solid has a regular geometric shape such as a cube or sphere, the volume can be calculated from the dimensions. You will work with an irregular solid and determine the volume another way.

PROCEDURE

1. Obtain a solid no. 1 rubber stopper from your desk or the stockroom.
2. Weigh a dry 50-mL beaker on a centigram or electronic balance and record the mass in Table 3.12.
3. Place the rubber stopper in the weighed beaker, weigh them together, and record the combined mass in Table 3.12.
4. Fill a 50- or 100-mL graduated cylinder approximately halfway with distilled water.
5. Read the water level in the cylinder (remember to estimate to 0.1 mL) and record the value in Table 3.12 in the blank labeled "cylinder reading without stopper."
6. Carefully put the stopper into the cylinder without splashing out any water. This is best done by tilting the cylinder and letting the stopper slide down the inside wall.
7. Tap the cylinder so the stopper sinks to the bottom. Swirl the water gently to remove any clinging air bubbles from the stopper.
8. Read the water level in the cylinder that now contains the stopper. Record the level in Table 3.12.

CALCULATIONS AND REPORT

A. The Graduated Cylinder

With rearrangement, Equation 3.1 can be written

$$v = \frac{m}{d}$$

Eq. 3.2

Thus, the volume of a liquid sample is equal to the mass of the sample divided by the liquid density. We will use this relationship and the measured masses of water samples to calculate the volume of each sample. These calculated volumes will then be compared with the volumes obtained by reading the volumetric equipment. In each calculation, the density of water will be assumed to have a value of 1.000 g/mL. Thus, the volume calculated in milliliters will be numerically equal to the mass measured in grams.

EXAMPLE 3.1

Use the following data to calculate the volume of water contained in a graduated cylinder on a reading basis and a mass basis. Calculate (1) the difference in volume between the mass and reading basis and (2) the percent difference.

Volume of water in cylinder	10.0 mL
Mass of beaker + sample	36.66 g
Mass of empty beaker	26.48 g

Solution

The volume of water contained on a reading basis is simply the cylinder reading of 10.0 mL. The volume on a mass basis is calculated using the water mass, the water density of 1.000 g/mL, and Equation 3.2. The mass of water is obtained by subtracting the mass of the empty beaker from the mass of beaker plus sample: 36.66 g – 26.48 g = 10.18 g. The volume of this sample is obtained by using Equation 3.2 and the water density:

$$v = \frac{m}{d} = \frac{10.18 \text{ g}}{1.000 \text{ g/mL}} = 10.18 \text{ mL}$$

Thus, the water volume in milliliters is equal to the water mass in grams.

The difference in volume between the mass and reading bases is 10.18 mL − 10.0 mL = 0.18 mL = 0.2, where the last value has been rounded to the correct number of significant figures.

The percent difference is calculated using Equation 3.3, given below in Step 5.

$$\% \text{ Difference} = \frac{\text{volume difference}}{\text{volume on a reading basis}} \times 100 = \frac{0.2 \text{ mL}}{10.0 \text{ mL}} \times 100 = 2\%$$

1. Record in Table 3.2 the volume of water delivered according to the readings taken from the graduated cylinder. Note that this will be ten mL in each case recorded using the proper number of significant figures.

2. Use Equation 3.2 and the water mass data in Table 3.1 to calculate the volume of water delivered (the volume on a mass basis). Note that the mass of the second water sample is obtained by subtracting the mass of the beaker containing the first sample from the mass of the beaker containing both samples. Record the calculated volumes in Table 3.2.

3. Determine the difference between the water volume on a mass basis and the volume on a reading basis for each sample. Record the difference in Table 3.2. If the reading basis volume is larger than the mass basis, this value of their difference will have a negative sign, which should be recorded.

4. Calculate and record the average difference in volume by averaging the two values obtained in Step 3. Remember to take into account any negative values recorded in Step 3.

5. Calculate the average percent difference between volumes on a mass and reading basis, using the following equation:

$$\text{Avg. } \% \text{ difference} = \frac{\text{avg. volume difference}}{\text{volume on a reading basis}} \times 100 \qquad \text{Eq. 3.3}$$

Note that if you obtained a negative value for your average in Step 4, the average percent difference will also be negative. This simply reflects the fact that your volumes on a reading basis were larger than those on a mass basis.

6. Record the average percent difference in Table 3.2.

B. The Pipet

Use the data in Table 3.3 and complete Table 3.4 as you did Table 3.2. Assume the pipet delivers exactly 10.00 mL when it is correctly filled.

C. The Buret

Use the data in Table 3.5 and complete Table 3.6, using the following equation for *each sample:*

$$\% \text{ Difference} = \frac{\text{volume difference}}{\text{volume on a reading basis}} \times 100 \qquad \text{Eq. 3.4}$$

Note that the volume of water delivered according to buret readings is equal to the final reading minus the initial reading for each sample. The average percent difference is obtained by averaging the two calculated percent differences.

D. The Volume of a Drop

Use the data recorded for the dropper experiments in Table 3.7 to calculate four values for the volume of a single drop of water. Do the calculation by dividing each recorded volume by the corresponding total number of drops. Record in Table 3.8 the calculated single-drop volumes together with the total number of drops used.

The data recorded in Table 3.7 for the drop-size experiment done with a buret will be used in two different ways to calculate the volume of a single water drop. Record in Table 3.9 the total number of drops delivered up to each point in the experiment, as recorded in Table 3.7, and the corresponding total volume of water delivered to each point. The total volume of water at each point is equal to the buret reading at that point minus the initial buret reading that is also recorded in Table 3.7. Calculate and record in Table 3.9 the volume of a single drop from the data at each point by dividing the total volume at each point by the corresponding total number of drops.

The data recorded in Table 3.9 will now be graphed, and the volume of a single drop will be determined from the graph. If you need a review of the techniques used to draw graphs from data, refer to Appendix A of this manual. Plot the total number of drops delivered at each point in the experiment along the *x* (horizontal) axis of the graph paper provided on the Data and Report Sheet. Plot the corresponding total volume of water delivered at each point on the *y* (vertical) axis. Remember, each value for the total number of drops and the corresponding total volume delivered will form one point on your graph. The resulting graph should be linear. Draw a straight line through the points and determine the slope of the resulting straight line by dividing the rise (vertical value) by the run (horizontal value). See Figure A.5 of Appendix A for an example of slope determination. The slope will be a value for the volume of a single drop. Record this graphical value of the volume of a single drop in Table 3.9.

E. Density of an Unknown Liquid

1. Record the unknown identification number in Table 3.11.
2. Record the volume of liquid (the pipet volume) used in each sample.
3. Use the data of Table 3.10 and determine the mass of each sample. Record the results in Table 3.11.
4. Use Equation 3.1 and the volume and mass obtained in Steps 2 and 3 and calculate the density of each sample. Record these results and the average of the two values in Table 3.11.

F. Density of a Solid

1. Use data from Table 3.12 and calculate the mass of the rubber stopper. Record the result in Table 3.13.
2. Use the two graduated cylinder readings of Table 3.12 to calculate the volume of water displaced by the stopper. Record this stopper volume.
3. Calculate and record the stopper density, using the mass and volume determined in Steps 1 and 2 and Equation 3.1.

G. Summary

Summarize the results of your volumetric measurements by completing Table 3.14. Record only the average percent difference for the buret measurements.

Experiment 3 ▪ Pre-Lab Review

The Use of Volumetric Ware and the Determination of Density

1. Are any specific safety alerts given in the experiment? List any that are given.

   _____

2. Are any specific disposal directions given in the experiment? List any that are given.

   _____

3. What is a meniscus? _____

4. Where is a liquid volume reading taken in relation to a meniscus? _____

   _____

5. Describe the steps followed when a pipet is used to measure a sample of liquid.

   _____

   _____

6. Describe the steps followed when a buret is used to measure a sample of liquid.

   _____

   _____

7. What type of weighing technique (direct weighing or weighing by difference) is used to determine the mass of water samples in this experiment?

   _____

8. Describe how to rinse the entire inside of a pipet with your density unknown in Part E.

   _____

   _____

9. Suppose you sequentially add 10 drops of water to a graduated cylinder and read the volume, add 10 more drops, read, and so on. How many total drops will have been added after you do this four times?

   _____

10. A water sample has a volume of exactly 12.0000 mL. How would this volume be recorded if it were measured from (a) a 50-mL graduated cylinder? (b) a buret?

    Graduated cylinder: _____   Buret: _____

11. Describe how the volume of an irregular solid can be determined. _____

    _____

    _____

Experiment 3 ▪ Data & Report Sheet

The Use of Volumetric Ware and the Determination of Density

A. The Graduated Cylinder

TABLE 3.1 (data)

Volume of water in cylinder	_____
Mass of beaker + both samples	_____
Mass of beaker + first sample	_____
Mass of empty beaker	_____

TABLE 3.2 (report)

	Sample 1	Sample 2
Water volume (reading basis)	_____	_____
Water volume (mass basis)	_____	_____
Volume difference	_____	_____
Average volume difference	_____	
Average % difference	_____	

B. The Pipet

TABLE 3.3 (data)

Pipet volume	_____
Mass of beaker + both samples	_____
Mass of beaker + first sample	_____
Mass of empty beaker	_____

TABLE 3.4 (report)

	Sample 1	Sample 2
Water volume (reading basis)	_____	_____
Water volume (mass basis)	_____	_____
Volume difference	_____	_____
Average volume difference	_____	
Average % difference	_____	

C. The Buret

TABLE 3.5 (data)

	Sample 1	Sample 2
Final buret reading	_____	_____
Initial buret reading	_____	_____
Mass of beaker + both samples		_____
Mass of beaker + first sample		_____
Mass of empty beaker		_____

TABLE 3.6 (report)

	Sample 1	Sample 2
Water volume (reading basis)	_____	_____
Water volume (mass basis)	_____	_____
Volume difference	_____	_____
% Difference	_____	_____
Average % difference	_____	

D. The Volume of a Drop

TABLE 3.7 (data)

Dropper Held Vertically		Dropper Held Horizontally	
Total number of drops	Grad. cyl. reading	Total number of drops	Grad. cyl. reading
_____	_____	_____	_____
_____	_____	_____	_____

Buret			
Initial buret reading	_____		
Total number of drops	_____	_____	_____
Buret reading	_____	_____	_____

TABLE 3.8 (report)

Dropper Held Vertically		Dropper Held Horizontally	
Total number of drops	Volume of a drop	Total number of drops	Volume of a drop
_____	_____	_____	_____
_____	_____	_____	_____

TABLE 3.9 (report)

Total number of drops	————	————	————	————
Total volume of water	————	————	————	————
Volume of a drop	————	————	————	————
Graphically determined volume of a single drop	————			

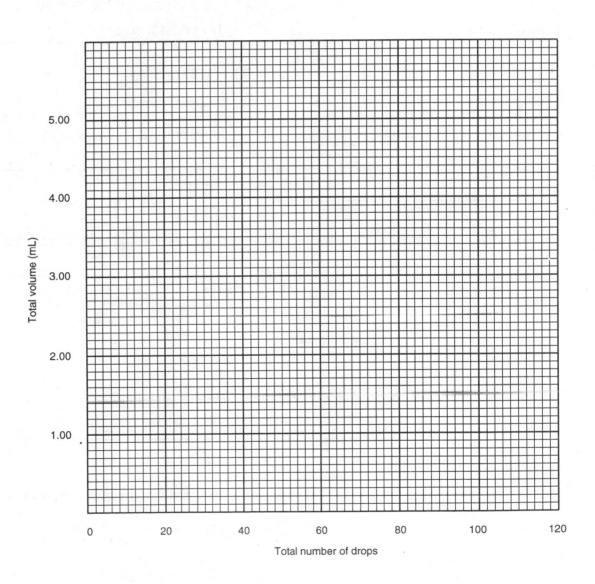

Total volume (mL) vs. Total number of drops

E. Density of an Unknown Liquid

TABLE 3.10 (data)

Unknown ID number _____

Pipet volume _____

Mass of beaker + both samples _____

Mass of beaker + first sample _____

Mass of empty beaker _____

TABLE 3.11 (report)

Unknown ID number _____

	Sample 1	Sample 2
Sample volume	_____	_____
Sample mass	_____	_____
Sample density	_____	_____
Average sample density	_____	

F. Density of a Solid

TABLE 3.12 (data)

Mass of beaker + stopper _____

Mass of empty beaker _____

Cylinder reading with stopper _____

Cylinder reading without stopper _____

TABLE 3.13 (report)

Stopper mass _____

Stopper volume _____

Stopper density _____

G. Summary

TABLE 3.14 (report)

	Average Volume Difference	Average Percent Difference
Graduated cylinder	_____	_____
Pipet	_____	_____
Buret		_____

QUESTIONS

1. Which of the following pieces of equipment is designed to deliver or contain a specific volume of liquid?

   a. Graduated cylinder   **b.** Pipet   **c.** Buret

   Explain your answer: _____

   _____

2. Suppose a 10-mL graduated cylinder is used that has calibration marks representing each mL and 0.1 mL. Accurate readings should be estimated and recorded to the nearest

   a. 1.0 mL.　　**b.** 0.1 mL.　　**c.** 0.01 mL.　　**d.** 0.001 mL.

   Explain your answer: _____

   _____

3. A student determined the volume of water from a pipet by the method described in Part B. The volume determined on a mass basis was consistently greater than 10.00 mL. Which of the following would account for this result?

   a. The meniscus was allowed to fall below the mark on the pipet before the sample was transferred to the weighed beaker.
   b. The remaining contents were blown out of the pipet into the beaker after the pipet had drained.
   c. Some of the water sample was spilled from the beaker before it was weighed.
   d. More than one response is correct.

   Explain your answer: _____

   _____

4. A student fills a buret with distilled water, adjusts the meniscus to a reading of 0.00 mL, and allows water to drain out until a reading of 10.00 mL is obtained. The water sample, weighed by difference, has a mass of 9.72 g. Which of the following experimental errors will account for the large difference between the volume according to readings and the volume according to mass?

   a. Initially the meniscus was above the 0.00-mL mark.
   b. The final meniscus was below the 10.00-mL mark.
   c. Air was not cleared from the buret tip before delivering the sample.
   d. Water leaked from the buret into the beaker after the final buret reading was taken.

   Explain your answer: _____

   _____

5. A student graphs the data collected in the drop-volume determination done with a buret. All the points except one fall on a straight line. One point is significantly above the straight line on the graph. Which of the following would account for this?

   a. The student miscounted the drops and actually delivered more drops than the number recorded in Table 3.7 for that point.
   b. The student miscounted the drops and actually delivered fewer drops than the number recorded in Table 3.7 for that point.
   c. The buret was misread at that point. The recorded value was less than the correct value.
   d. More than one response is correct.

   Explain your answer: _____

   _____

6. A student determines the density of an unknown liquid by weighing a sample of known volume. The reported density is found by the instructor to be significantly higher than it should be. Which of the following experimental errors could account for this result?

   a. The pipet used was filled to a level above the calibration mark.
   b. The balance was read incorrectly, and the recorded sample mass was lower than the true sample mass.
   c. A part of the sample was spilled from the beaker before it was weighed.

   Explain your answer: _____

   _____

7. A student determines the density of a rubber stopper by putting it into a graduated cylinder of water as you did in Part F. The stopper sinks to the bottom of the cylinder. The reported density of the stopper is 0.12 g/mL. This result

   a. might be correct.   b. is probably correct.   c. cannot possibly be correct.

   Explain your answer: _____

   _____

Physical and Chemical Changes

In this experiment, you will

- Cause changes to occur in several substances by various methods.
- Test and observe the substances before and after you change them.
- Interpret the test results and decide if

composition changes took place in the original substances.

- Classify the changes that took place as physical or chemical.

INTRODUCTION

Most changes that occur in substances can be classified as physical or chemical changes. **Physical changes** occur without any accompanying changes in composition. **Chemical changes** result in composition changes; that is, a substance is converted into one or more new substances.

In this experiment, you will subject a number of substances to conditions that will cause changes to occur. You will observe the changes and test the substances before and after the changes have occurred. You will use the observations and test results to classify the changes as physical or chemical. The tests are not exhaustive, but they will provide you with sufficient information to decide whether or not changes in composition have taken place. However, you must be sure to record everything you see.

The three most important things to look for are: color changes, the formation of precipitates, and the evolution of gases (fizzing or bubbling). When precipitates form, the amount produced influences what you will see. Light precipitates will cause a previously-clear liquid to become cloudy, whereas heavy precipitates will settle out of the liquid and collect on the bottom of the container in the form of clumps or finely-divided solid. Any observation from one of these extremes to the other should be interpreted as precipitate formation.

EXPERIMENTAL PROCEDURE

A. Solution Formation

Solutions result when one or more substances, called **solutes,** are dissolved in another substance called the **solvent.** In liquid solutions, the solutes may be gases, liquids, or solids; the solvent is a liquid. Some solid solutes can be recovered from liquid solutions by evaporating away the solvent.

In this part of the experiment, you will test sodium chloride before and after it is dissolved in water. You will use the test results to decide the type of change (physical or chemical) that takes place in sodium chloride when it is dissolved in water.

PROCEDURE

1. Place about 0.2 g of solid sodium chloride (NaCl) into each of two dry small (10-cm) test tubes.
2. Test the sample in one tube as follows: Moisten the end of a clean stirring rod with distilled water. Dip the moist rod into the sample so that a few crystals of solid adhere to it. Put the stirring rod with salt crystals into the flame of your burner.
3. Note and record in Table 4.1 (under the "Original Solute" heading) the color of the flame that results in Step 2.
4. Test the remaining solid in the test tube as follows. Add 20 drops of distilled water and agitate the tube until the solid dissolves completely. Add 2 drops of 0.1 M silver nitrate solution (AgNO₃) to the clear liquid in the test tube and agitate to mix. Record the results of this test under the "Original Solute" heading of Table 4.1.
5. Form a solution of sodium chloride in water by adding 20 drops of distilled water to the test tube containing the second sample of solid that you prepared in Step 1. Agitate the test tube to completely dissolve the solid. You have now changed the solid by dissolving it to form a solution.

FIGURE 4.1
Evaporation of liquids

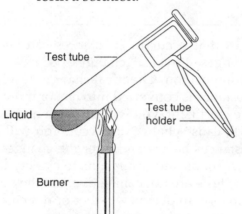

Test tube

Liquid

Test tube holder

Burner

6. Recover the dissolved solid solute from the solution by evaporating away the water solvent. This can be done by tilting the test tube and gently heating it above the liquid level, while gently agitating the tube to bring the liquid into contact with the hot portion. Figure 4.1 illustrates the technique, which will also be demonstrated by your instructor.

SAFETY ALERT

This must be done correctly or hot liquid might spatter out of the test tube. Do not point the mouth of the test tube at anyone while you do this evaporation.

7. Test a small portion of the recovered solid solute in a flame as you did in step 2. Record the results of this test under the "Recovered Solute" heading of Table 4.1.
8. Test the remainder of the recovered solid solute in the test tube by following the directions given in Step 4. Record the test results in Table 4.1 under the "Recovered Solute" heading.

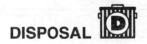

DISPOSAL

Tested solute solution in container labeled "Exp. 4, Used Chemicals"

B. Heating Copper Carbonate

Upon being heated, some substances undergo chemical changes, some are changed physically, and others do not change at all. In this part and Part C, you will investigate the effect of heating substances.

PROCEDURE

1. Place approximately 0.1 g of solid copper carbonate ($CuCO_3$) into each of two small (10-cm), dry test tubes. Note the color of the solid and record it in Table 4.3 under the "Unheated Sample" heading.
2. Tap the tubes on the bench top so any sample clinging to the sides falls to the bottom.
3. Test one sample by adding 15 drops of dilute (6 M) hydrochloric acid (HCl). Agitate the tube until the sample completely dissolves. Record the test results in Table 4.3.

SAFETY ALERT

Hydrochloric acid is very corrosive. If you get any on your skin, wash it off immediately with cold water and inform your instructor.

FIGURE 4.2
Test tube mounted for heating

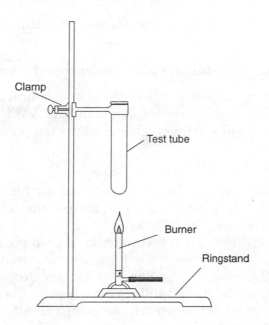

Clamp

Test tube

Burner

Ringstand

4. Mount the other test tube on a ringstand as shown in Figure 4.2 and heat the sample very strongly with a burner for a minimum of 5 minutes.
5. Allow the tube to cool for at least 10 minutes. Record the color of the heated sample in Table 4.3 under the "Heated Sample" heading.
6. Test the sample, which has been changed by heating, by following the directions given in Step 3. Record the results of the test in Table 4.3.

DISPOSAL Tested copper carbonate samples in container labeled "Exp. 4, Used Chemicals"

C. Melting Iron Chloride

PROCEDURE

1. Place approximately 0.1 g of solid iron chloride ($FeCl_3$) into each of two dry, small (10-cm) test tubes.
2. Add 30 drops of distilled water to one solid sample and shake until the solid dissolves completely. Record the color of the resulting solution in Table 4.5.
3. Make three test solutions by putting 10 drops of the solution prepared in Step 2 into each of two other test tubes.
4. Add 2 drops of 0.1 M silver nitrate ($AgNO_3$) solution to one test solution. Mix well, and record the results in Table 4.5 under the "Unmelted Sample" heading.
5. Add 2 drops of 0.1 M ammonium thiocyanate (NH_4SCN) to the second test solution, mix well, and record the results.

6. Add 2 drops of 0.1 M potassium ferrocyanide ($K_4Fe(CN)_6$) to the third test solution, mix well, and record the results.

7. Heat the other solid sample of $FeCl_3$ *very gently* until it just melts. This is best done by moving the test tube back and forth through a low burner flame. Do not overheat the sample.

8. Allow the melt to cool for about 5 minutes. The sample might remain in the liquid form even when cool. You have now changed the sample by melting it.

9. Add 30 drops of distilled water to the cool sample and shake until the sample dissolves completely. Note and record the color of the resulting solution in Table 4.5 under the "Melted Sample" heading.

10. Test this solution made from the melted sample the same way you did the solution made from the unmelted sample by repeating Steps 3-6. Record the results of the tests in Table 4.5.

DISPOSAL  Tested iron chloride samples in container labeled "Exp. 4, Used Chemicals"

D. Sodium Bicarbonate plus Hydrochloric Acid ▭

Substances may also undergo physical or chemical changes when brought together. The formation of a solution in Part A was an example of such a process; here you will investigate another example.

PROCEDURE

1. Place approximately 0.1 g of solid sodium bicarbonate ($NaHCO_3$) into each of two dry, small (10-cm) test tubes. Record the color of this unreacted solid in Table 4.7.

2. Add 30 drops of distilled water to one sample and agitate until the solid dissolves completely. If the sample doesn't dissolve completely after 2 minutes of agitation, add 10 more drops of distilled water and agitate again. If solid still remains, discard the sample into the sink and weigh out another sample, being careful not to exceed 0.1 g.

3. Test the solution formed in Step 2 by adding 6 drops of 0.1 M calcium nitrate solution ($Ca(NO_3)_2$). Mix well and record the results in Table 4.7.

4. You will attempt to change the second sample by reacting it with hydrochloric acid. Slowly and carefully add 15 drops of dilute (6 M) hydrochloric acid (HCl) to the second solid sample. No solid should remain after the reaction stops. If necessary, add 5 more drops of HCl to finish dissolving the solid.

5. Evaporate the resulting solution from Step 4 and recover the solid solute by the method described in Step 6 of Part A and illustrated in Figure 4.1. Record the color of the recovered solid in Table 4.7.

6. Allow the test tube to cool. Then add 30 drops of distilled water to the recovered solid and shake until it dissolves completely.

7. Test the resulting solution by adding 6 drops of 0.1 M calcium nitrate solution. Mix well and record the test results in Table 4.7 under the "Recovered Solid" heading.

DISPOSAL  Tested sodium bicarbonate samples in container labeled "Exp. 4, Used Chemicals"

E. Combustion of Magnesium

Combustion is a familiar process to most of us. It is a source of energy used to move vehicles and generate most of the electricity we use.

PROCEDURE

1. Obtain two 5-cm strips of magnesium ribbon (Mg). Record the appearance of the metal in Table 4.9 under the "Before Combustion" heading.
2. Place one strip in a small (10-cm) test tube and cautiously add 10 drops of dilute (6 M) HCl one drop at a time. Record your observations of the reaction that takes place.
3. Grasp one end of the other strip with your crucible tongs and hold the strip in the flame of your burner until the magnesium ignites.

SAFETY ALERT

> Do not look directly at the burning metal

4. Collect the combustion product, being very careful to exclude any unburned metal. Record the appearance of the product in Table 4.9 under the "After Combustion" heading.
5. Put the product in a small test tube and test it with dilute HCl as you did in Step 2. Record the results of this test.

DISPOSAL   Tested magnesium samples in container labeled "Exp. 4, Used Chemicals"

REPORT

Review the observations and test results recorded in Tables 4.1, 4.3, 4.5, 4.7, and 4.9. Decide which of these is different before and after the attempt was made to change the original substance. For example, in Table 4.1, you would compare the flame color recorded under the "Original Solute" heading with the flame color recorded under the "Recovered Solute" heading. Similarly, in Table 4.1, you would compare the recorded results of adding $AgNO_3$ to a solution of the original solute and the recovered solute. Any differences in the two results or observations you compared indicate that a composition change has taken place in the original substance. Decide which pairs of test results or observations indicate composition changes and record your conclusions in Tables 4.2, 4.4, 4.6, 4.8, and 4.10 by writing yes or no in the blanks under the heading "Is a Composition Change Indicated?" Then, in Table 4.11, classify the change that took place in each process as physical or chemical.

Experiment 4 ▪ Pre-Lab Review

Physical and Chemical Changes

1. Are any specific safety alerts given in the experiment? List any that are given.

   _____

   _____

2. Are any specific disposal directions given in the experiment? List any that are given.

   _____

   _____

3. What three important behaviors will you look for and record in this experiment?

   _____

   _____

   _____

4. What is the difference between a physical change and a chemical change?

   _____

   _____

   _____

5. Are solid samples weighed accurately or approximately in this experiment? How can you tell?

   _____

   _____

6. Describe the process used to recover solid solutes from liquid solutions.

   _____

   _____

   _____

7. In Part D, what do you do if your 0.1g sample of sodium bicarbonate does not dissolve in 30 drops of water?

_____

_____

_____

8. In Part E, what special precaution is given about how to collect the combustion product?

_____

_____

_____

Experiment 4 ■ Data & Report Sheet

Physical and Chemical Changes

A. Solution Formation

TABLE 4.1 (data)

	Original Solute	Recovered Solute
Flame Color	_____	_____
Result of adding $AgNO_3$ to dissolved sample	_____	_____

TABLE 4.2 (report)

Test or Observation	Is a Composition Change Indicated?
Flame color	_____
Addition of $AgNO_3$	_____

B. Heating Copper Carbonate

TABLE 4.3 (data)

	Unheated Sample	Heated Sample
Sample Color	_____	_____
Result of adding HCl to sample	_____	_____

TABLE 4.4 (report)

Test or Observation	Is a Composition Change Indicated?
Sample color	_____
Addition of HCl	_____

C. Melting Iron Chloride

TABLE 4.5 (data)

	Unmelted Sample	Melted Sample
Solution color	_____	_____
Result of adding $AgNO_3$ to test solution	_____	_____
Result of adding NH_4SCN to test solution	_____	_____
Result of adding $K_4Fe(CN)_6$ to test solution	_____	_____

TABLE 4.6 (report)

Test or Observation	Is a Composition Change Indicated?
Solution color	_____
Addition of $AgNO_3$	_____
Addition of NH_4SCN	_____
Addition of $K_4Fe(CN)_6$	_____

D. Sodium Bicarbonate plus Hydrochloric Acid

TABLE 4.7 (data)

	Unreacted Solid	Recovered Solid
Color of Sample	_____	_____
Result of adding Ca(NO₃)₂ to dissolved sample	_____	_____

TABLE 4.8 (report)

Test or Observation	Is a Composition Change Indicated?
Sample color	_____
Addition of Ca(NO₃)₂	_____

E. Combustion of Magnesium

TABLE 4.9 (data)

	Before Combustion	After Combustion
Appearance of sample	_____	_____
Results of adding HCl to sample	_____	_____

TABLE 4.10 (report)

Test or Observation	Is a Composition Change Indicated?
Sample appearance	_____
Addition of HCl	_____

TABLE 4.11 (report)

Process Studied	Type of Change
A. Solution Formation	_____
B. Heating Copper Carbonate	_____
C. Melting Iron Chloride	_____
D. Sodium Bicarbonate plus Hydrochloric Acid	_____
E. Combustion of Magnesium	_____

QUESTIONS

1. Certain popular antacids are dissolved in water (with much fizzing) and are taken in the form of a solution. When the antacid and water are mixed, what kind of change occurs?

   a. Physical   b. Chemical   c. Can't tell from the information given.

   Explain your answer: _____

   _____

2. A sample of solid potassium chlorate is heated strongly for 30 minutes. The sample melts, and bubbles of gas escape during the heating. Upon cooling, the heated sample becomes solid. The solid is white in color before and after heating. A sample of unheated solid is dissolved in water and 10 drops of silver nitrate are added. Nothing happens. A sample of the heated solid is also dissolved in water and treated with 10 drops of silver nitrate. The solution becomes very cloudy. What classification should be given to the change that occurs when potassium chlorate is heated?

    a. Physical   b. Chemical   c. Can't be determined from the observations and tests.

    Explain your answer: _____

    _____

3. Oxygen gas dissolved in water can be used by fish, which remove it from the water by means of their gills. Classify the change that occurs when oxygen dissolves in water.

    a. Physical   b. Chemical   c. Can't tell from the information given.

    Explain your answer: _____

    _____

4. In Part D of the experiment, a large amount of fizzing took place when HCl was added to solid $NaHCO_3$. Is this observation consistent with the type of change listed for Part D in Table 4.11?

    a. Yes   b. No   c. Can't be determined from the observation.

    Explain your answer: _____

    _____

5. Seawater can be made into drinking water by boiling the seawater, collecting the steam, and condensing the steam to liquid water. Classify the changes in water that occur when it is boiled to steam, then condensed to water.

    a. Both are physical.   b. One is physical, and one is chemical.
    c. Both are chemical.   d. Can't be determined from the information given.

    Explain your answer: _____

    _____

    _____

Experiment 5

Separations and Analysis

In this experiment, you will

- Use physical properties to separate the components of mixtures.
- Use chemical properties to verify that separation took place.

- Use paper chromatography to separate and identify the components in an unknown.

INTRODUCTION

The analysis of materials is one of the most well-known and widely used applications of chemistry. Sometimes analysis is used to determine the identity of a substance. This type of determination is called **qualitative analysis**. **Quantitative analysis** is used when the amount of a substance present in a sample is determined. It is sometimes necessary to separate the components of a mixture from each other before either qualitative or quantitative analysis can be done.

In this experiment, you will use physical properties to separate several simple mixtures into their components. You will use chemical tests to do qualitative analysis of some of the separated substances in order to verify that separation took place.

You will use plastic droppers quite often in this experiment, and it is necessary that they be clean when used. One way to do this is to put about 150 mL of distilled water into a 400-mL or larger beaker. Then, squeeze the bulb of each dropper, put the tip under the water in the beaker, and release the bulb. This will draw distilled water into each dropper. Leave the droppers stored this way until needed. When a clean dropper is needed, remove a dropper from the storage beaker and expel the water from it into the sink. Then shake the dropper to remove the last few drops of water. When you are finished using a dropper, expel the liquid it contains into an appropriate container or the sink (if appropriate). Then squeeze the bulb, put the tip under the water in the storage beaker, release the bulb, and allow it to partially fill with water. Without squeezing the bulb, remove the dropper from the storage beaker, shake it so water rinses the entire inside of the bulb, and expel the water into the sink. Repeat this rinsing procedure twice, and then store the dropper in the beaker as described earlier.

EXPERIMENTAL PROCEDURE

A. Sublimation

Sublimation is the process in which a solid changes directly to the gaseous state without first melting to a liquid. Solid carbon dioxide (dry ice) is a familiar example of a solid that undergoes this process.

In this part of the experiment, you will use sublimation to separate the components of a mixture that contains solid calcium carbonate and solid ammonium chloride. One of the solids sublimes when it is heated. You will heat the mixture, condense a part of the sublimed substance back into a solid, and test the collected solid to identify it as calcium carbonate or ammonium chloride

PROCEDURE

1. Use a weighing dish and a centigram balance to weigh out approximately 1.0 g of a mixture that contains solid calcium carbonate ($CaCO_3$) and ammonium chloride (NH_4Cl). Return to your work area and put the weighed sample into a 50-mL beaker.

FIGURE 5.1
Sublimation apparatus

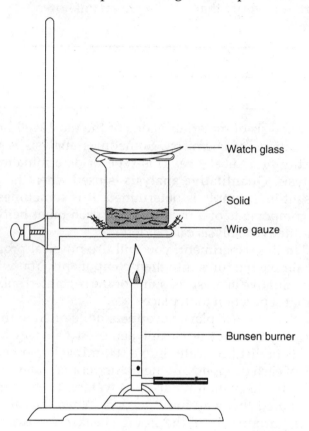

Watch glass

Solid

Wire gauze

Bunsen burner

2. Assemble a ringstand, burner, and your beaker that contains the sample, as shown in Figure 5.1. Be sure to cover the mouth of the beaker with a watch glass as shown.
3. Heat the beaker with a moderate flame from your burner for a minimum of 5 minutes, or until a significant amount of solid collects on the bottom of the watch glass that covers the beaker. While the heating is going on, you can continue with Steps 4, 5, and 6.
4. Go to the reagent area and use your spatula to put small samples of the following solids into separate dry, small (10 cm) test tubes: calcium carbonate, ammonium chloride, and the mixture of calcium carbonate and ammonium chloride. Each sample should cover up the bottom of the test tube to a depth of about 2 mm.
5. Test each of the three solid samples by cautiously adding 10 drops of dilute (6 M) hydrochloric acid (HCl) to each test tube. Observe what happens to each sample while the acid is being added.

SAFETY ALERT

Hydrochloric acid is very corrosive. If you get any on your skin, wash it off immediately with cold water and inform your instructor.

6. Record in Table 5.1 the behavior of each solid sample as the acid was added.

7. After the heating of the beaker is completed, allow the beaker and watch glass to cool for a minimum of 10 minutes.

8. When the cooling is completed, carefully remove the watch glass from the beaker and use your spatula to scrape the solid from the watch glass onto a clean piece of filter paper. Transfer the collected solid to a dry, small (10 cm) test tube and test it by the method described in Step 5. Record the test results in Table 5.1.

DISPOSAL All test tube samples in sink
Surplus solids in wastebasket

B. Solubility

The **solubility** of a substance in a specific solvent depends on a number of factors such as temperature, the nature of the solute, and the nature of the solvent. In this part of the experiment, you will use the difference in solubility of two solid substances in water and the technique of filtration to separate the substances. Chemical tests will be used to differentiate between the substances.

PROCEDURE

1. Use a weighing dish and a centigram balance to weigh out approximately 1.0 g of a mixture that contains solid calcium carbonate ($CaCO_3$) and solid copper sulfate ($CuSO_4$).

2. Put the sample into a 50-mL beaker and use a graduated cylinder to add about 10 mL of distilled water. Swirl to mix the materials, then set the beaker aside while you go on to Step 3.

3. Fold a filter paper twice and open it into a cone as illustrated in Figure 5.2. Place the filter paper into a funnel that is mounted on a ringstand (Figure 5.2).

FIGURE 5.2
Preparation of filter paper and funnel for filtration.

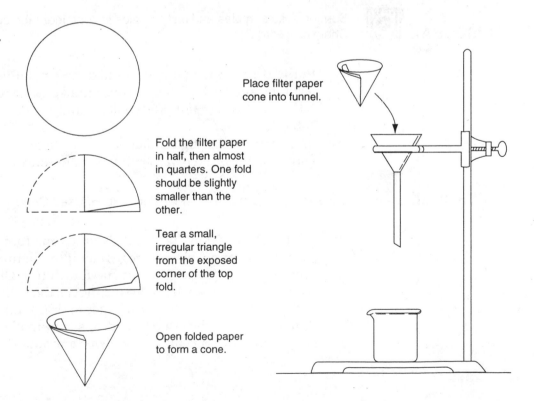

Fold the filter paper in half, then almost in quarters. One fold should be slightly smaller than the other.

Tear a small, irregular triangle from the exposed corner of the top fold.

Open folded paper to form a cone.

Place filter paper cone into funnel.

4. Swirl the mixture of water and solid in the 50-mL beaker until it appears that one of the solids has dissolved completely. Moisten the filter paper in the funnel with a little distilled water from your wash bottle. Swirl the beaker again to suspend the solid, then quickly pour the contents into the center of the filter paper in the funnel. Collect the liquid that passes through the filter in a 100 mL-beaker. The liquid is called the filtrate.

5. While the filtrate is draining from the funnel, take three pieces of filter paper to the reagent area and use your spatula to obtain small individual samples of solid calcium carbonate ($CaCO_3$), solid copper sulfate ($CuSO_4$), and the mixture that contains the two solids. The samples should cover an area about the size of a 10-cent coin, and should not be a pile. Be careful not to get samples that are too large.

6. Return to your work area and divide each of the solid samples into two approximately equal smaller samples. Put each of the smaller samples into separate small (10-cm) test tubes. Add 3 drops of dilute (6 M) hydrochloric acid (HCl) to one sample of each of the solids. Record your observations in Table 5.3. To the remaining sample of each solid, add 10 drops of distilled water. Agitate each test tube to dissolve or suspend the solid, add 4 drops of concentrated aqueous ammonia (NH_3 or NH_4OH), and agitate the test tube to mix the contents.

SAFETY ALERT

> Both dilute hydrochloric acid and concentrated aqueous ammonia are very corrosive. If any gets on your skin, wash it off immediately with cold water and inform your instructor.

Record your observations in Table 5.3.

DISPOSAL Surplus solid samples and test samples in container labeled "Exp. 5, Used Chemicals Part B."

7. Put 10 drops of the filtrate collected in Step 4 into a small (10-cm) test tube and add 4 drops of concentrated aqueous ammonia (NH_3 or NH_4OH). Mix the test tube contents and record your observations in Table 5.3.

8. Empty the beaker that contains the filtrate into the proper container and rinse the beaker well with tap water followed by distilled water.

DISPOSAL Collected filtrate in container labeled "Exp. 5, Used Chemicals Part B."

9. Put the clean 100-mL beaker back under the funnel, and wash the solid in the funnel by spraying water into the funnel from your wash bottle until the funnel is about one-fourth full. Direct the stream of water from your wash bottle so it moves the solid toward the center of the funnel. Allow the wash water to drain until it is almost all drained. Collect the last 5 to 10 drops in a small clean test tube and add 4 drops of concentrated aqueous ammonia. Note the results of

this test, but do not record it. Wash the solid once again with water from your wash bottle, and test the last 5 to 10 drops that drain out with concentrated aqueous ammonia. If the test indicates that all of the dissolved material has not been washed from the solid in the funnel, repeat the washing and testing of filtrate one more time.

10. Use your spatula to remove a small sample of moist solid from the funnel. Put the sample on a watch glass. Test the sample by adding 3 drops of dilute (6 M) hydrochloric acid (HCl) to it. Record your observations in Table 5.3.

DISPOSAL

Filter paper with solid in wastebasket
Tested samples in container labeled "Exp. 5, Used Chemicals Part B."

C. Paper Chromatography

The separation technique known as **chromatography** was first used in the early part of the twentieth century. The word *chromatography* comes from two Greek words that mean "colored graph," a good description of the earliest applications of the technique to separate colored plant pigments.

All chromatographic methods involve the use of the same general principle. A mixture of dissolved solutes that is to be separated is passed over a stationary material. The moving phase in which the solutes are dissolved is called the **mobile phase,** and the stationary material is called the **stationary phase.** Separation occurs when the dissolved solutes have different affinities for the mobile and stationary phases. A solute that is strongly attracted to the stationary phase spends more time attached to that phase than a solute that is strongly attracted to the mobile phase. As a result, the solute with a stronger affinity for the stationary phase spends less time in the moving mobile phase and does not move along as rapidly as the solute that is attracted more strongly to the mobile phase. Thus, the solutes move different distances and are separated during the time the mobile phase is moving through the stationary phase.

In this part of the experiment, the stationary phase will be paper, and the mobile phase will be acetone with a little hydrochloric acid and nitric acid added.

The extent of separation of solutes in a chromatogram is expressed by the R_f value of each solute. The R_f (retention factor) value is calculated by dividing the distance traveled by the mobile phase into the distance traveled by the solute:

$$R_f = \frac{\text{Solute distance}}{\text{Mobile phase distance}}$$

Eq. 5.1

The distances used in Equation 5.1 must be measured from the same starting point, which is usually the point at which the solute sample was applied to the paper. These distances are illustrated in Figure 5.3.

FIGURE 5.3
Measurement of
distances traveled during
chromatography

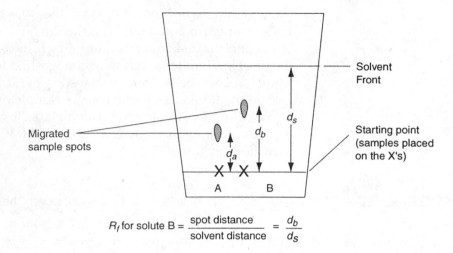

$$R_f \text{ for solute B} = \frac{\text{spot distance}}{\text{solvent distance}} = \frac{d_b}{d_s}$$

In this part of the experiment, you will separate the ions Fe^{3+}, Co^{2+}, and Cu^{2+} that are produced in solution when the salts $FeCl_3$, $CoCl_2$, and $CuCl_2$, respectively, are dissolved in water. You will also determine which of the three ions are present in an unknown solution.

PROCEDURE

1. Make six spotting capillaries by heating the middle of glass melting-point capillaries in the flame of a burner until the glass softens. Then stretch the softened glass to form a narrow neck. Bend the narrowed glass until it breaks to form the tip of the spotting capillary. Your instructor will demonstrate the technique.

SAFETY ALERT

> Handle the hot glass carefully to avoid burning yourself.

2. Obtain a sheet of chromatography paper from the reagent area.
3. Obtain a ruler from the stockroom, measure the chromatography paper, then trim it with a pair of scissors so the dimensions match those given in Figure 5.4(a). Note that the paper is wider at the top than at the bottom.

FIGURE 5.4
Preparation of
chromatography paper
(not actual size)

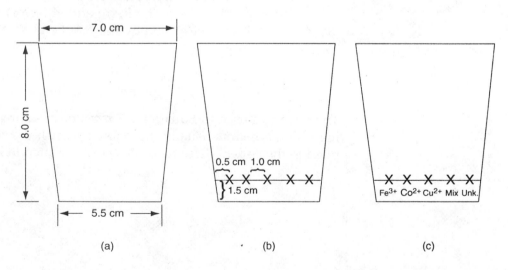

4. Use a pencil (not a pen) to draw a line across the sheet of paper 1.5 cm from the bottom as shown in Figure 5.4(b).

5. Measure from the edge of the paper and use a pencil to mark an X on the line of the paper at the following distances from the edge: 0.5 cm, 1.5 cm, 2.5 cm, 3.5 cm, and 4.5 cm. Note that this can be easily done by measuring 0.5 cm from the edge for the first X, then measuring 1.0 cm from that X to the next one, and so on.

6. Take a clean glass plate or a plastic weighing dish to the reagent area. Put 1-drop samples of the following solutions on separate parts of the glass plate or weighing dish. Make certain none of the samples mix with neighboring samples. 0.2 M $FeCl_3$, 0.2 M $CoCl_2$, 0.2 M $CuCl_2$, and a solution labeled *mixed ions* that contains all three compounds. Identify the location of various samples so you can remember which sample is which.

7. Use one of the spotting capillaries you prepared in Step 1 and practice spotting samples by dipping the narrow end of the capillary into a sample of distilled water formed by putting 2 or 3 drops on a glass plate or plastic weighing dish. After dipping the capillary into the water sample, touch the narrow tip of the capillary to a piece of filter paper. Water will flow from the capillary to the filter paper. Practice until you can consistently form spots that are 3-4 mm in diameter. Be sure they are *no* more than 4 mm in diameter.

8. Put one of the remaining five capillaries near each of the four samples on your glass plate or plastic weighing dish obtained in Step 6. You will use a separate and different capillary for each sample when you spot the chromatography paper in Step 10.

9. Obtain a sample of unknown solution from the stockroom. Record the identification number of the unknown in Table 5.6. You will use the remaining capillary with this unknown.

10. Use a pencil to label under the X's on the sheet of chromatography paper, as shown in Figure 5.4(c). Spot the appropriate sample on each X of the paper, being sure to use a different capillary for each sample you spot. Make the spots 3-4 mm in diameter. Allow the spots to dry for at least 5 minutes.

11. While the sample spots are drying, prepare a chromatography chamber as follows: Use your 10-mL graduated cylinder to measure 3.0 mL of acetone. Put the acetone into a clean, dry 250-mL beaker and cover the top of the beaker with a watch glass.

SAFETY ALERT

> Acetone is very flammable. Do not allow an open flame to come near the acetone-containing beaker.

12. Add to the acetone in the beaker 10 drops of dilute (6 M) nitric acid (HNO_3) and 5 drops of dilute (6 M) hydrochloric acid (HCl). Swirl the beaker to mix the contents and replace the watch glass on the top.

13. Put the beaker into a location where it can remain undisturbed. Carefully put the spotted paper into the beaker with the spotted end down. The paper should rest evenly on the bottom of the beaker. It might be necessary to push the paper down into the beaker to position it properly. Do this carefully and quickly so the liquid is not splashed onto the paper, but begins to move up the paper uniformly. The liquid level in the beaker should be below the line on the chromatography paper. Replace the watch glass. Your chamber and paper should now look like Figure 5.5, except the migrating solute spots will be very faint and difficult to see.

FIGURE 5.5
A developing
chromatogram

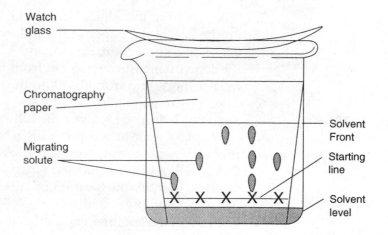

14. Allow the chromatogram to develop undisturbed until the solvent front gets to within about 1 cm of the top of the paper. This will take 10 to 15 minutes. When the solvent reaches the appropriate level, remove the paper from the beaker and use a pencil to mark the position of the solvent front. Dispose of the chromatography solvent.

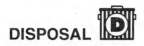

DISPOSAL Chromatography solvent in container labeled "Exp. 5, Used Chromatography Solvent."

15. The spots on the your chromatogram are made visible by exposing them to gaseous ammonia. This is to be done ***under a hood*** by pouring about 10 mL of concentrated aqueous ammonia (NH_3 or NH_4OH) into a 100-mL beaker, then laying your chromatogram across the top of the beaker. Each of the three ions forms a characteristic color when exposed to ammonia. However, some will fade quickly when removed from the ammonia; so after all spots become visible, be sure to circle each one with a pencil. It might be necessary to move the chromatogram around on the top of the beaker in order to get all the spots exposed to the gaseous ammonia.

16. After the spots are made visible and marked, use a ruler to measure the distance from the starting line to the center of each spot and from the starting line to the line that marks the final location of the solvent front. Record these distances in Table 5.5, using the proper number of significant figures. Note that your unknown might not contain all three ions and so will not produce three spots. Use only as many blanks for the unknown sample in Table 5.5 as you need.

DISPOSAL Chromatography samples and unknown in container labeled "Exp. 5, Used Chromatography Unknowns".

DISPOSAL Used spotting capillaries should be put into a special broken-glass container. Check with your instructor.

DISPOSAL Used concentrated aqueous ammonia in container labeled "Exp. 5, Used Ammonia."

A. Sublimation

1. Review the observations you recorded in Table 5.1 and decide which of the two solids in the mixture (calcium carbonate or ammonium chloride) sublimed when it was heated. Record your conclusion in Table 5.2.
2. In the blanks provided in Table 5.2, briefly explain how you used the observations recorded in Table 5.1 to arrive at your conclusion about the identity of the material that sublimed.

B. Solubility

1. Review the observations you recorded in Table 5.3 and decide which of the two solids in the mixture (calcium carbonate or copper sulfate) was soluble in water. Record your conclusion in Table 5.4.
2. In the blanks provided in Table 5.4, briefly explain how you used the observations recorded in Table 5.3 to identify the soluble solid and the insoluble solid.

C. Paper Chromatography

1. Use the distances recorded in Table 5.5 and Equation 5.1 to calculate the R_f value for each spot on the chromatogram, including those for the unknown. Record the calculated R_f values in Table 5.6.
2. Use the calculated R_f values to determine which ions are present in your unknown. Record your conclusion in Table 5.6.
3. Attach your chromatogram to the report sheet in the space provided.

Experiment 5 ▪ Pre-Lab Review

Separations and Analysis

1. Are any specific safety alerts given in the experiment? List any that are given.

   _____

   _____

2. Are any specific disposal directions given in the experiment? List any that are given.

   _____

   _____

   _____

3. Describe how to maintain a supply of clean droppers available for use in the experiment.

   _____

   _____

   _____

   _____

4. What is the approximate size of the sample of mixture you obtain for separation in Parts A and B?

   _____

5. What is the minimum time you should heat the sample you are to separate in Part A?

   _____

6. What is the name of the technique used to separate the substances that have different solubilities in water?

   _____

7. How many spotting capillaries do you prepare for use in Part C? _____

8. What size sample spots should be put on the chromatography paper in Part C?

   _____

9. Describe how you make the spots on your chromatogram visible in Part C.

_____

_____

10. What do you do with the chromatogram in Part C after it has been developed and the distances measured?

_____

_____

Experiment 5 ▪ Data & Report Sheet

Separations and Analysis

A. Sublimation

TABLE 5.1 (data)

Sample Tested	Behavior when HCl was added
Solid CaCO$_3$	_____
Solid NH$_4$Cl	_____
Mixture	_____
Collected solid	_____

TABLE 5.2 (report)

Identity of solid
that sublimed _____

Explanation _____

_____

_____

_____

_____

B. Solubility

TABLE 5.3 (data)

Sample Tested	Observations	
	Add HCl	Add NH$_3$
Solid CaCO$_3$	_____	_____
Solid CuSO$_4$	_____	_____
Mixture	_____	_____
Filtrate	_____	_____
Solid from funnel	_____	_____

TABLE 5.4 (report)

Identity of water-soluble solid _____

Explanation _____

_____

_____

C. Paper Chromatography

TABLE 5.5 (data)

Sample	Distance
Fe^{3+}	
Co^{2+}	
Cu^{2+}	
Mixture	
Unknown	
Solvent	

TABLE 5.6 (report)

Sample	R_f
Fe^{3+}	
Co^{2+}	
Cu^{2+}	
Mixture	
Unknown	
ID number of unknown	
Ions present in unknown	

Attach Chromatogram Below

QUESTIONS

1. A student performs the HCl tests on the three solid samples as described in Step 5 of Part A and gets the same result for the solid $CaCO_3$ sample and the mixture. Which of the following statements is true concerning these results?

    a. The results are to be expected.
    b. The results are not expected and could have been caused by using contaminated HCl.
    c. The results are not expected and could have been caused by using contaminated samples.

    Explain your answer: _____

    _____

    _____

2. In Part B, chemical tests were used to differentiate between the two substances in the mixture. Which of the following physical properties could also have been used conveniently?

    a. Density   b. Color   c. Hardness

    Explain your answer: _____

    _____

    _____

3. A student performs Part B, but instead of discarding the filtrate, the student heats it carefully and eventually ends up with a small amount of solid in the beaker. What is the solid?

    a. $CaCO_3$ b. $CuSO_4$ c. A mixture of $CaCO_3$ and $CuSO_4$

    Explain your answer: _____

    _____

    _____

4. According to the chromatogram you prepared in Part C, which of the three ions you separated had the stronger affinity for the paper stationary phase?

    a. Fe^{3+} b. Co^{2+} c. Cu^{2+}

    Explain your answer: _____

    _____

    _____

5. A student collects chromatographic data as you did in Part C and calculates an R_f value of 1.4 for one of the solute spots. Which of the following is true?

   a. The calculated R_f value might be correct.
   b. The calculated R_f value is probably incorrect.
   c. The calculated R_f value is definitely incorrect.

   Explain your answer: _____

   _____

   _____

6. A student separated an unknown mixture of ions using the procedure of Part C. When the spots were made visible, a faint spot corresponding to Cu^{2+} was seen. However, the student later learned that the unknown did not contain Cu^{2+}. Which of the following could cause this result?

   a. The same spotting capillary was used to spot the Cu^{2+} sample and the unknown sample.
   b. The Cu^{2+} sample was contaminated with Co^{2+} and Fe^{3+}.
   c. The chromatography paper was not trimmed to the correct dimensions.
   d. The position of the solvent front was incorrectly marked.

   Explain your answer: _____

   _____

   _____

Classification of Chemical Reactions

In this experiment, you will

- Cause chemical reactions of the following types to take place: combination, decomposition, single replacement, and double replacement.

- Identify the products of some of the reactions.
- Write balanced equations for the reactions.

INTRODUCTION

The study of chemical reactions is made easier by classifying the reactions according to some specific characteristics. Numerous classification schemes are possible. For example, reactions could be classified according to the physical state (gas, g; liquid, l; or solid, s) of the reacting materials. In the scheme used in this experiment, reactions are classified according to relationships that exist between the reactants and products. On this basis, reactions are classified as combinations, decompositions, single replacements, or double replacements.

In order to properly classify the reactions, it is necessary to make observations of the reactions as they proceed. Observations are the things we see, hear, smell or otherwise detect by using our senses. Thus, colors, odors, the evolution of gases (fizzing, bubbling, etc.), and flammability can all be determined on the basis of observations. As you do this experiment and make observations of the reactions, try to observe as many behaviors, etc. as you can and avoid the common mistake of focusing on just a single one. For example, when the glowing splint tests are done in Part B, some students will record only that behavior as their observations and ignore other observations that are equally as important.

EXPERIMENTAL PROCEDURE

A. Combination Reactions

Combination reactions are also called addition or synthesis reactions because the characteristic relationship between reactants and products is that two or more reactants (written on the left of chemcial equations) react to form a single product (written on the right of chemical equations). The general form of a combination reaction is

$$A + B \rightarrow C \qquad \text{Eq. 6.1}$$

A specific example is the reaction that occurs when sulfur burns. The melted sulfur combines with gaseous oxygen from the air to form gaseous sulfur dioxide:

$$S(l) = O_2(g) \rightarrow SO_2(g) \qquad \text{Eq. 6.2}$$

PROCEDURE: Magnesium Metal

1. Obtain a short strip of magnesium ribbon (Mg). Record the appearance of the metal in Table 6.1.
2. Grasp one end of the metal strip with your crucible tongs and hold the strip in the flame of your burner until the magnesium ignites.

SAFETY ALERT

Do not look directly at the burning metal. This can injure your eyes.

3. Collect the combustion product on a piece of filter paper (exclude any unburned metal) and record its appearance in Table 6.1.

DISPOSAL

Burned and unburned metal in wastebasket

PROCEDURE: Hydrogen Chloride and Ammonia

Liquid hydrochloric acid and aqueous ammonia commonly used in the laboratory are actually water solutions of hydrogen chloride gas (HCl) and ammonia gas (NH_3), respectively. A significant amount of gas escapes from concentrated solutions and can be detected easily because both gases cause choking and have unpleasant orders.

SAFETY ALERT

Because of the strong odors and hazards associated with breathing these gases, most of the following experiment will be done in a fume hood. The necessary chemicals will be found in the hood.

1. In a fume hood, put 5 drops of concentrated hydrochloric acid (HCl) into a 7.5-cm test tube (your smallest test tube).
2. In a fume hood, put 5 drops of concentrated aqueous ammonia (NH_3 or NH_4OH) into a second 7.5-cm test tube.
3. For a short time, remove the test tubes from the hood. Initially hold the test tubes about 1 foot apart, then bring them together so the test tube mouths are as close as possible to each other and tilted toward each other.

SAFETY ALERT

Do not mix the liquid contents of the test tubes. A rapid hazardous reaction results when they are mixed.

Observe what happens when the mouths are close together and record your observations in Table 6.1. Return the test tubes to the hood.

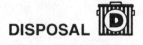

DISPOSAL

The liquid contents of each test tube should be poured separately into a sink with the cold water running. If possible, the sink should be located inside a hood.

B. Decomposition Reactions

The relationship between reactants and products in **decomposition reactions** is that a single reactant forms two or more products. The general form of a decomposition reaction is

$$A \rightarrow B + C \qquad \text{Eq. 6.3}$$

A specific example is the reaction used to produce calcium oxide (CaO) (also called lime) by heating calcium carbonate ($CaCO_3$) (limestone).

$$CaCO_3(s) \rightarrow CaO(s) + CO_2(g) \qquad \text{Eq. 6.4}$$

The CO_2 product of Equation 6.4 is a gaseous material, while $CaCO_3$ and CaO are solids. Gaseous products will be formed in both reactions of this part of the experiment. A simple test will help you identify two common gaseous products, oxygen (O_2) and carbon dioxide (CO_2). A glowing wooden splint inserted into the mouth of a test tube containing the gas will behave differently depending on the identity of the gas. The splint is lighted in a flame, then carefully blown out to leave a glowing end. The glowing end is then slowly inserted about 3 to 4 cm into the test tube. If the gas is oxygen, the splint will glow brighter, might burst into flame or pop. If the gas is carbon dioxide, the splint will go out immediately upon being inserted.

PROCEDURE: Potassium Chlorate

1. Put about one-half gram (centigram balance) of solid potassium chlorate ($KClO_3$) into a dry 10-cm test tube. Pure $KClO_3$ is white, but the material you will use has some red-orange or black iron oxide added to speed up the decomposition reaction. The iron oxide is not changed by the process that occurs (it acts as a catalyst), and it is not included in the balanced equation for the reaction.

FIGURE 6.1
Mounted test tube

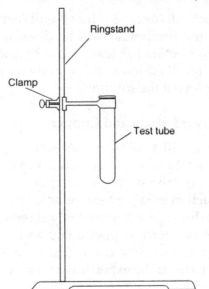

Ringstand

Clamp

Test tube

2. Mount the test tube containing the solid sample on a ringstand as shown in Figure 6.1.

3. Heat the test tube steadily with a moderate flame of a burner. The mixture will begin to agitate and then will melt. Wait about 10 seconds after the melting occurs, then continue the heating and test the gas produced with a glowing wooden splint. Record the results of the glowing splint test in Table 6.2 along with other observations. Continue the heating for 1 minute after you get results with the glowing splint that will let you correctly identify the gas.

SAFETY ALERT

Handle the hot test tube with care to avoid burning yourself.

DISPOSAL

Used $KClO_3$ in sink. Allow the test tube to cool, then half-fill it with tap water, and use a stirring rod or spatula to dislodge and suspend the solid. Pour the resulting suspension into the sink and flush it down the drain with water.

PROCEDURE: Hydrogen Peroxide

1. Put 7 drops of yeast suspension into a clean 10-cm test tube. The yeast will act as a catalyst in the decomposition reaction of hydrogen peroxide and will not be included in the balanced equation for the reaction.
2. Add 10 drops of 3% hydrogen peroxide to the test tube and agitate to mix the contents.
3. Allow the reaction to proceed for at least 1 minute, then test the evolved gas with a glowing splint. Note, the glowing splint test will generally work correctly only once for this reaction. So, if you want to repeat the test, you should begin with Step 1.
4. Record the results of the glowing splint test along with other observations in Table 6.2.

DISPOSAL 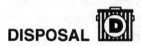 Used yeast and hydrogen peroxide mixture in sink

C. Single-Replacement Reactions

Single-replacement reactions are sometimes called substitution reactions because one element reacts with a compound and in the process displaces another element from the compound and takes its place. This results in a different element and compound as the reaction products. The general reaction is

$$A + BX \rightarrow B + AX \qquad \text{Eq. 6.5}$$

In this reaction, A and B are elements, and BX and AX are compounds. A specific example is the reaction used to produce iron (Fe) from its ore (Fe_2O_3):

$$3C(s) + 2Fe_2O_3(s) \rightarrow 4Fe(l) + 3CO_2(g) \qquad \text{Eq. 6.6}$$

In the reactions you will observe, a metal will displace another metal from a compound, but the compound will be dissolved in water. You will be able to verify that a reaction has taken place by changes you observe in the appearance of the metal added to the solution of compound in water and the appearance (color) of the solution.

PROCEDURE: Silver Nitrate and Copper

1. Put 10 drops of 0.10 M silver nitrate solution ($AgNO_3$) into a 7.5-cm test tube and set the test tube in a small beaker or your test tube block.
2. Put a small piece (about 2 cm) of copper wire (Cu) into the test tube.
3. Allow the reaction to take place for at least 30 minutes. Look at it several times during the 30 minutes and record the observed changes in the appearance of the copper metal and the solution (color) in Table 6.3. Observe solution colors against a white background. You can do other parts of the experiment while the reaction proceeds.

DISPOSAL 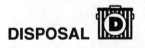 Test tube contents in container labeled "Exp. 6, Used Chemicals Part C."

PROCEDURE: Copper Nitrate and Zinc

1. Put 10 drops of 0.10 M copper nitrate solution ($Cu(NO_3)_2$) into a 7.5-cm test tube and set the test tube in a small beaker or your test tube block.
2. Put a small piece of zinc metal (Zn) into the test tube.

3. Allow the reaction to take place for at least 30 minutes, while observing it several times. Record any observed changes in the metal and the solution in Table 6.3. You can go on to Part D while the reaction proceeds.

DISPOSAL  Test tube contents in container labeled "Exp. 6, Used Chemicals Part C."

D. Double-Replacement Reactions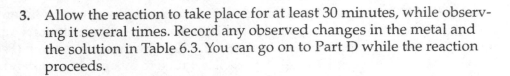

Double-replacement reactions (also called metathesis reactions) often take place between substances that are dissolved in water. The reactions can be described as partner swapping because the reacting substances trade partners to form the products. The general reaction is

$$AX(aq) + BY(aq) \rightarrow AY(aq) + BX(aq) \qquad \text{Eq. 6.7}$$

In this reaction, the *aq* in parentheses indicates that the substances are dissolved in water (aqueous). A specific example is the reaction of dissolved barium chloride ($BaCl_2$), with dissolved sodium sulfate (Na_2SO_4), to give insoluble barium sulfate ($BaSO_4$), which settles out of the solution as a solid, and dissolved sodium chloride (NaCl):

$$BaCl_2(aq) + Na_2SO_4(aq) \rightarrow BaSO_4(s) + 2NaCl(aq) \qquad \text{Eq. 6.8}$$

The double-replacement reactions you will study in this part of the experiment all take place between compounds dissolved in water. Because the reactants are all dissolved in water, it is difficult to be certain that a reaction takes place unless one or both of the products can be detected. Detection is possible if one or more of the products is insoluble in water (like $BaSO_4$ in Equation 6.8). When a reaction product is insoluble in water, it forms a solid (a precipitate) that makes the solution cloudy when the reactants are mixed. Some solids will settle from the liquid and become visible on the bottom of the container. Others are gel-like and may remain in suspension or be a little difficult to see even when they have settled to the bottom of the container.

PROCEDURE

1. Put 4 drops of 1 M sodium hydroxide solution (NaOH) into each of three separate, clean 7.5-cm test tubes. Number the test tubes 1, 2, and 3.
2. Add 4 drops of 0.10 M copper nitrate solution ($Cu(NO_3)_2$) to test tube 1, 4 drops of 0.10 M silver nitrate solution ($AgNO_3$) to test tube 2, and 4 drops of 0.10 M iron (III) nitrate solution ($Fe(NO_3)_3$) to test tube 3.
3. Agitate each test tube to mix the contents and observe the behavior of each mixture. Record your observations in the appropriate blanks of Table 6.4. Be especially careful to look for the formation of solids (precipitates). Note that the blanks in Table 6.4 are numbered to correspond to the numbers assigned to your test tubes and are arranged in a grid that indicates the contents of each test tube. Thus, Blank 3 (test tube 3) will contain a mixture of NaOH solution and $Fe(NO_3)_3$ solution.
4. Put 4 drops of 1 M sodium chloride solution (NaCl) into each of three separate, clean 7.5-cm test tubes. Number the test tubes 4, 5, and 6.

5. Add 4 drops of 0.10 M copper nitrate solution ($Cu(NO_3)_2$) to test tube 4, 4 drops of 0.10 M silver nitrate solution ($AgNO_3$) to test tube 5, and 4 drops of 0.10 M iron nitrate solution ($Fe(NO_3)_3$) to test tube 6.

6. Agitate each test tube to mix the contents and observe the behavior of each mixture. Record your observations in the appropriate blanks of Table 6.4.

7. Put 4 drops of 1 M sodium nitrate solution ($NaNO_3$) into each of three separate, clean 7.5-cm test tubes. Number the test tubes 7, 8, and 9.

8. Add 4 drops of 0.10 M copper nitrate solution ($Cu(NO_3)_2$) to test tube 7, 4 drops of 0.10 M silver nitrate solution ($AgNO_3$) to test tube 8, and 4 drops of 0.10 M iron nitrate solution ($Fe(NO_3)_3$) to test tube 9.

9. Agitate the tubes to mix the contents and record your observations in Table 6.4

DISPOSAL Test tube contents in container labeled "Exp. 6, Used Chemicals Part D." This is most easily done by first emptying the test tubes into a small beaker and emptying the beaker into the labeled container after Part D is completed.

REPORT

A. Combination Reactions

1. The product formed when magnesium metal burned was magnesium oxide (MgO). In Blank 1 in Table 6.1, write a balanced equation for the combination reaction that took place when magnesium burned. Remember, the correct formula for oxygen gas is O_2.

2. The product formed when gaseous HCl and NH_3 reacted was ammonium-chloride (NH_4Cl). In Blank 2 in Table 6.1, write a balanced equation for the combination reaction that took place between gaseous HCl and NH_3.

B. Decomposition Reactions

1. Refer to the information in Table 6.2 and identify the gas that was given off when potassium chlorate ($KClO_3$) was decomposed by heating. The other product of the decomposition was potassium chloride (KCl). In Blank 1 in Table 6.2, write a balanced equation for the decomposition reaction that took place.

2. Refer to the information in Table 6.2 and identify the gas that was given off when hydrogen peroxide (H_2O_2) decomposed. The other product of the reaction was water (H_2O). In Blank 2 in Table 6.2, write a balanced equation for the decomposition reaction that took place.

C. Single-Replacement Reactions

The single-replacement reactions of this part of the experiment took place between an element (the copper and zinc metals) and a compound dissolved in water (the silver nitrate and copper nitrate solutions). Both silver nitrate and copper nitrate are ionic compounds that dissociate to form ions when they are dissolved. Thus, a solution of silver nitrate does not contain $AgNO_3$ molecules but does contain Ag^+ and NO_3^- ions in the water solution. Such ions in solution will be designated by (aq) following their symbols. Balanced equations for reactions that involve dissolved ionic compounds can be written in several ways. Full (or molecular) equations contain the correct formulas for all compounds or elements involved in the reaction. In total ionic equations, all dissolved ionic materials are represented by the ions they form when they dissolve; materials that do not dissolve are represented by their correct formulas. In net ionic equations, any ions that appear on both sides of a total ionic equation are called spectator ions and are subtracted from both sides of the equation. All balanced equations must contain the same number of atoms of each kind on each side of the equation. Balanced total ionic or net ionic equations must also have the same total electrical charge on each side of the equation. These charges arise from the charges of the ions included in the equation.

When aluminum metal is put into a solution of zinc nitrate, a dark coating of zinc metal is deposited on the aluminum. This imples that the aluminum has replaced the zinc in the zinc nitrate compound. Thus, the reactants (left side of the equation) for this single-replacement reaction are aluminum metal and dissolved zinc nitrate. The products (right side of the equation) are zinc metal and aluminum nitrate. The blanced equation for this reaction is written in all three forms below:

Full equation: $2Al(s) + 3Zn(NO_3)_2(aq) \rightarrow 3Zn(s) + 2Al(NO_3)_3(aq)$

Total ionic: $2Al(s) + 3Zn^{2+}(aq) + 6NO_3^-(aq) \rightarrow 3Zn(s) + 2Al^{3+}(aq) + 6NO_3^-(aq)$

Net ionic: $2Al(s) + 3Zn^{2+}(aq) \rightarrow 3Zn(s) + 2Al^{3+}(aq)$

Notice that six NO_3^- ions appear on both sides of the total ionic equation. The NO_3^- ions are thus spectator ions and are left out of both sides of the equation when it is written in net ionic form. Also note that the total charges on the left and right sides of the total ionic and net ionic equations balance. In the total ionic form, the left side has 6 + charges (from the three Zn^{2+} ions) and 6 − charges (from the six NO_3^- ions). The right side also has 6 + charges (from the two Al^{3+} ions) and 6 − charges (from the six NO_3^- ions). In the net ionic form, the charges are also balanced because of the presence of three Zn^{2+} ions on the left and two Al^{3+} ions on the right.

Ionic charges are also useful for determining the formulas of ionic compounds such as $Al(NO_3)_3$ that are used to write full equations. In such compounds, the ions combine in numbers that will give an equal number of positive and negative charges in the final formula. Thus, three NO_3^- ions (a total of 3 − charges) will combine with one Al^{3+} ion (a total of 3 + charges) to give the formula of $Al(NO_3)_3$ that is used in the equation. The formulas of all ions that are involved in this part of the experiment and in Part D are given in Table 6.6.

1. Refer to the observations recorded in Table 6.3. The deposit that formed on the copper wire was metallic silver, Ag. The other reaction product was $Cu(NO_3)_2$. Use this information to write a balanced equation for the reaction in full, total ionic and net ionic forms in Blanks 1–3 in Table 6.3. See Table 6.6 for the formulas and charges of the ions involved.

2. Refer to the observations recorded in Table 6.3. The deposit that formed on the piece of zinc was metallic copper, Cu. The other reaction product was zinc nitrate, $Zn(NO_3)_2$. Use this information to write a balanced equation for the reaction in full, total ionic and net ionic forms in Blanks 4–6 in Table 6.3. See Table 6.6 for the formulas and charges of the ions involved.

D. Double-Replacement Reactions

The partner swapping of double-replacement reactions is possible because the dissolved compounds exist in solution in the form of ions. When Equation 6.8 is written in ionic form, $BaCl_2$ becomes $Ba^{2+} + 2Cl^-$, Na_2SO_4 becomes $2Na^+ + SO_4^{2-}$, and $2NaCl$ becomes $2Na^+ + 2Cl^-$. The $BaSO_4$ product is still written $BaSO_4$ because it is an insoluble solid that does not dissolve and dissociate. Thus, Equation 6.8 (which is written as a full equation) can be written as follows in total ionic and net ionic forms.

Total ionic: $Ba^{2+}(aq) + 2Cl^-(aq) + 2Na^+(aq) + SO_4^{2-}(aq) \rightarrow BaSO_4(s) + 2Na^+(aq) + 2Cl^-(aq)$

Net ionic: $Ba^{2+}(aq) + SO_4^{2-}(aq) \rightarrow BaSO_4(s)$

Note that the Na^+ and Cl^- ions in the total ionic equation were spectator ions that did not appear in the net ionic equation.

1. Refer to the observations recorded in Table 6.4. Note and record in Table 6.5 those pairs of reactants that formed a solid (precipitate).
2. For each pair of reactants noted in Step 1, write each reactant in the form of the ions it forms when dissolved in water. For example, if one of the reactants was copper nitrate, $Cu(NO_3)_2$, you would write it as $Cu^{2+} + 2NO_3^-$.
3. Keep in mind that no positive metal ions form insoluble compounds with nitrate ions, NO_3^-, and write the formulas of the solids that formed in each reaction. Do this by combining the positive metal ion from one of the reactants with the negative ion from the other reactant. Remember that the ions will combine in numbers such that the resulting formula will contain an equal number of positive and negative charges. Refer to Table 6.6 for formulas and charges of the ions involved.
4. In the blanks in Table 6.5, write full, total ionic and net ionic equations for the reactions that formed each solid whose formula you wrote in Step 3.

Experiment 6 ▪ Pre-Lab Review

Classification of Chemical Reactions

1. Are any specific safety alerts given in the experiment? List any that are given.

2. Are any specific disposal directions given in the experiment? List any that are given.

3. In which part of the experiment do you react two gases with each other? What are the gases?

4. When gases are liberated in Part B, what simple test allows you to differentiate between oxygen gas and carbon dioxide gas?

5. In which reactions are catalysts used to speed the reactions? List the catalyst for each reaction.

6. Which reactions must be allowed to go on for at least 30 minutes? _____

7. In which part of the experiment is the formation of solids (precipitates) used as an indication that a reaction has taken place?

8. In which part of the experiment are you instructed to label your test tubes with numbers?

Experiment 6 ▪ Data & Report Sheet

Classification of Chemical Reactions

A. Combination Reactions

TABLE 6.1 (data)

Samples Reacted	Observations
Mg (combustion)	
HCl and NH$_3$	
Blank 1 (report)	
Blank 2 (report)	

B. Decomposition Reactions

TABLE 6.2 (data)

Samples Reacted	Observations
KClO$_3$	
H$_2$O$_2$	
Blank 1 (report)	
Blank 2 (report)	

C. Single-Replacement Reactions

TABLE 6.3 (data)

Samples Reacted	Observations
$AgNO_3$ and Cu	
$Cu(NO_3)_2$ and Zn	
Blank 1 (report)	
Blank 2 (report)	
Blank 3 (report)	
Blank 4 (report)	
Blank 5 (report)	
Blank 6 (report)	

D. Double-Replacement Reactions

TABLE 6.4 (data)

	NaOH		NaCl		$NaNO_3$	
$Cu(NO_3)_2$	1.		4.		7.	
$AgNO_3$	2.		5.		8.	
$Fe(NO_3)_3$	3.		6.		9.	

TABLE 6.5 (report)

Reactants That Formed a Solid	Ionic Form of Reactants	Formula of Solid That Formed
_____	_____	_____
_____	_____	_____
_____	_____	_____
_____	_____	_____

Blank 1 (report) _____

Blank 2 (report) _____

Blank 3 (report) _____

Blank 4 (report) _____

Blank 5 (report) _____

Blank 6 (report) _____

Blank 7 (report) _____

Blank 8 (report) _____

Blank 9 (report) _____

Blank 10 (report) _____

Blank 11 (report) _____

Blank 12 (report) _____

TABLE 6.6 (report)

Ion Formula	Ion Name	Ion Formula	Ion Name
Ag^+	Silver ion	Na^+	Sodium ion
Cu^{2+}	Copper ion	Cl^-	Chloride ion
Fe^{3+}	Iron ion	OH^-	Hydroxide ion
Zn^{2+}	Zinc ion	NO_3^-	Nitrate ion

QUESTIONS

1. A student carries out a reaction in which a single reactant gives products. Which of the following statements is true?

   a. The reaction might be a combination reaction.
   b. The reaction might be a decomposition reaction.
   c. The reaction might be a single-replacement reaction.
   d. The reaction might be a double-replacement reaction.
   e. More than one response is correct.

   Explain your answer: _____

   _____

   _____

2. Solid copper carbonate, $CuCO_3$, is heated strongly in a test tube. The solid is observed to agitate vigorously and change color from gray-green to black during the heating. A glowing splint immediately goes out when inserted into the mouth of the test tube during the heating. Which of the following equations for the decomposition reaction that took place is consistent with the observations?

   a. $CuCO_3(s) \rightarrow CuCO(s) + O_2(g)$
   b. $2CuCO_3(s) \rightarrow 2Cu(s) + 2C(s) + 3O_2(g)$
   c. $CuCO_3(s) \rightarrow CuO(s) + CO_2(g)$

   Explain your answer: _____

   _____

   _____

3. In the following total ionic equation, which ions are spectator ions?

   $$Cu^{2+}(aq) + 2NO_3^-(aq) + 2Na^+(aq) + S^{2-}(aq) \rightarrow CuS(s) + 2NO_3^-(aq) + 2Na^+(aq)$$

   a. Cu^{2+} b. Na^+ c. NO_3^- d. S^{2-} e. More than one response is correct.

   Explain your answer: _____

   _____

   _____

4. Suppose you had an unknown solution that contained either dissolved NaOH, NaCl, or NaNO₃. You added 4 drops of $AgNO_3$ solution to 4 drops of the unknown solution and observed that no solid formed. What can be concluded about the unknown solution?

    a. It contains NaOH.　　**b.** It contains NaCl.　　**c.** It contains $NaNO_3$.
    d. No conclusion can be made based on the observations.

    Explain your answer: _____

    _____

    _____

5. Suppose you tried to carry out a double-replacement reaction by mixing together equal volumes of a solution that contains dissolved NaOH and a solution that contains dissolved NaCl. What would you expect to happen when the two solutions were mixed?

    a. A solid would form.　　**b.** A solid would not form.

    Explain your answer: _____

    _____

    _____

    _____

Experiment 7

Analysis Using Decomposition Reactions

In this experiment, you will

- Heat a hydrate salt to drive off the water and use the collected data to determine the formula of the original hydrate.
- Decompose a sample of impure potassium chlorate by heating and use the collected data to determine the percentage of potassium chlorate in the original sample.

INTRODUCTION

Many solid, crystalline, ionic compounds, called **hydrates,** contain definite amounts of water as a part of their crystal structures. This water of hydration can be driven from a solid by heating. When appropriate weighings are done, the mass and number of moles of water driven off can be calculated and the formula of the hydrate determined.

Some crystalline solids do not contain water of hydration but will still undergo decomposition reactions when heated. When one of the products of decomposition is a gas, appropriate weighings allow the mass of gas given off to be calculated. This mass and the equation for the decomposition reaction can be used to calculate the mass of solid that was decomposed. This result can then be used to calculate the percentage of the solid present in a mixture that contains other solids.

EXPERIMENTAL PROCEDURE

A. Determining the Formula of a Hydrate

The water of hydration is included in the formula of solid hydrates as follows: $MgCl_2 \cdot 6H_2O$. In this formula, the dot between the $MgCl_2$ and the $6H_2O$ indicates that the water is a part of the crystal structure of the solid. The 6 preceding the H_2O indicates that 6 water molecules are present in the crystal for every formula unit of $MgCl_2$ or that 6 moles (mol) of water are present for every mole of $MgCl_2$.

In this part of the experiment, you will drive the water from a sample of barium chloride hydrate whose formula will be represented by $BaCl_2 \cdot nH_2O$. Your task will be to determine the value of n in the formula.

PROCEDURE

1. Use soapy water and a test tube brush to clean a porcelain crucible and cover. Permanent stains that cannot be removed will not influence your results and can be ignored. Rinse the cleaned crucible and cover with small amounts of distilled water.

FIGURE 7.1
Heating a crucible

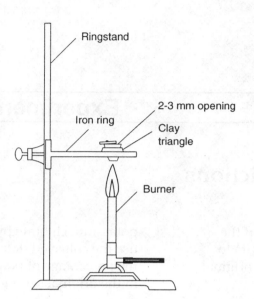

Ringstand

2-3 mm opening

Iron ring

Clay triangle

Burner

2. Place the crucible with its cover on a clay triangle mounted on a ringstand and located over a burner as shown in Figure 7.1.
3. Heat the crucible and cover gently for 1 to 2 minutes, then heat them strongly with the hottest part of your burner flame for another 5 minutes.
4. After heating the crucible strongly for 5 minutes, use crucible tongs to remove the crucible and lid from the clay triangle. Put them on a wire gauze and allow them to cool for a minimum of 10 minutes.

SAFETY ALERT

The crucible, lid, ringstand, ring, and clay triangle get very hot during parts of this experiment. Be aware of this and avoid touching them while they are hot.

While the crucible and lid cool, use a centigram balance and a weighing dish to weigh out a sample of barium chloride hydrate with an approximate mass in the range of 0.4 to 0.5 g.

5. After the crucible and lid are cool, weigh them together on an electronic balance with a sensitivity of no less than 1 mg. Record their combined mass in Table 7.1.
6. Add the sample of barium chloride hydrate you weighed in Step 4 to the crucible.
7. Weigh the crucible, its cover, and contained-sample together on the same balance used in Step 5. Record their combined mass in Table 7.1.
8. Return the crucible, lid, and sample to the clay triangle and assemble the apparatus again as shown in Figure 7.1. Heat the crucible with a moderate flame from your burner for 2 to 3 minutes, then heat it strongly for 5 minutes.
9. After the heating is completed, set the lid of the crucible so it covers the crucible opening completely and allow the crucible, lid, and contents to cool for a minimum of 10 minutes.
10. Reweigh the cool crucible, lid, and contents on the same balance used in Steps 5 and 7. Record the combined mass in Table 7.1.

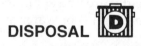

DISPOSAL

Used barium chloride sample in container labeled "Exp. 7, Used Barium Chloride."

B. Determining the Percentage of $KClO_3$ in a Mixture

When potassium chlorate ($KClO_3$) is heated, it decomposes into oxygen gas and potassium chloride (KCl). If a mixture containing potassium chlorate and iron oxide is heated, only the potassium chlorate decomposes. The oxygen gas that is produced leaves the mixture. If the mixture is weighed before and after heating, the loss in mass is equal to the mass of the oxygen that was given off during heating. In this part of the experiment, you will heat a sample of an unknown mixture that contains potassium chlorate and iron oxide. The masses of the sample will be determined before and after heating, and the data used to calculate the percentage of potassium chlorate in the mixture.

PROCEDURE

1. Obtain a sample of an unknown from the stockroom and record the identification number in Table 7.4.
2. Use a centigram balance and weighing dish to weigh out a sample of the unknown with an approximate mass in the range of 0.4 to 0.5 g.
3. Weigh a clean, dry 10-cm test tube on an electronic balance with a sensitivity of no less than 1 mg. Record the mass of the empty test tube in Table 7.3.
4. Add the sample of unknown you weighed in Step 2 to the test tube and weigh them together on the same balance used in Step 3. Record their combined mass in Table 7.3.
5. Mount the test tube that contains the sample on a ringstand as shown in Figure 6.1 of Experiment 6. Rotate the clamp so the test tube is not vertical but is tilted at about a 45-degree angle.
6. Use a moderate flame from your burner to heat the test tube gently. The solid will agitate and might melt as the decomposition reaction takes place. Control the heating carefully so the reaction does not get rapid enough to expel reactants from the mouth of the test tube. If this happens, you will have to start this part of the experiment over.

SAFETY ALERT

> The hot mixture in the test tube can cause severe burns to the skin and may ignite any combustible material it contacts, including paper. If any of the hot mixture is expelled, clean it up with a cloth soaked in cold water. Any that gets on the skin should be washed off immediately with cold water, and your instructor should be informed.

7. After the reaction appears to be completed (after the agitation stops and the mixture appears to be a solid again), increase the intensity of the heating and heat the test tube strongly for about 2 minutes.
8. After the heating is completed, allow the test tube and contents to cool for a minimum of 15 minutes. Then, weigh the test tube and contents on the same balance used in Steps 3 and 4. Record the mass in Table 7.3.

DISPOSAL

Heated sample in sink. Rinse sample from test tube and flush down drain with cold water.
Unused unknown in container labeled "Exp. 7, Used Potassium Chlorate."

≡ CALCULATIONS AND REPORT

A. Determining the Formula of a Hydrate

The decomposition reaction of solid $BaCl_2 \cdot nH_2O$ that takes place when the solid is heated can be represented by the following equation:

$$BaCl_2 \cdot nH_2O(s) \rightarrow BaCl_2(s) + nH_2O(g) \qquad \text{Eq. 7.1}$$

We see from this equation that n moles of gaseous water (water vapor) are produced for every mole of anhydrous (without water) $BaCl_2$ produced by the decomposition. Thus, the ratio of moles H_2O/moles $BaCl_2$ will give the value of n in the hydrate sample.

1. Refer to Table 7.1 and calculate the mass of hydrate sample that was heated by subtracting the mass of the crucible and lid from the mass of crucible, lid, and hydrate sample before heating. Record this mass in Table 7.2.

2. Calculate the mass of the hydrate sample left after the heating by subtracting the mass of the crucible and lid from the mass of crucible, lid, and hydrate sample after heating. Note that this is the mass of anhydrous $BaCl_2$ produced by the reaction. Record this mass in Table 7.2.

3. Calcuate the mass of water lost as a result of the heating by subtracting the mass calculated in Step 2 from the mass calculated in Step 1. Record this mass in Table 7.2.

4. Note that the mass (in grams) of water lost can be converted to moles of water by multiplying the mass by the following conversion factor: 1.00 mol H_2O/18.0 g H_2O. Convert the mass of water lost that you calculated in Step 3 into moles of water lost and record the value in Table 7.2.

5. Note that the mass (in grams) of anhydrous $BaCl_2$ produced by the reaction can be converted to moles of anhydrous $BaCl_2$ by multiplying the mass by the following conversion factor: 1.00 mol $BaCl_2$/ 208 g $BaCl_2$. Convert the mass of anhydrous $BaCl_2$ produced by the reaction (calculated in Step 2) to moles of anhydrous $BaCl_2$ and record the value in Table 7.2.

6. Divide the number of moles of water lost by the number of moles of anhydrous $BaCl_2$ produced and record the value of the ratio in Table 7.2.

7. Round the ratio calculated in Step 6 to the nearest whole number and record the number in Table 7.2. This is the n value for the hydrate.

8. Record the determined formula for the hydrate in Table 7.2.

B. Determining the Percentage of $KClO_3$ in a Mixture ▭

When potassium chlorate is heated, it undergoes a decomposition reaction that is represented by the following balanced equation:

$$2KClO_3(s) \rightarrow 2KCl(s) + 3O_2(g) \qquad \text{Eq. 7.2}$$

At the high temperature used, the solid $KClO_3$ melts. We see from the balanced equation that 3 mol O_2 are produced when 2 mol $KClO_3$ are decomposed. Thus, the mass of oxygen lost can be used to calculate the number of moles of oxygen lost, which in turn can be used to calculate the number of moles and mass of $KClO_3$ that were decomposed. This mass can be compared to the mass of mixture that was heated and the percentage of $KClO_3$ in the mixture can be calculated. This last calculation assumes that no other component of the mixture decomposes when it is heated. The other component of the mixture used in this part of the experiment is iron oxide, which does not decompose when it is heated.

1. Refer to Table 7.3 and calculate the mass of the sample of unknown that was heated by subtracting the mass of the empty test tube from

the mass of the test tube and sample before heating. Record this mass in Table 7.4.

2. Calculate the mass of oxygen lost by the reaction by subtracting the mass of the test tube and sample after heating from the mass of the test tube and sample before heating. Record this mass in Table 7.4.

3. Note that the mass (in grams) of oxygen lost can be converted to moles by multiplying the mass by the following conversion factor: 1.00 mol O_2/32.0 g O_2. Convert the mass of oxygen lost that you calculated in Step 2 into moles of oxygen lost and record the number in Table 7.4.

4. According to Equation 7.2, the decomposition of 2 mol $KClO_3$ produces 3 mol O_2. Thus, the number of moles of $KClO_3$ that were decomposed can be calculated by multiplying the number of moles of oxygen produced by the following conversion factor: 2 mol $KClO_3$/ 3 mol O_2. Calculate the number of moles of $KClO_3$ that were decomposed and record the value in Table 7.4. Note that this is the number of moles of solid $KClO_3$ that were in the heated sample of unknown mixture.

5. Calculate the number of grams of $KClO_3$ that were decomposed by multiplying the number of moles of $KClO_3$ calculated in Step 4 by the following conversion factor: 123 g $KClO_3$/1.00 mol $KClO_3$. Record the calculated mass of $KClO_3$ in Table 7.4.

6. Use the mass of the sample of unknown you calculated in Step 1 and the mass of $KClO_3$ you calculated in Step 5 to calculate the percentage of $KClO_3$ in the sample of unknown. This is done by using the following equation:

$$\% \text{ KClO}_3 = \frac{\text{KClO}_3 \text{ mass}}{\text{unknown sample mass}} \times 100 \qquad \text{Eq. 7.3}$$

Record the calculated percentage in Table 7.4.

Experiment 7 ▪ Pre-Lab Review

Analysis Using Decomposition Reactions

1. Are any specific safety alerts given in the experiment? List any that are given.

    _____

    _____

2. Are any specific disposal directions given in the experiment? List any that are given.

    _____

    _____

    _____

3. Two different containers are used to heat samples in this experiment. What are the two containers, and in which parts of the experiment are they used?

    _____

    _____

4. What is the allowable mass range for the samples used in Parts A and B?

    _____

5. Which type of balance is used to weigh out samples with approximate masses in Parts A and B?

    _____

6. Are the masses that are determined on the balance of Question 5 recorded on the experiment sheet? Explain why or why not.

    _____

    _____

7. After heating is completed, how long should the samples be allowed to cool before they are weighed in Parts A and B?

    _____

    _____

Experiment 7 ▪ Data & Report Sheet

Analysis Using Decomposition Reactions

A. Determining the Formula of a Hydrate

TABLE 7.1 (data)

Mass of crucible and lid   _____

Mass of crucible, lid, and hydrate
  sample before heating   _____

Mass of crucible, lid, and hydrate
  sample after heating   _____

TABLE 7.2 (report)

Mass of hydrate sample before
heating   _____

Mass of hydrate sample after
  heating (mass of anhydrous
  $BaCl_2$ produced)   _____

Mass of water lost   _____

Moles of water lost   _____

Moles of anhydrous $BaCl_2$
  produced   _____

Ratio: Moles of water lost/moles
  of anydrous $BaCl_2$ produced   _____

Whole-number ratio   _____

Formula of hydrate   _____

B. Determining the Percentage of KClO₃ in a Mixture

TABLE 7.3 (data)

Mass of empty test tube	_____
Mass of test tube and sample before heating	_____
Mass of test tube and sample after heating	_____

TABLE 7.4 (report)

ID number of unknown	_____
Mass of sample that was heated	_____
Mass of oxygen lost	_____
Moles of oxygen lost	_____
Moles of KClO₃ decomposed (moles of KClO₃ in sample)	_____
Mass of KClO₃ decomposed (mass of KClO₃ in sample)	_____
Percentage of KClO₃ in sample	_____

QUESTIONS

1. Suppose in Step 8 of Part A of the experiment that a student fails to heat the crucible and hydrate sample long enough to drive off all of the water. How would this mistake influence the value of n determined for the hydrate if everything else was done correctly in Part A?

   a. Make it greater than the correct value.
   b. Make it less than the correct value.
   c. Would not have an influence on the value of n.

   Explain your answer: _____

   _____

   _____

2. Suppose in Step 5 of Part A of the experiment that a student misreads the balance and records a mass for the crucible and lid that is greater than the correct mass. If all other measurements are done correctly, how would this mistake influence the value of n determined for the hydrate?

    a. Make it greater than the correct value.
    b. Make it less than the correct value.
    c. Would not have an influence on the value of n.

    Explain your answer: _____

    _____

    _____

3. Suppose in Step 6 of Part A of the experiment that a student spills a small amount of sample from the weighing dish while attempting to put it into the crucible. The student does not realize that the spill occurred and does the rest of Part A according to the directions. How will this influence the value of n determined for the hydrate?

    a. Make it greater than the correct value.
    b. Make it less than the correct value.
    c. Would not influence the value of n.

    Explain your answer: _____

    _____

    _____

4. Suppose in Step 2 of Part B of the experiment that a student weighs out a sample that has a mass of 0.51 g. All other steps in this part of the experiment are done according to the directions. How will this influence the determined percentage of $KClO_3$ in the sample?

    a. Make it greater than the correct value.
    b. Make it less than the corrrect value.
    c. Would not influence the determined percentage.

    Explain your answer: _____

    _____

    _____

5. Suppose in Step 3 of Part B that a student weighs a test tube that is clean but not dry. The mass is recorded, and the student then realizes the mistake and dries the test tube but does not reweigh it. How will this influence the determined percentage of $KClO_3$ in the sample if all other steps are done according to the directions given?

    a. Make it greater than the correct value.
    b. Make it less than the correct value.
    c. Would not influence the determined value.

    Explain your answer: _____

    _____

    _____

6. Suppose in Step 6 of Part B that the heating is done too strongly and a small amount of the re-action mixture spurts out of the test tube onto the desk top. The spill is cleaned up according to the directions given, but the student involved continues the experiment without starting over. All other steps are done correctly. How would this mistake influence the determined percentage of $KClO_3$ in the sample?

   a. Make it greater than the correct value.
   b. Make it less than the correct value.
   c. Would not influence the determined value.

   Explain your answer: _____

   _____

   _____

Experiment 8

Gas Laws

In this experiment, you will

- Verify Boyle's law, the pressure–volume behavior for a gas.
- Verify Charles's law, the temperature–volume behavior for a gas.
- Use temperature–volume data for a gas to determine the value of absolute zero.
- Verify Graham's law of diffusion.

INTRODUCTION

In 1662 Robert Boyle, an Irish chemist, reported that he had discovered a quantitative relationship between the pressure and volume of a gas sample maintained at constant temperature. The mathematical expression of this relationship is called Boyle's law in honor of his discovery. More than a century later, Jacques Charles, a French scientist, reported on his studies of the behavior of gas samples kept at constant pressure but whose temperature and volume were varied. The mathematical expression of this temperature–volume behavior is called Charles's law in honor of its discoverer. In 1828 Thomas Graham, a British scientist, discovered that the diffusion rate of a gas is inversely proportional to the square root of its molecular weight. The mathematical expression that relates the relative diffusion rates of two gases to their molecular weights is known as Graham's law.

EXPERIMENTAL PROCEDURE

A. Boyle's Law

According to **Boyles law,** the volume of a fixed amount (mass) of gas is inversely proportional to the pressure of the gas, provided the gas temperature is kept constant. Thus, the volume of a gas sample decreases as the pressure on the sample increases. Mathematically, this is expressed as

$$P = \frac{k}{V} \qquad \text{Eq. 8.1}$$

$$PV = k \qquad \text{Eq. 8.2}$$

In these equations, P is the pressure, V is the volume, and k is a constant number that depends on the temperature and quantity of gas in the sample. The units of k depend on the units used to express P and V.

You will collect pressure and volume readings for a gas sample (air) and verify that the product PV is equal to a constant number. The data collection for this part of the experiment is easier if you work with a partner. Check with your instructor for permission to work with a partner.

PROCEDURE

1. Assemble or obtain from the stockroom the apparatus shown in Figure 8.1.

SAFETY ALERT

If you assemble the apparatus yourself, be very careful when you put the capillary tube into the rubber stopper. Be sure the tube is lubricated with glycerin and your hands are protected by being wrapped in a towel.

FIGURE 8.1
Test tube assembly

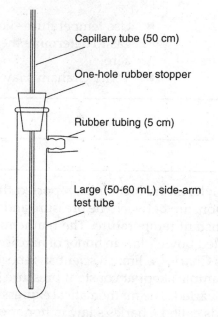

Capillary tube (50 cm)

One-hole rubber stopper

Rubber tubing (5 cm)

Large (50-60 mL) side-arm test tube

If you have any questions about the procedure to follow, check with your lab instructor.

2. Remove the stopper and capillary from the test tube, put your finger over the end of the rubber tubing, and fill the test tube to the top with tap water. Insert the capillary and stopper snugly into the top of the test tube. Do this over a sink because the water will overflow.

3. Keep your finger over the side-arm rubber tubing and remove the capillary and stopper from the test tube.

4. Measure the volume of water that remains in the test tube by pouring it carefully into a 50- or 100-mL graduated cylinder. If a 50-mL cylinder is used, you might have to fill it to 50, empty it, and refill it. Record this test tube volume in Table 8.1 of the Data and Report Sheet.

5. Use a 10-mL graduated cylinder to measure 5.0 mL of distilled water. Add 2 drops of food coloring to the water and put it into the side-arm test tube, then clamp it to a ringstand.

FIGURE 8.2
Boyle's law apparatus

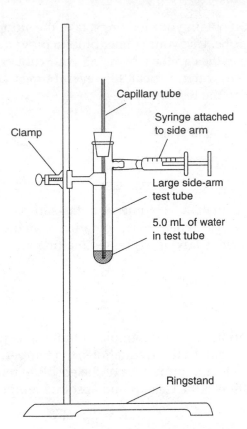

Capillary tube

Syringe attached
to side arm

Clamp

Large side-arm
test tube

5.0 mL of water
in test tube

Ringstand

6. Make sure all water has been removed from the capillary tube (use a rubber bulb to blow it out if necessary) and put the capillary tube and its stopper into the side-arm test tube. This step is very important, so make sure you do it properly.

7. Obtain a 5- or 10-cc (mL) plastic hypodermic syringe from the stockroom.

8. Adjust the plunger of the syringe to a reading of 5.0 mL and attach the syringe to the rubber tubing on the side of the test tube with the syringe markings toward you for easier reading. Your assembled apparatus should look like that shown in Figure 8.2.

9. Check the syringe and adjust it if necessary to read 5.0 mL. Use a ruler with 0.1-cm scale divisions to read the height of the water in the capillary tube above the level of water in the test tube (see Figure 8.3). Read this height only to the nearest 0.1 cm and record the height and syringe readings in Table 8.1.

10. Push the syringe plunger in about 1 mL, making sure it is exactly on one of the scale divisions. Measure the height of the water in the capillary above the water level in the test tube and record the height and syringe reading in Table 8.1. If the water level in the capillary is inside the stopper where it cannot be read, adjust the syringe until the level is slightly above or below the stopper.

FIGURE 8.3
Measurement of water
height in capillary

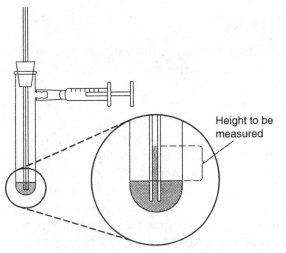

Height to be
measured

11. Continue pushing the syringe plunger in and measuring the height of water in the capillary until you have at least five readings. Record the values in Table 8.1. Be certain the water is not pushed out the top of the capillary on any measurement.

12. Obtain three or four more syringe readings and capillary water heights by pulling the syringe plunger out in increments and measuring the corresponding water heights in the capillary tube. Record these values in Table 8.1.

13. Obtain a reading of the barometric pressure in the laboratory in cm Hg and record the value in Table 8.1. The value will probably be written on the blackboard. If no value seems to be available, ask your instructor for it.

14. After you obtain your readings, take the stopper and capillary from the test tube. Use your rubber bulb to blow the water from the capillary. Clean the capillary by using your rubber bulb to draw and expel clean water. Repeat this several times. Discard the water that remains in the test tube.

DISPOSAL  Used water in test tube in sink

B. Charles's Law

Charles found that the volume of a gas sample, maintained at constant pressure, increased when its temperature was increased. This direct relationship (**Charles's law**) is expressed mathematically as:

$$V = k'T \qquad\qquad \text{Eq. 8.3}$$

$$\frac{V}{T} = k' \qquad\qquad \text{Eq. 8.4}$$

In these equations, V is the sample volume, T is the sample temperature, and k' is a constant. In these equations, the temperature must be expressed in kelvins, the temperature unit of the absolute temperature scale. The relationship between the Celsius and absolute temperature scales is:

$$K = {}^{\circ}C + 273 \qquad\qquad \text{Eq. 8.5}$$

According to this equation, a temperature given in Celsius degrees is converted to kelvins by adding 273.

You will collect temperature and volume data for a gas sample and use the data to verify Charles's law and to determine the value of absolute zero. The pressure of the sample will be the atmospheric pressure in the lab, which is constant.

The data for this part of the experiment can be collected conveniently by individuals or pairs. Check with your instructor.

PROCEDURE

1. Obtain a 5- or 10-cc (mL) plastic hypodermic syringe and a thermometer from the stockroom or your locker. The syringe used in Part A will be satisfactory.
2. Put enough water into a 400-mL beaker to come above the 5.0-cc mark on the syringe when it is held vertically in the beaker with its tip near the beaker bottom.
3. Add hot water to the beaker or heat it with a burner until the water temperature is about 40°C.
4. Set the plunger of the syringe to read 4.5 cc (mL) and put the syringe, tip down, into the beaker of warm water. Hold the syringe under the water (don't let it float) for at least 5 minutes.
5. After 5 minutes, keep the syringe under water and carefully pull the plunger back until water enters the syringe barrel. Then, push the

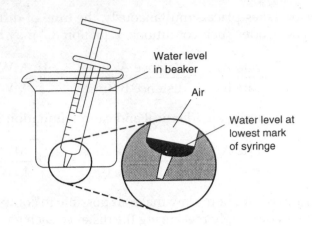

FIGURE 8.4
Adjustment of water level
in syringe

Water level
in beaker

Air

Water level at
lowest mark
of syringe

plunger in carefully
until the water is even
with the lowest mark
on the syringe barrel
(see Figure 8.4). At this
point, note the location
of the plunger in the
syringe. This reading
(estimate to 0.1-mL) is
the air volume in the
syringe. Measure the
temperature of the
water bath (estimate to
0.1°C), which is also the temperature of the air inside the syringe.
Record the volume and temperature of the air sample in Table 8.5.

6. Without removing the syringe from the water, cool the water to
    about 30°C by slowly adding small pieces of ice. Stir the water so the
    temperature is uniform.

7. Allow the syringe to remain in the water at this new temperature for
    about 5 minutes.

8. Carefully adjust the air volume by pushing in the syringe plunger
    until the water is again even with the lowest syringe mark. (NOTE: If
    any air is pushed out of the syringe tip at this or any future point,
    the experiment must be started over.)

9. Read the air volume and temperature as before and record the val-
    ues in Table 8.5.

10. Repeat Steps 6–9 and obtain air-sample volumes and temperatures
    at bath temperatures of about 20°C, 10°C and 0°C (use a bath of ice
    and water).

C. Graham's Law

Graham found that the rate of diffusion of a gas is inversely proportional
to the square root of the molecular weight of the gas. When the diffusion
rates of two gases are compared mathematically by forming them into a
ratio, **Graham's law** results in the following mathematical forms:

$$\frac{\text{diffusion rate A}}{\text{diffusion rate B}} = \sqrt{\frac{\text{molecular weight B}}{\text{molecular weight A}}} \qquad \text{Eq. 8.6}$$

or in an abbreviated form

$$\frac{\text{rate A}}{\text{rate B}} = \sqrt{\frac{MW_B}{MW_A}} \qquad \text{Eq. 8.7}$$

In each of these equations, A and B represent the two different gases.

When a gas is allowed to move down a tube by diffusion, the rate of
diffusion is equal to the distance the gas moves divided by the time it takes
for the movement to occur. In this part of the experiment you will allow
two gases to diffuse simultaneously in a glass tube. The design of the experiment will allow you to measure the distance each gas moves. Since the

diffusion takes place simultaneously, the time of diffusion is the same for each gas. Under such conditions, Equation 8.7 may be written as follows:

$$\frac{\text{rate A}}{\text{rate B}} = \frac{\text{distance A}/\text{time}}{\text{distance B}/\text{time}} = \sqrt{\frac{MW_B}{MW_A}} \qquad \text{Eq. 8.8}$$

Because the times are identical and cancel, Equation 8.8 becomes:

$$\frac{\text{rate A}}{\text{rate B}} = \frac{\text{distance A}}{\text{distance B}} = \sqrt{\frac{MW_B}{MW_A}} \qquad \text{Eq. 8.9}$$

This form of Graham's law makes it possible to compare diffusion rates for two gases by simply measuring the distance each gas travels in a tube during a fixed diffusion time.

The two gases you will allow to diffuse are hydrogen chloride (HCl) and ammonia (NH_3). Both of these gases are available in the laboratory as water solutions called hydrochloric acid and aqueous ammonia (or ammonium hydroxide). The respective gases readily escape from the concentrated solutions as evidenced by the strong odors of these solutions. When the two gases meet, a reaction occurs that produces finely divided white solid ammonium chloride. The reaction is:

$$HCl(g) + NH_3(g) \rightarrow NH_4Cl(s) \qquad \text{Eq. 8.10}$$

It will be necessary to work in pairs on this part of the experiment.

PROCEDURE

1. Obtain a 50-cm piece of glass tubing from the stockroom for each team.
2. Make certain the glass tubing is completely dry. If it is not, check with your instructor for directions on how to dry it. Obtain two small pieces of cotton, form them into small plugs and insert a plug into each end of the glass tube. Each plug should extend about 5-6 mm into the tube.

SAFETY ALERT

> Concentrated hydrochloric acid and concentrated aqueous ammonia will be used in steps 3,4 and 5 of this part of the experiment. They both have strong odors, and release hazardous fumes, so the steps must be done in a fume hood. The necessary chemicals will be found in the hood.

3. Take the plugged tube to a fume hood that contains concentrated (12 M) hydrochloric acid (HCl) and concentrated aqueous ammonia (NH_3 or NH_4OH).
4. At this point it is necessary to simultaneously moisten the cotton plug on one end of the tube with concentrated HCl, and the plug on the other end with concentrated NH_3. Draw some concentrated HCl into one dropper and some concentrated NH_3 into another dropper. Each team member take one dropper and hold it near the cotton plugs at the opposite ends of the tube. Simultaneously touch the droppers to the respective plugs and carefully squeeze enough liquid onto the plugs to just moisten them. No liquid should be visible in the glass tube beyond the cotton plugs. If liquid is visible, the glass tube must be cleaned and dried, and the experiment repeated.

5. Allow the tube to remain in the hood undisturbed until a distinct cloudy white deposit forms on the tube where the two gases meet and react.
6. The distance each gas moved during diffusion is the distance from the plug that contained the aqueous solution of the gas to the middle of the white deposit. When the deposit has formed, remove the tube from the hood and use a ruler to measure the distance from the inside end of each cotton plug to the middle of the white deposit on the glass. Record the measured distances in the appropriate blanks of Table 8.7.
7. After the measurements are completed, allow tap water to flow into the plugged tube, remove the plugs and discard them in the wastebasket, and continue to rinse the tube with tap water until all of the white deposit is removed. Give the tube a final rinse with distilled water, and return it to the stockroom.

CALCULATIONS AND REPORT

A. Boyle's Law

The total volume of air in the sample is equal to the volume of the test tube, minus the volume of water added to the test tube, plus the volume in the syringe attached to the side-arm test tube:

$$V = V_t - V_w + V_s \qquad \text{Eq. 8.11}$$

In this equation, V is the volume of the air sample, V_t is the measured volume of the side-arm test tube, V_w is the volume of water added to the test tube (5.0 mL in this experiment), and V_s is the volume of air in the syringe.

The pressure on the air in the sample is equal to the atmospheric pressure (barometric pressure) plus the pressure represented by the height of the water column in the capillary tube:

$$P = P_b + P_w \qquad \text{Eq. 8.12}$$

In this equation, P_b is the barometric pressure (in cm Hg), and P_w is the pressure represented by the water column in the capillary tube. Before P_b and P_w can be added, they must be expressed in the same units. Since mercury is 13.6 times as dense as water, a pressure represented by a 1.0-cm column of mercury is equivalent to a pressure represented by a 13.6-cm column of water. Thus, the water column heights must be divided by 13.6 (or multiplied by 1/13.6 = 0.0735) before they can be added to the barometric pressure to get the gas sample pressure.

1. Record in Table 8.2 the test tube volume (V_t), volume of water used V_w (5.0 mL), and the syringe reading V_s (volume of air in syringe) for each trial done. Use Equation 8.11 to calculate the total air volume for each measurement. Record these values in Table 8.2.
2. Record in Table 8.3 the barometric pressure and height of water in the capillary for each trial done. Convert the capillary tube water heights to their equivalents in cm Hg and record the values in Table 8.3. Add the barometric pressure and mercury equivalent of the capillary water heights to give the total pressure values of the gas sample in each trial. Record the values in Table 8.3.

3. Record the total air volumes from Table 8.2 and the total air pressures from Table 8.3 for each trial into Table 8.4. Multiply the total volume and total pressure for each trial to get k values. Record the values in Table 8.4.
4. Calculate the average k value and record it in Table 8.4.

B. Charles's Law

According to Equation 8.3, the volume of a gas sample shrinks to zero when the sample is cooled to 0 K. This is one way to theoretically define absolute zero (0 K). Of course, real gases become liquids at temperatures above absolute zero, so their volumes can't become zero. However, graphical data can be extrapolated (extended) from measured values back to zero volume, and the temperature corresponding to absolute zero can be estimated.

1. Record the temperatures and volumes of Table 8.5 in Table 8.6.
2. Convert the Celsius temperature to kelvins (Equation 8.5) and record them in Table 8.6.
3. Use Equation 8.4 and calculate k' for each trial. Record the result in Table 8.6.
4. Calculate the average value for k' and record the value in Table 8.6.
5. Graph the T (in kelvins) and V values of Table 8.6 on the graph paper provided. Draw a straight line through the points and extend it to the left until it intersects the vertical temperature axis. The point of intersection is the temperature at which the gas volume would theoretically be equal to zero; this is absolute zero. Record your value of absolute zero from the graph in Table 8.6.

C. Graham's Law

The experimentally-determined diffusion differences for the two gases will be used along with Equation 8.9 to calculate an experimental value for the ratio (rate NH_3)/(rate HCl). A theoretical value for the ratio will also be calculated using Equation 8.9 and the molecular weights of NH_3 and HCl.

1. Use atomic weights given in Appendix C of this manual and calculate the molecular weights of NH_3 and HCl in atomic mass units (u). Record the calculated values in Table 8.8.
2. Use the distances recorded in Table 8.7 and Equation 8.9 to calculate an experimental value for the ratio (rate NH_3)/(rate HCl). Record the calculated value in Table 8.8 in the blank labeled *Rate ratio (exp.)*.
3. Use the molecular weights calculated in Step 1 and Equation 8.9 to calculate a theoretical value for the ratio (rate NH_3)/(rate HCl). Record the calculated value in Table 8.8 in the blank labeled *Rate ratio (theor.)*.

Experiment 8 ▪ Pre-Lab Review

Gas Laws

1. Are any specific safety alerts given in the experiment? List any that are given.

   _____

   _____

2. Are any specific disposal directions given in the experiment? List any that are given.

   _____

3. What characteristic of a gas sample stays constant according to Boyle's law?

   _____

4. What characteristic of a gas sample stays constant according to Charles's law?

   _____

5. How closely should the readings be estimated of the height of water in the capillary tube in Part A?

   _____

6. In Part A, what quantity should you ask your instructor for if it doesn't seem to be available?

   _____

7. In Part B, what mistake will require that you start the experiment over?

   _____

8. In Part B, what temperature scale must be used? How are Celsius temperature readings converted to this scale?

   _____

9. In Part A, what must be done before P_b and P_w can be added together?

   _____

   _____

   _____

10. Explain how an estimate of absolute zero is made in Part B. _____

_____

_____

11. What are the identities and the sources of the two gases you will allow to diffuse in Part C of the experiment?

_____

_____

_____

Experiment 8 ▪ Data & Report Sheet

Gas Laws

A. Boyle's Law

TABLE 8.1 (data)

Test tube volume, V_t (mL) _____

Barometric pressure, P_b (cm Hg) _____

Trial	Syringe Reading (mL)	Height of Water in Capillary (cm)
1	_____	_____
2	_____	_____
3	_____	_____
4	_____	_____
5	_____	_____
6	_____	_____
7	_____	_____
8	_____	_____
9	_____	_____

TABLE 8.2 (report)

Trial	Test Tube Volume (mL) V_t	Volume of Water Added (mL) V_w	Syringe Reading (mL) V_s	Total Air Volume (mL) V
1	_____	_____	_____	_____
2	_____	_____	_____	_____
3	_____	_____	_____	_____
4	_____	_____	_____	_____
5	_____	_____	_____	_____
6	_____	_____	_____	_____
7	_____	_____	_____	_____
8	_____	_____	_____	_____
9	_____	_____	_____	_____

TABLE 8.3 (report)

Trial	Barometric Pressure (cm Hg) P_b	Water Height in Capillary (cm)	Hg Equiv. of Water Height (cm Hg) P_w	Total Air Pressure (cm Hg) P
1	_____	_____	_____	_____
2	_____	_____	_____	_____
3	_____	_____	_____	_____
4	_____	_____	_____	_____
5	_____	_____	_____	_____
6	_____	_____	_____	_____
7	_____	_____	_____	_____
8	_____	_____	_____	_____
9	_____	_____	_____	_____

TABLE 8.4 (report)

Trial	Total Air Volume (mL) V	Total Air Pressure (cm Hg) P	k $P \times V$
1	_____	_____	_____
2	_____	_____	_____
3	_____	_____	_____
4	_____	_____	_____
5	_____	_____	_____
6	_____	_____	_____
7	_____	_____	_____
8	_____	_____	_____
9	_____	_____	_____
		Average value	_____

B. Charles's Law

TABLE 8.5 (data)

Trial	Sample Temperature (°C)	Sample Volume (mL)
1	_____	_____
2	_____	_____
3	_____	_____
4	_____	_____
5	_____	_____

TABLE 8.6 (report)

Trial	Sample Temperature (°C)	Sample Volume (mL)	Sample Temp. in kelvins (K)	k'
1	_____	_____	_____	_____
2	_____	_____	_____	_____
3	_____	_____	_____	_____
4	_____	_____	_____	_____
5	_____	_____	_____	_____

Absolute zero value from graph (K) _____   Average k' value _____

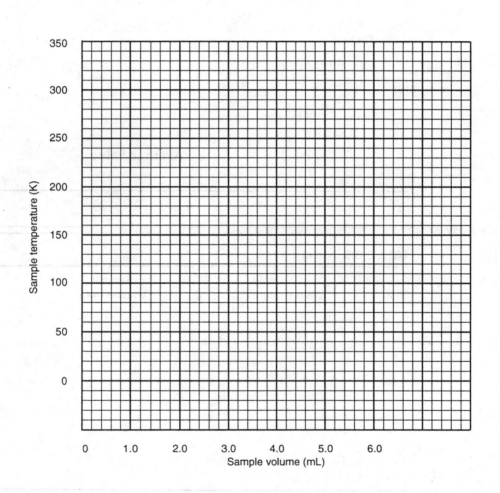

C. Graham's Law

TABLE 8.7 (data)

Gas	Distance
NH_3	_____
HCl	_____

TABLE 8.8 (report)

MW of NH_3	_____
MW of HCl	_____
Rate ratio (exp.)	_____
Rate ratio (theor.)	_____

QUESTIONS

1. A student calculates an average value of 4289 for a Boyle's law constant k. Based on the number of significant figures in the measured quantities, how many significant figures should this result have after proper rounding?

    a. 1   **b.** 2   **c.** 3   **d.** 4

    Explain your answer: _____

    _____

    _____

    _____

2. Use the unrounded k value given in Question 1 and calculate the total volume of a gas sample if the total sample pressure is 64.3 cm Hg.

    a. 0.0150 mL   **b.** 15.0 mL   **c.** 0.067 mL   **d.** 66.7 mL

    Explain your answer: _____

    _____

    _____

3. Suppose some water was pushed out the top of the capillary and lost in Step 2 of the Boyle's law experiment. How would this influence the values of k obtained by the readings taken in Step 13?

    a. Increase k.   **b.** Decrease k.   **c.** Have no effect on k.

    Explain your answer: _____

    _____

    _____

    _____

4. Did the results you obtained for k in Table 8.4 verify Boyle's law?

    a. Yes.   **b.** No.   **c.** Hard to tell.

    Explain your answer: _____

    _____

    _____

    _____

5. Suppose the thermometer you used in your Charles's law experiment consistently gave low temperature readings. How would that influence the value of k' you would obtain?

    a. Make it too high.
    b. Make it too low.
    c. Would not influence it.

    Explain your answer: _____

    _____

    _____

6. Suppose in Step 8 of your Charles's law experiment you push some of the air sample out the tip of the syringe. How would this influence k' values determined from the data collected after the mistake if the experiment were continued?

    a. Make them too high.
    b. Make them too low.
    c. Would not influence them.

    Explain your answer: _____

    _____

    _____

7. Did the k' values you obtained and recorded in Table 8.6 verify Charles's law?

    a. Yes.   **b.** No.   **c.** Hard to tell.

    Explain your answer: _____

    _____

    _____

8. Suppose in Step 8 of the Graham's law experiment one of the partners moistened the cotton on the NH_3 end of the tube several seconds before the cotton on the other end was moistened. How would this mistake influence the calculated experimental value of the ratio (rate NH_3)/(rate HCl)?

    a. Increase it   **b.** Decrease it   **c.** Would have no effect

    Explain your answer: _____

    _____

    _____

    _____

Solution Formation and Characteristics

In this experiment, you will

- Test the influence of temperature on solubility.
- Attempt to dissolve various solutes in water and classify each solute as soluble or insoluble.

- Test for the dissociation of soluble solutes.
- Attempt to dissolve a variety of solutes in a polar solvent and in a nonpolar solvent.

INTRODUCTION

Chemicals are often used in the form of homogeneous mixtures called **solutions**. Generally, a solution contains a relatively large amount of one component (the **solvent**) in which smaller amounts of one or more other components (**solutes**) are dissolved.

The behavior of a solute towards a solvent depends on the nature of both the solute and the solvent. Thus, polar covalent solutes will dissolve in polar solvents and either remain in the form of molecules or dissociate to form ions as represented by the following reactions, in which water is the solvent:

$$C_2H_5OCl(l) + H_2O(l) \rightarrow C_2H_5OCl(aq) \qquad \text{Eq. 9.1}$$

$$HCl(g) + H_2O(l) \rightarrow H_3O^+(aq) + Cl^-(aq) \qquad \text{Eq. 9.2}$$

Ionic solutes dissolve in polar solvents to form ions in solution:

$$CaCl_2(s) + H_2O(l) \rightarrow Ca^{?+}(aq) + 2Cl^-(aq) \qquad \text{Eq. 9.3}$$

Factors other than the nature of the solvent and solute also affect solubility. Temperature is an important one. An increase in solvent temperature increases the solubility of most solutes, but not always by the same amount. In some cases, such as solutions that contain gaseous solutes, an increase in solvent temperature decreases the solute solubility. This can be easily verified by opening containers of carbonated soft drinks kept at different temperatures.

EXPERIMENTAL PROCEDURE

A. Influence of Temperature on Solubility

The solubility of most solutes increases with an increase in the temperature of the solvent. The extent of the increase varies from solute to solute. In this exercise, you will investigate such a variation for two solutes.

PROCEDURE

1. Use a centigram balance to accurately weigh 1.50 g samples of potassium nitrate (KNO_3) and sodium chloride (NaCl).
2. Place each sample into a separate 15-cm test tube containing 3.0 mL of distilled water measured carefully with a 10-mL graduated cylinder. Mark the tubes so you know which solute each contains.
3. Swirl the contents of each tube for 30 seconds and record the appearance of the mixtures in Table 9.1.

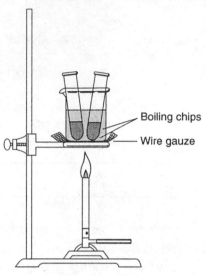

FIGURE 9.1
Heating test tubes in a water bath

Boiling chips

Wire gauze

4. Heat both tubes for 5 minutes in a boiling water bath constructed from a 250-mL beaker (see Figure 9.1).
5. Swirl each test tube two or three times during the heating.
6. Note and record the appearance of both mixtures after the 5 minutes of heating.
7. Remove the test tubes from the water bath and let them cool for 15 minutes. Swirl each tube occasionally while they cool. You can go on to other parts of the experiment while the samples cool.
8. After cooling, swirl each tube; note and record the appearance of each mixture in Table 9.1.

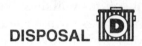

DISPOSAL

Test tube contents in container labeled "Exp. 9, Used Chemicals Part A."
Swirl each test tube to suspend the solids before pouring the contents into the container.

B. Solubility and Dissociation of Chlorine-Containing Solutes

The amount of solute that dissolves in a specified amount of solvent is called the **solubility** of that solute and given in specific units at a specific temperature. For example, the solubility of potassium sulfate (K_2SO_4), at 25°C, is 12.7 g per 100 g of water. It is often useful to express solubility in a less specific way. Thus, if the solubility of a solute is less than a specific amount, it is sometimes the practice to simply say the solute is insoluble, or not soluble in the solvent. If the solubility is greater than the specific amount, the solute is said to be soluble, with no specific amount mentioned. In this part of the experiment, and in Parts C, D, and E solubility will be determined in a nonspecific way by attempting to dissolve 0.1 g of solid solutes or 3 drops of liquid solutes in 2.0 mL of solvent. If all the solute dissolves, it will be classified as soluble. If any of the solute does not dissolve, it will be classified as insoluble.

This definition of solubility makes it very important that the samples of solid solute be weighed quite carefully even though the masses will not be recorded in any data table. Also, the amounts of liquid solvent and liquid solute must be carefully measured either by counting drops or using an appropriate graduated cylinder.

After the solute and solvent are mixed, it is important to attempt to dissolve the solute using the following technique. A piece of plastic kitchen wrap about 2 inches wide is folded in half twice along the longest dimension to form a pad that is at least four thicknesses thick. This pad is placed over the mouth of the test tube containing the solvent and solute. The pad serves as a stopper for the test tube. You should then grasp the test tube in one hand with the thumb of that hand holding the pad securely in place. The test tube should then be shaken very vigorously for a minimum of 10 seconds. The shaking should be done in an up and down direction so that the liquid in the tube is caused to forcefully impact repeatedly with the bottom of the test tube and then with the pad at the top of the test tube. After each sample is shaken, the pad can be reused after the part that came in contact with the liquid is rinsed off with distilled water.

When the technique described in the preceding paragraph is used, soluble solid solutes will form a clear (although not necessarily colorless) solution in which no undissolved solute will be visible. However, the vigorous shaking will cause some air bubbles to form, which will rise to the top and disappear. Do not confuse the air bubbles with undissolved solute. Undissolved solid solutes will be visible in the bottom of the test tube after any air bubbles have risen to the top. Some undissolved solids will also cause the resulting liquid to appear cloudy. Undissolved liquid solutes will be dispersed in the solvent as fine droplets that will cause the liquid to appear uniformly cloudy for 15 seconds or more following the shaking. Once again, it is important to not confuse air bubbles with the cloudiness resulting from undissolved liquid solutes. You might want to shake a test tube that contains only water to get an idea of what the air bubbles look like.

The chloride ion (Cl⁻) is easily detected in solutions formed when soluble solutes dissolve in water. This is done by adding silver ions (Ag⁺) in the form of a solution of silver nitrate (AgNO₃). The chloride ion and silver ion react to form an insoluble white solid precipitate of silver chloride (AgCl):

$$Cl^-(aq) + Ag^+(aq) \rightarrow AgCl(s) \qquad \text{Eq. 9.4}$$

Because of this simple test, it is convenient to use chlorine-containing solutes to illustrate solute dissociation.

PROCEDURE

1. Put 2.0 mL of distilled water into each of five clean 10-cm test tubes.
2. Test the solubility of each solute listed in Table 9.3 of the Data and Report Sheet by adding 0.1 g (if a solid) or 3 drops (if a liquid) to one of the test tubes of distilled water. Remember to measure carefully. Use a 10-mL graduated cylinder to measure the 2.0 mL of water and a centigram or electronic balance and weighing dish to weigh the solids by difference.

SAFETY ALERT

Monochloroacetic acid is a strong solid acid that will attack the skin. If you contact it, wash it off immediately with tap water and inform your instructor. 1,1,1-Trichloroethane is toxic. Avoid inhaling any fumes from the bottle. If you contact it, wash it off immediately with soap and water and inform your instructor.

3. Stopper and shake each test tube for 10 seconds as described earlier.
4. Observe each test tube and determine whether or not the solute dissolved.
5. Record your conclusions in Table 9.3. Indicate solubility with a + and insolubility with a –.
6. In each test tube in which the solute is classified as soluble, test for the presence of chloride ions by adding 3 drops of 0.1 M silver nitrate solution to the test tube. Mix the test tube contents by agitating the test tube and record the test results in Table 9.3. Use a + if chloride ions are present as indicated by the formation of a significant amount of precipitate or use a – if chloride ions are absent.
7. Dispose of the contents of the test tubes.

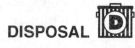

DISPOSAL All test tube contents except the one containing 1,1,1-trichloroethane in container labeled "Exp. 9, Used Chemicals Part B." The test tubes that contain insoluble solutes should be shaken vigorously to suspend the insoluble solute, then poured into the container.

DISPOSAL Test tube containing 1,1,1-trichloroethane in container labeled "Exp. 9, Used 1,1,1-trichloroethane Part B."

C. Solubility and Dissociation of Hydroxyl-Containing Solutes

Solutes that contain hydroxyl groups (OH) may dissociate and form hydroxide ions (OH⁻) when the solute dissolves. Hydroxide ions can be detected by measuring the pH of a solution. The pH concept is discussed in more detail in a future experiment, but for our purposes here, a solution pH significantly greater than the pH of the pure solvent (distilled water) indicates that the solute has dissociated to produce OH⁻ ions.

PROCEDURE

1. Put 2.0-mL samples of distilled water into each of four clean 10-cm test tubes. Put a 10.0-mL sample of distilled water into a clean 15-cm test tube.
2. Test the solubility of each solute listed in Table 9.5, as you did in Part B, with the exception of sodium hydroxide (NaOH). To avoid having to crush a solid NaOH pellet, put one pellet into the test tube that contains 10.0 mL of distilled water. Set that tube aside and swirl it occasionally for a few minutes until the pellet dissolves.

SAFETY ALERT Sodium hydroxide is a very caustic material that can damage the skin. If you contact it, wash it off immediately with tap water and inform your instructor.

3. Use the + and – notation and record the solubility test results in Table 9.5.
4. For each soluble solute, test the resulting solution with long-range (1 to 11) pH paper. Get a strip of paper about 3-cm long and test each solution by dipping in a glass stirring rod and touching the rod to the pH paper. Compare the color of the resulting spot on the paper to the chart attached to the paper dispenser. One 3-cm piece of paper

should be sufficient for all the pH tests in this part of the experiment. Rinse the stirring rod with distilled water and dry it between each test.

5. Record the measured pH values in Table 9.5.
6. In a similar way, measure the pH of the distilled water sample that contains no solute. Record this value.

DISPOSAL

Test tube contents in container labeled "Exp. 9, Used Chemicals Part C." The test tubes containing insoluble solutes should be shaken to suspend the solutes before they are emptied into the container.

D. Solubility and Dissociation of Acids

Solutes that dissociate and produce hydronium ions (H_3O^+) when dissolved in water are called acids. The strength of an acid is measured by the percentage of dissolved molecules that dissociate. In strong acids, more than 90% of the dissolved molecules dissociate, whereas in very weak acids only 0.1% or less may dissociate. Hydronium ions also can be detected by using pH paper. However, in this case, a pH value lower than that of distilled water indictes the contribution of H_3O^+ ions to the solution from a dissolved solute. The lower the pH, the greater the contribution of H_3O^+ ions by the solute, and thus the stronger the solute is as an acid.

PROCEDURE

1. Prepare mixtures as you did in Parts B and C, but use the solutes listed in Table 9.7.

SAFETY ALERT

> Phosphoric acid, acetic acid, and valeric acid will attack the skin. If you contact any of these substances, wash the contacted area immediately with tap water and inform your instructor.

2. Determine and record in Table 9.7 the solubility of each solute, using the + and − notation.
3. For each soluble solute, measure and record the pH of the resulting solution using short-range (1.4 to 2.8) pH paper.
4. Measure and record the pH of distilled water with long-range paper.

DISPOSAL

Test tube contents in container labeled "Exp. 9, Used Chemicals Part D." Shake the test tubes that contain insoluble solutes to suspend the solutes before putting them in the container.

E. Influence of Solvent and Solute Nature on Solubility

The polar nature of solvent and solute molecules exerts a significant influence on solution formation. The statement "like dissolves like" reflects this influence in a general way, where the "like" property being compared is polarity.

PROCEDURE

1. Put 2.0 mL of distilled water (a polar solvent) into each of five clean 10-cm test tubes.

2. Test the solubility of the five solutes listed in Table 9.9, as you did before. Use 0.1-g solid samples and 3-drop liquid samples. Make certain the paraffin wax sample is in the form of small thin shavings. If not, allow extra time for the solution process to take place before you classify it as insoluble.
3. Record the solubility results in Table 9.9.
4. Use five clean, *dry* test tubes and repeat the solubility tests, but use as the solvent 2.0 mL of toluene, a nonpolar solvent. If necessary, dry your test tubes and graduated cylinder used to measure the solvent by rinsing them with a little acetone, followed by air drying. Use a new, dry pad of kitchen wrap.

SAFETY ALERT

Both toluene and acetone are quite flammable. Check your area of the laboratory to make certain no open flames are nearby before you do this part of the experiment. If you are in doubt, ask your lab instructor.

5. Record the solubility test results in Table 9.9, using the + and − notation.

DISPOSAL

Test tube contents in container labeled "Exp. 9, Used Chemicals Part E." Shake the test tubes that contain insoluble solutes to suspend them before putting them into the container.
Used kitchen-wrap pads in wastebasket.

REPORT

A. Influence of Temperature on Solubility

1. On the basis of the observations in Table 9.1, decide which solute has a solubility most influenced by temperature.
2. Record your choice and the reasons for the choice in Table 9.2.

B. Solubility and Dissociation of Chlorine-Containing Solutes

1. Refer to Table 9.3 where you classified each solute as soluble (+) or insoluble (−). Transcribe the classifications to Table 9.4.
2. Use data in Table 9.3 and classify each soluble solute as dissociated (+) or nondissociated (−). Record the classifications in Table 9.4.

C. Solubility and Dissociation of Hydroxyl-Containing Solutes

1. Transcribe the solubility classifications of Table 9.5 to Table 9.6.
2. Compare the solution pHs and the pH of pure water recorded in Table 9.5 and classify each soluble solute as dissociated (+) or nondissociated (−). Record the classifications in Table 9.6.

D. Solubility and Dissociation of Acids ⊂▭▭▭▭▭⊃

1. Transcribe the solubility classifications of Table 9.7 to Table 9.8.
2. Use the pH data in Table 9.7 and label the soluble solutes in order of decreasing strength. Label the strongest 1, the next strongest 2, and so forth.
3. Record these results in Table 9.8.

E. Influence of Solvent and Solute Nature on Solubility ⊂▭▭▭⊃

1. Assume that polar and ionic solutes are soluble in polar solvents and that nonpolar solutes are soluble in nonpolar solvents.
2. On the basis of the assumption in Step 1 and the data in Table 9.9, classify each solute as polar/ionic (P) or nonpolar (N).
3. Record the classifictions in Table 9.10.

Experiment 9 ▪ Pre-Lab Review

Solution Formation and Characteristics

1. Are any specific safety alerts given in the experiment? List any that are given.

   _____

   _____

   _____

   _____

2. Are any specific disposal directions given in the experiment? List any that are given.

   _____

   _____

   _____

   _____

3. What is the mass of the solid samples used in Part A of the experiment?

   _____

4. What is the mass of the solid samples used in Parts B to E of the experiment?

   _____

5. Why is it important to carefully measure the amounts of materials used in all parts of the experiment, even though the amounts are not recorded anywhere in the experiment?

   _____

   _____

   _____

6. What observations indicate that a solid or liquid is insoluble in a solvent?

   _____

   _____

   _____

7. What test is used to detect Cl⁻ ions in Part B of this experiment? _____

   _____

   _____

8. How are OH⁻ ions detected in Part C of this experiment? _____

_____

9. How are H_3O^+ ions detected in Part D of this experiment? _____

_____

10. Describe how to correctly use pH paper. _____

_____

_____

11. What is different about the pH paper used in Parts C and D of the experiment?

_____

12. What special directions are given about the paraffin wax sample used in Part E of the experiment?

_____

_____

Experiment 9 ▪ Data & Report Sheet

Solution Formation and Characteristics

A. Influence of Temperature on Solubility

TABLE 9.1 (data)

Sample	Appearance
KNO_3 + water before heating	
NaCl + water before heating	
KNO_3 + water after heating	
NaCl + water after heating	
KNO_3 + water after cooling	
NaCl + water after cooling	

TABLE 9.2 (report)

Solute with solubility most influenced by temperature _____

Reasons for choice _____

B. Solubility and Dissociation of Chlorine-Containing Solutes

TABLE 9.3 (data)

Solute	Solubility	Ag^+ Test
Ammonium chloride (NH_4Cl)		
1,1,1-Trichloroethane ($C_2H_3Cl_3$)		
Copper (I) chloride (CuCl)		
Monochloroacetic acid ($C_2H_3O_2Cl$)		
Nickel (II) chloride ($NiCl_2$)		

TABLE 9.4 (report)

Solute	Solubility Classification	Dissociation Classification
NH_4Cl		
$C_2H_3Cl_3$		
CuCl		
$C_2H_3O_2Cl$		
$NiCl_2$		

C. Solubility and Dissociation of Hydroxyl-Containing Solutes

TABLE 9.5 (data)

Solute	Solubility	pH
Glucose (C₅H₆(OH)₅CHO)	_____	_____
Isopropyl alcohol (C₃H₇OH)	_____	_____
Octyl alcohol (C₈H₁₇OH)	_____	_____
Sodium hydroxide (NaOH)	_____	_____
Distilled water		_____

TABLE 9.6 (report)

Solute	Solubility Classification	Dissociation Classification
$C_5H_6(OH)_5CHO$	_____	_____
C_3H_7OH	_____	_____
$C_8H_{17}OH$	_____	_____
NaOH	_____	_____

D. Solubility and Dissociation of Acids

TABLE 9.7 (data)

Solute	Solubility	pH
Acetic acid	_____	_____
Citric acid	_____	_____
Phosphoric acid	_____	_____
Stearic acid	_____	_____
Valeric acid	_____	_____
Distilled water		_____

TABLE 9.8 (report)

Solute	Solubility Classification	Acid Strength
Acetic acid	_____	_____
Citric acid	_____	_____
Phosphoric acid	_____	_____
Stearic acid	_____	_____
Valeric acid	_____	_____

E. Influence of Solvent and Solute Nature on Solubility

TABLE 9.9 (data)

Solute	Solubility in Water	Solubility in Toluene
Cooking oil	_____	_____
Ethylene glycol	_____	_____
Sodium chloride	_____	_____
Sucrose	_____	_____
Paraffin wax	_____	_____

TABLE 9.10 (report)

Solute	Classification
Cooking oil	_____
Ethylene glycol	_____
Soidum chloride	_____
Sucrose	_____
Paraffin wax	_____

QUESTIONS

1. Two students test the solubility and dissociation of a chlorine-containing solute as described in Part B. Both students find a specific solute to be soluble in the solvent. However, one student gets no precipitate upon adding $AgNO_3$, while the second detects a distinct clouding. Which of the following might explain the difference in results?

   a. The distilled water in the lab was contaminated with Cl^- ions.
   b. The second student used tap water for the solvent instead of distilled water.
   c. The $AgNO_3$ solution was contaminated with Cl^-.

   Explain your answer: _____

   _____

   _____

   _____

2. According to your data in Table 9.5, which of the following characteristics or tests would be useful in differentiating between isopropyl alcohol and octyl alcohol?

   a. pH of solutions   c. Addition of $AgNO_3$ to solutions
   b. Solubility in water   d. Color of solutions

   Explain your answer: _____

   _____

   _____

   _____

3. Gasoline is drawn from the fuel tank of an auto through an inlet tube that is attached to a float that rides on the surface of the gasoline. Which of the following characteristics of water and gasoline might account for this arrangement?

   a. Water is quite soluble in gasoline.
   b. Water is insoluble in gasoline and is more dense than gasoline.
   c. Water is insoluble in gasoline and is less dense than gasoline.

   Explain your answer: _____

   _____

   _____

   _____

4. For unexplained reasons, the distilled water in a lab has a pH of about 6. A solid is added to a sample of the water and dissolves. The pH of the resulting solution is about 6. Which of the following best describes the solute?

   a. Soluble and dissociated to give OH⁻
   b. Soluble and dissociated to give H_3O^+

   c. Soluble but undissociated
   d. Soluble and may or may not be dissociated

   Explain your answer: _____

   _____

   _____

   _____

   _____

Colligative Properties of Solutions

In this experiment, you will

- Demonstrate the movement of solvent through a membrane as a result of osmotic pressure.
- Measure the depression in freezing point for water resulting from the addition of a solute and evaluate the freezing-point constant for water.

- Measure the elevation in boiling-point for water resulting from the addition of a solute, and evaluate the boiling-point constant for water.

INTRODUCTION

Certain properties of solvents change when a solute is dissolved in the solvent to form a solution. The freezing point of saltwater, for example, is lower than the freezing point of pure water. Properties of this type, called **colligative properties,** also include the vapor pressure, boiling point, and osmotic pressure of solutions. All colligative properties depend on the concentration of solute particles present in a solution. For example, a 1 molar (*M*) solution of potassium chloride (KCl) has a lower freezing point than a 1 *M* solution of sugar. This difference occurs because 1 mol of KCl dissociates and produces 2 mol of solute particles (K^+ and Cl^- ions), whereas 1 mol of sugar produces only 1 mol of undissociated sugar molecules upon dissolving.

EXPERIMENTAL PROCEDURE

A. Osmosis and Osmotic Pressure

If two solutions of different concentration are separated by a **semipermeable membrane** through which solvent but not solute can flow, solvent will flow from the dilute solution into the concentrated solution. The pressure necessary to prevent solvent flow between such a pair of solutions is called the **osmotic pressure** of the system and depends on the solute concentration difference between the two solutions. In this part of the experiment, the magnitude of the osmotic pressure will not be measured, but the associated flow of solvent through membranes will be illustrated.

PROCEDURE

1. Prepare a sugar solution by dissolving about 10 g of sucrose (table sugar) in about 90 mL of distilled water contained in a 125-mL flask.
2. Mix the resulting solution thoroughly by swirling the flask.

3. Obtain two 8-cm lengths of dialysis tubing from the stockroom. Moisten each piece in water and open it.
4. Tie one end of each piece tightly closed with string to form a tube.
5. Put 5 mL of the sugar solution prepared in Step 1 into one tube, then tie the tube tightly closed while excluding most of the air. Make sure no solution can leak out.
6. Put 5 mL of distilled water into the other tube and tie it closed, again making sure that it doesn't leak.
7. Identify the tubes in some way to remind you of their contents.
8. Rinse both tubes under running water for a short time, then pat them dry carefully with a towel. Do not attempt to completely dry the strings.
9. Weigh each tube on a centigram or electronic balance.
10. Record the mass of the water-filled tube in Table 10.1 and the mass of the sugar-solution-filled tube in Table 10.2.
11. Put the remaining sugar solution from Step 1 into a 250-mL beaker.
12. Put about 100 mL of distilled water into another 250-mL beaker.
13. Label the beakers so you know what each contains.
14. Put the water-filled dialysis tube into the beaker of sugar solution and the sugar-solution-filled tube into the beaker of water.
15. Obtain two dried prunes (nonpitted), rinse them in water, and wipe off the excess water with a towel.
16. Weigh each prune on the same balance used in Step 9.
17. Record the mass of one prune in Table 10.1, then put the prune into the beaker containing the sugar solution.
18. Record the mass of the second prune in Table 10.2, then put the prune into the beaker containing distilled water.
19. Let all samples remain in the beakers for at least 1 hour. A longer time will improve your results. Go on to other parts of the experiment while the samples soak in the beakers.
20. After at least 1 hour, remove all samples, being careful to note which beaker they were in.
21. Rinse the samples in water, wipe off excess water, and weigh them on the same balance used for the earlier weighings.
22. Record the masses of the samples from the beaker containing sugar solution in Table 10.1 and those from the beaker containing distilled water in Table 10.2.

DISPOSAL   Beaker contents and dialysis-tubing contents in sink
Used prunes in wastebasket

B. Freezing-Point Depression

The freezing point of a solution is lower than the freezing point of the pure solvent. The difference in temperature, Δt_f, between the pure solvent and solution freezing points is

$$\Delta t_f = mk_f \qquad \text{Eq. 10.1}$$

In this expression, m is the molal concentration of solute in the solution and k_f is the freezing-point constant of the solvent. Numerical values of k_f are different for each solvent and must be determined experimentally. For example, k_f is 4.90 for benzene and 3.90 for acetic acid.

The molal concentration (m) used in Equation 10.1 is defined below and can be calculated as follows:

m = the number of moles of solute dissolved in 1 kg of solvent

$$m = \frac{\text{moles of solute}}{\text{kilograms of solvent}}$$ Eq. 10.2

In this part of the experiment, you will collect data that will enable you to estimate the value of k_f for water.

PROCEDURE

1. Weigh a clean, dry, empty 100-mL beaker on a centigram or electronic balance. Record the mass in Table 10.4 in the blank corresponding to sodium chloride as the solute.
2. Add enough crushed ice to fill the beaker ½ full.
3. Dry the outside of the beaker and weigh it on the same balance used in Step 1. Record the mass in Table 10.4 in the blank corresponding to sodium chloride solute.
4. Use your graduated cylinder and add 25.0 mL of distilled water to the beaker. Record the volume in Table 10.4.
5. Use a thermometer and measure the lowest temperature resulting in the stirred ice-and-water mixture (estimate to the nearest 0.1°C). Record this temperature, which is the freezing point of pure water.
6. Accurately weigh about 5 g of sodium chloride on a centigram or electronic balance. Use the difference technique and record the empty-container (or paper) mass and the container-plus-sample mass in Table 10.4.
7. Add the salt to the ice-and-water mixture in the beaker.
8. Put your thermometer into the mixture. Stir the mixture with a glass stirring rod while observing the temperature. Continue stirring the mixture at intervals and observing the temperature until the last piece of ice just melts. The temperature at that point is the solution freezing point. Record the freezing point in Table 10.4. The process of melting might take up to 10 minutes, but be patient and get a good temperature reading.

DISPOSAL  Beaker contents in sink

9. Empty, rinse, and dry the beaker. Repeat Steps 1 to 5, then set the mixture aside and proceed with Steps 10 and 11. Record the data in Table 10.4 under the ethylene glycol solute heading.
10. Weigh an empty, dry 50-mL beaker on the centigram or electronic balance used in the repeated Steps 1 and 2. Record the mass.
11. Using a dry graduated cylinder, put about 4.5 mL of ethylene glycol into the 50-mL beaker, reweigh, and record the combined mass.
12. Add the ethylene glycol sample to the ice-and-water mixture from Step 9. You might have to rinse the last traces of ethylene glycol from the weighing beaker, using a little of the water from the ice–water mixture. Pour any such rinsings back into the ice–water mixture. Stir the resulting mixture while observing the temperature, as you did before. Record the temperature of the stirred mixture

at the time the last piece of ice melts. This is the solution freezing point.

 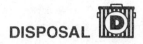

C. Boiling-Point Elevation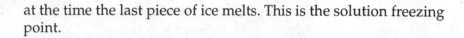

The boiling point of a solution is also different than the boiling point of the pure solvent. However, the boiling point of the solution is higher, and the difference, Δt_b, is given by Equation 10.3, where m is the molal concentration of solute in the solution and k_b is the boiling-point constant for the solvent:

$$\Delta t_b = mk_b \qquad \text{Eq. 10.3}$$

We see that the Δt values for boiling and freezing points of solutions are calculated using similar equations—only the k's are different. The boiling point constant, k_b, can be evaluated experimentally by procedures similar to those used to evaluate k_f.

PROCEDURE

1. Put about 25 mL of distilled water into a 100-mL beaker and add 1 or 2 boiling chips.
2. Heat until the water boils steadily.
3. Measure the boiling point by submerging the thermometer bulb in the boiling water, without the thermometer bulb touching the bottom of the beaker.
4. Record in Table 10.6 the maximum temperature reading from the thermometer (the boiling point of pure water). Once again, estimate the thermometer reading to 0.1°C.
5. Empty and dry the beaker.
6. Use your graduated cylinder and put exactly 25.0 mL of distilled water into the beaker. Add 1 or 2 boiling chips. Record the volume of water used in Table 10.6 under the sucrose solute heading.
7. On a centigram balance, accurately weigh by difference a sample of sucrose (table sugar) equal to about 12 g. Record the appropriate data in Table 10.6.
8. Add the sucrose sample to the water in the beaker and stir well.
9. Stirring carefully, heat the mixture until it boils steadily, then measure and record the boiling point as before. Measure the temperature as soon as you can to avoid boiling solvent from the mixture.
10. Empty, rinse, and dry the beaker.

11. Repeat Steps 5 to 9, but substitute 10 mL of ethylene glycol for the sucrose. Using a 50-mL beaker as a container, weigh the ethylene glycol by difference. Rinse the last traces of ethylene glycol from the weighing beaker by pouring a little of the water sample into the weighing beaker, swirling, and returning the rinse material to the rest of the water sample in the 100-mL beaker. Record the data in Table 10.6 under the ethylene glycol solute reading.

 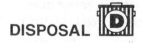

CALCULATIONS AND REPORT

A. Osmosis and Osmotic Pressure

1. Use the masses recorded in Tables 10.1 and 10.2 to determine whether each sample gained or lost mass. Indicate your results in Table 10.3, with a + for a gain and a − for a loss.
2. Use the results of Step 1 and determine the direction of solvent flow for each sample: into the sample (+) or out of the sample (−). Record your conclusions.
3. On the basis of these conclusions, decide whether the solution concentration inside the sample was higher (+), or lower (−), or about equal (0) to that of the surrounding liquid. Remember that, during osmosis, solvent flows through membranes from the dilute solution into the concentrated solution. Record your conclusions.

B. Freezing-Point Depression

1. The total mass of water in each sample equals the mass of ice plus the mass of water added from the graduated cylinder. Determine this total mass from data in Table 10.4 by assuming the water poured from the graduated cylinder had a density of 1.00 gram per milliliter (g/mL). Record the total mass of the water sample in Table 10.5.
2. Calculate and record the mass of the sodium chloride solute sample used.
3. One mole of sodium chloride weighs 58.5 g. Use this value and the sodium chloride sample mass to calculate the moles of sodium chloride in the sample (Equation 10.4). Record the result in Table 10.5.

$$\text{moles NaCl} = \text{mass NaCl} \times \frac{1 \text{ mol NaCl}}{58.5 \text{ g NaCl}} \qquad \text{Eq. 10.4}$$

4. Calculate and record Δt_f from the freezing points of the ice and water mixture and the ice water and salt mixture.
5. Express the total mass of the water sample in kilograms and record the result.
6. Calculate the molality of the solution by dividing the number of moles of sodium chloride by the number of kilograms of water in the sample (Equation 10.2). Record the result.
7. Note that Equation 10.1 can be rearranged into the form

$$\frac{\Delta t_f}{m} = k_f \qquad \text{Eq. 10.5}$$

Use this equation, the calculated Δt_f, and the calculated molality of the solution to evaluate k_f. Record the result of this calculation in Table 10.5.

8. Repeat the above calculations, using the data collected in Table 10.4 for ethylene glycol (1 mol = 62.0 g). Record the results.

C. Boiling-Point Elevation

1. The mass of water in each sample is obtained from the volume used (25.0 mL) and the water density of 1.00 g/mL. Use data from Table 10.6 and calculate and record this mass in Table 10.7.

2. Using data from Table 10.6, calculate and record the mass of sucrose solute added to the water.
3. One mole of sucrose weighs 342.3 g. Use this value and the mass of the sucrose sample to calculate the number of moles of sucrose in the sample (see Equation 10.4). Record this result.
4. Express the mass of water in the sample in kilograms and record the result.
5. Calculate the molality of the solution by using Equation 10.2.
6. Rearrange Equation 10.3 to allow you to evluate k_b from Δt_b and m. Then calculate k_b. Record the result.
7. Repeat the above calculations, using the data collected in Table 10.6 for ethylene glycol (1 mol = 62.0 g). Record the result in Table 10.7.

Experiment 10 ▪ Pre-Lab Review

Colligative Properties of Solutions

1. Are any specific safety alerts given in the experiment? List any that are given.

   _____

2. Are any specific disposal directions given in the experiment? List any that are given.

   _____

   _____

3. List the steps used to weigh a sample of material using the difference technique.

   _____

   _____

4. What special precautions should be taken when you prepare dialysis tubes full of liquid in Part A?

   _____

   _____

5. In Part A, why is it important to note which centigram or electronic balance is used when the liquid-filled dialysis tubes and the prunes are first weighed?

   _____

   _____

6. To what extent should you estimate your thermometer readings when making temperature measurements in this experiment?

   _____

7. In Part B, why must the 100-mL beaker used in Step 1 be dry? _____

   _____

8. How do you know what temperature to record as a freezing point in Part B?

   _____

   _____

9. What precaution is taken in this experiment to make sure that boiling occurs smoothly without spattering?

   _____

10. In Parts B and C, how do you make certain that all of the weighed ethylene glycol sample is added to the water solution?

_____

_____

11. How do you know what temperature to record as a boiling point in Part C?

_____

_____

Experiment 10 ▪ Data & Report Sheet

Colligative Properties of Solutions

A. Osmosis and Osmotic Pressure

TABLE 10.1   Samples Placed in Sugar Solution (data)

Sample	Mass Before Soaking	Mass After Soaking
Dialysis tubing containing water	_____	_____
Prune	_____	_____

TABLE 10.2   Samples Placed in Water (data)

Sample	Mass Before Soaking	Mass After Soaking
Dialysis tubing containing sugar solution	_____	_____
Prune	_____	_____

TABLE 10.3 (report)

Sample	Surrounding Liquid	Change in Sample Mass	Direction of Solvent Flow	Conc. of Solution (inside sample relative to surrounding liquid)
Dialysis tubing + water	Sugar solution	_____	_____	_____
Dialysis tubing + sugar solution	Pure water	_____	_____	_____
Prune	Sugar solution	_____	_____	_____
Prune	Pure water	_____	_____	_____

B. Freezing-Point Depression

TABLE 10.4 (data)

	Sodium Chloride Solute	Ethylene Glycol Solute
Mass of beaker + ice	_____	_____
Mass of empty beaker	_____	_____
Volume of water added	_____	_____
Freezing point of pure water	_____	_____
Mass of solute sample + container	_____	_____
Mass of empty container	_____	_____
Freezing point of solution	_____	_____

TABLE 10.5 (report)

	Sodium Chloride Solute	Ethylene Glycol Solute
Total water mass	_____	_____
Mass of solute	_____	_____
Moles of solute	_____	_____
Δt_f	_____	_____
Water mass in kg	_____	_____
Solution molality	_____	_____
k_f	_____	_____

C. Boiling-Point Elevation

TABLE 10.6 (data)

Boiling point of pure water _____

	Sucrose Solute	Ethylene Glycol Solute
Volume of water used	_____	_____
Mass of solute + container	_____	_____
Mass of empty container	_____	_____
Boiling point of solution	_____	_____

TABLE 10.7 (report)

	Sucrose Solute	Ethylene Glycol Solute
Total water mass	_____	_____
Mass of solute	_____	_____
Moles of solute	_____	_____
Δt_b	_____	_____
Water mass in kg	_____	_____
Solution molality	_____	_____
k_b	_____	_____

QUESTIONS

1. Suppose that in Part A of the experiment you confused the dialysis tube samples and put them into the wrong beakers. What results would have been observed?

   a. Both samples would increase in mass.
   b. Both samples would decrease in mass.
   c. One sample would increase and the other would decrease in mass.
   d. Both samples would remain essentially constant in mass.

   Explain your answer: _____

   _____

2. In the calculations used to obtain the k_f values tabulated in Table 10.5, one fact has not been taken into account. This will cause a significant difference between the k_f value obtained using NaCl solute and the value obtained using ethylene glycol solute. What is this fact?

   a. NaCl dissociates in solution while ethylene glycol does not.
   b. NaCl is solid while ethylene glycol is a liquid.
   c. NaCl is less soluble in water than ethylene glycol.
   d. NaCl has a higher melting point than ethylene glycol.

   Explain your answer: _____

   _____

   _____

3. How should the k_f value obtained with NaCl as the solute and recorded in Table 10.5 be modified before it is equivalent to the value obtained using ethylene glycol solute?

   a. It should be multipled by 2.   b. It should be divided by 2.
   c. No modification is needed.

   Explain your answer: _____

   _____

   _____

4. Suppose in Part C that a student ignores the direction in Step 9 to measure the boiling point as soon as possible. As a result, a significant amount of the solvent (water) boils away and is lost. How would this mistake influence the measured boiling point compared to the correct value it would have if the mistake was not made?

   a. Measured value would be lower than correct value.
   b. Measured value would be higher than correct value.
   c. Measured value would not be different from correct value.

   Explain your answer: _____

   _____

5. Suppose you want to purchase a syrup containing water and sugar and you have two brands from which to choose. If you want to purchase on the basis of highest sugar content, which should you buy?

   a. The syrup with the highest freezing point.
   b. The syrup with the highest boiling point.

   Explain your answer: _____

   _____

   _____

6. How would the freezing and boiling points of distilled water and seawater compare?

   a. Both the freezing and boiling points of seawater would be lower.
   b. Both the freezing and boiling points of seawater would be higher.
   c. The seawater freezing point would be lower and the seawater boiling point higher.
   d. The seawater freezing point would be higher and the seawater boiling point lower.

   Explain your answer: _____

   _____

   _____

7. Ethylene glycol is the active ingredient in most commercial automobile antifreeze. Why wouldn't sodium chloride be used in a similar way?

   a. Sodium chloride solute does not lower the freezing point of water.
   b. Sodium chloride does not dissolve in water.
   c. Sodium chloride lowers the freezing point of water but has other undesirable characteristics.
   d. Sodium chloride is a solid while ethylene glycol is a liquid.

   Explain your answer: _____

   _____

   _____

Reaction Rates and Equilibrium

In this experiment, you will

- Demonstrate the influence of reactant concentration on reaction rates.
- Demonstrate the influence of temperature on reaction rates.
- Demonstrate the influence of catalysts on reaction rates.
- Demonstrate Le Chatelier's principle.

INTRODUCTION

It is often important and useful to find answers to the following questions about chemical reactions: (1) How fast does a reaction take place under specific conditions, (2) what factors influence the rate of a chemical reaction, and (3) what factors influence the equilibrium position for a reaction?

The benefits derived from answering these questions are illustrated by the development of the Haber process for "fixing" (reacting) nitrogen. Nitrogen gas, which comprises nearly 80% of the atmosphere, is quite inert and does not react readily to form compounds. However, it is the nitrogen compounds that are useful to most living organisms. Prior to 1905, the nitrogen compounds used to manufacture fertilizer and explosives came primarily from natural nitrate-containing deposits. In 1905 ammonia (NH_3) was first synthesized from atmospheric nitrogen and hydrogen, using the Haber process. The reaction involved appears to be quite simple:

$$N_2(g) + 3H_2(g) \rightleftharpoons 2NH_3(g) \qquad \text{Eq. 11.1}$$

However, the process was developed and made practical only after an understanding was gained of the factors that influence the reaction rate and equilibrium position of the reaction in Equation 11.1.

The rates of chemical reactions depend on a number of factors including the nature of the reactants, reactant concentrations, reaction temperature, and the presence of catalysts. We will examine the last three factors in this experiment.

EXPERIMENTAL PROCEDURE

A. Influence of Concentration on Reaction Rates

Reaction rates normally increase with an increase in reactant concentrations. For example, wood (a reactant) burns more rapidly in pure oxygen (a reactant) than in air that contains only about 20% oxygen.

Both the concentration and temperature effects will be studied using the following slow reaction:

$$3HSO_3^-(aq) + IO_3^-(aq) \rightarrow 3HSO_4^-(aq) + I^-(aq) \qquad \text{Eq. 11.2}$$

To study such a reaction, it is necessary to follow the rate at which products are formed or reactants are consumed. In this experiment, the rate of consumption of the bisulfite ion (HSO_3^-) will be followed. This is possible because of the following reactions that take place rapidly as long as reactants are available:

$$5I^-(aq) + IO_3^-(aq) + 6H^+(aq) \rightarrow 3I_2(aq) + 3H_2O(l) \qquad \text{Eq. 11.3}$$

$$3I_2(aq) + 3H_2O(l) + 3HSO_3^-(aq) \rightarrow 3HSO_4^-(aq) + 6I^-(aq) + 6H^+(aq) \quad \text{Eq. 11.4}$$

We see that as soon as the HSO_3^- is depleted, I_2 formed by the reaction of Equation 11.3 will not be changed to I^- by the reaction of Equation 11.4. The I_2 will quickly accumulate and form a dark blue complex with starch, which is added as an indicator:

$$\text{Starch} + I_2 \rightarrow \text{blue color} \qquad \text{Eq. 11.5}$$

Thus, the time that elapses between mixing the reactants and the appearance of a blue color in the solution is a measurement of the rate at which HSO_3^- is reacted. The shorter the time, the greater the rate.

PROCEDURE

1. Work in pairs on this part and the temperature-effect part of the experiment. Each partner is to record the data, do the calculations, and submit a separate report.

2. Take two clean, dry 10-cm test tubes and two clean, dry 10-mL graduated cylinders to the reagent storage area. Use one of the cylinders to measure 7 mL of 0.03 M potassium iodate solution (KIO_3). Pour the solution into one of the test tubes and identify it in some way so you can remember what it contains. Use the other graduated cylinder to measure 5 mL of 0.05 M sodium bisulfite solution ($NaHSO_3$). Pour the solution into the other test tube and identify it so you can remember the contents.

3. Return to your work area and put a clean, dry plastic dropper into each of the test tubes. Put distilled water into a clean 10-cm test tube, identify the tube so you can remember its contents, and put a plastic dropper into it. The droppers must be used with only the liquids in which they are now located during all of the trials done in this part of the experiment and Part B, so be very careful not to mix up the droppers.

4. Refer to Table 11.1 and perform the three trials outlined there. For each trial, use the appropriate dropper to put the indicated number of drops of 0.03 M KIO_3 solution into a clean, dry 10-cm test tube. Add the indicated number of drops of distilled water to the same test tube and agitate the mixture to mix the liquids. The KIO_3 solution is the source of the IO_3^- ion of Equation 11.2.

5. Use the appropriate dropper to put the indicated number of drops of $NaHSO_3$ solution into a clean, dry 15-cm test tube. The $NaHSO_3$ solution is the source of the HSO_3^- ion of Equation 11.2. This solution also contains the starch indicator. *It is very important that the number of drops of each reactant be exact.* If you lose count or make some other kind of counting error, begin that trial again using clean, dry test tubes. If it is necessary to wash test tubes before they are used, be careful to dry them completely on the inside, using a cloth or paper towel or an acetone rinse followed by air drying.

   NOTE: The contents of the two test tubes are to be mixed and a stopwatch or a watch with second-counting capability is to be used to measure the time that elapses between mixing and the appearance of a blue color. Read Steps 6 to 9 completely before you do any trials.

6. Quickly pour the contents of the small test tube into the larger test tube.
7. Begin timing immediately upon mixing the contents of the test tubes.
8. During the first 5 seconds after mixing, agitate the large test tube rapidly to make certain the solutions are well mixed.
9. When a blue color first appears, stop the stopwatch or otherwise note the elapsed time to the nearest second. Record the elapsed time in Table 11.1 for each trial.

DISPOSAL    Test tube contents into container labeled "Exp. 11, Used Chemicals Parts A and B"

B. Influence of Temperature on Reaction Rates

Most chemical reactions speed up when the temperature is increased. Chemists sometimes use the rough rule of thumb that a 10°C increase in temperature will double the rate of a reaction. The influence of temperature on a reaction rate will be studied using the same reactions that were used in Part A. Reaction temperatures will be established and maintained by water baths.

PROCEDURE

1. For each trial listed in Table 11.3, measure the KIO_3 solution and distilled water into a clean, dry 10-cm test tube as you did in Part A. Measure the $NaHSO_3$ solution into a clean, dry 15-cm test tube as you did in Part A.
2. Before mixing the reactants, being them both close to the same temperature by placing them in a water bath made from a 250-mL beaker containing about 150-mL of water. The water is adjusted to the appropriate temperature. Leave them in the bath for at least 5 minutes before mixing.
   NOTE: The reactants are to be mixed and the reaction timed as in Part A. However, the baths must be set to proper temperatures before the test tubes are put in. Read Steps 3 to 5 completely before you proceed.
3. Carry the reactions out at three temperatures: room temperature (no bath required), about 10°C below room temperature (use cold tap water or ice to adjust the bath temperature), and about 10°C above room temperature (use hot tap water to adjust the bath temperature).
4. In each trial, mix the reactants and leave the larger test tube in the bath until the reaction is completed.
5. After the reaction is completed, insert a thermometer into the bath. Record the observed temperature in Table 11.3. The room temperature should be obtained from your thermometer after it has been exposed to the room conditions for at least 5 minutes.
6. Record the elapsed time to the nearest second for each reaction in Table 11.3.

DISPOSAL    Test tube contents and unused reactants in container labeled "Exp. 11, Used Chemicals Parts A and B"

C. Influence of Catalysts on Reaction Rates

Some substances, known as **catalysts,** have the ability to speed up chemical reactions without being consumed in the process. Catalysts operate in a variety of different ways. Some provide an alternate, faster pathway for a reaction by forming intermediate compounds with one or more of the reactants (**homogeneous catalysis**). Others provide surfaces on which the reactants adsorb and react more easily (**heterogeneous catalysis**). The catalytic exhaust converter of automobiles is an example of a heterogeneous surface catalyst. Whatever mechanism is involved, catalysts are very useful and important because of their rate-enhancing abilities.

The effect of catalysts will be studied as a part of this experiment by observing the rate of decomposition of a solution of hydrogen peroxide:

$$H_2O_2(aq) \rightarrow 2H_2O(l) + O_2(g) \qquad \text{Eq. 11.6}$$

This reaction can be followed easily by noting the rate of evolution of gaseous oxygen bubbles from the solution.

PROCEDURE

1. The remainder of the experiment (this part and Part D) should be performed on an individual basis.
2. Use a clean graduated cylinder and obtain about 10 mL of 3% hydrogen peroxide.
3. Put 20 drops of hydrogen peroxide into each of eight clean 10-cm test tubes.
4. Test the catalytic activity of the substances listed in Table 11.5 by adding 2 drops of solutions or about 0.1 g of solids to separate test tubes containing hydrogen peroxide.
5. Mix well and look for the immediate rapid evolution of gas. If a rapid evolution of gas is seen, note that fact as described in Step 6.
6. Observe the other test tubes for 5 minutes. Look for the evolution of gas bubbles or the collection of gas bubbles on the side of the test tube. Record your observations in Table 11.5 in terms of the rate of production of oxygen gas (+++ rapid; ++ slow; + very slow; – none).

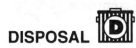

DISPOSAL Test tube contents and unused hydrogen peroxide solution in container labeled "Exp. 11, Used Chemicals Part C." The test tubes that contain the solid metal oxides should be shaken vigorously to suspend the solid before being emptied into the container.

D. Equilibrium and Le Chatelier's Principle

In 1884 Henri Le Chatelier stated a very important principle concerning equilibrium systems. According to Le Chatelier's principle, *any stress applied to a system at equilibrium will cause the position of equilibrium to shift in a way that will relieve the stress.* Consider, for example, the following hypothetical reaction:

$$A + B \rightleftharpoons C + \text{heat} \qquad \text{Eq. 11.7}$$

According to this expression, the reactants A and B react to produce product C and liberate a quantity of heat. Reactions that give off heat as a product are called exothermic. Reactions that absorb heat (heat is a reactant) are called endothermic. At equilibrium, the rate at which A and B react to form C and liberate heat is equal to the rate at which C absorbs heat, and decomposes to form A and B. In short, at equilibrium the rates of the forward and reverse reactions are identical. Under such equilibrium conditions, the concentrations of A, B, and C in the reaction mixture must remain constant with time. Now, suppose some A were added to the equilibrium system. The resulting stress can be relieved by eliminating the excess A. This is accomplished when some B reacts with a part of the added A to form C. This process will occur until the rates of the forward and reverse reactions are once again equal. The new equilibrium will be characterized by a higher concentration of A and C, and a lower concentration of B than the original; thus, the equilibrium has shifted toward the right. If the original equilibrium system were heated (heat added), the resulting stress would be relieved by using up some of the excess heat. This is accomplished when C reacts with heat to form A and B. The equilibrium will therefore shift to the left, and the new equilibrium will be characterized by higher A and B concentrations, and a lower C concentration than the original.

In this experiment, you will set up two equilibrium systems, put stresses on the systems, and observe the resulting changes. You will then determine whether the involved reactions are exothermic or endothermic. The two systems to be studied are represented by Equations 11.8 and 11.9.

PROCEDURE

$$Fe^{3+}(aq) \; + \; SCN^-(aq) \; \rightleftharpoons \; FeSCN^{2+}(aq)$$
$$\text{light brown} \quad\quad \text{colorless} \quad\quad\quad \text{red}$$

Eq. 11.8

1. Put 1 drop of 0.1 M iron (III) nitrate solution ($Fe(NO_3)_3$) into a clean 10-cm test tube. Add 1 drop of 0.1 M potassium thiocyanate solution (KSCN) to the test tube. Add 5 mL of distilled water to the test tube and mix well.

2. Put 20 drops of the solution prepared in Step 1 into each of five clean 10-cm test tubes.

3. Use one of the test tubes as a color reference that represents the initial equilibrium condition. Note its orange color that results from the equilibrium mixture containing both the light-brown-colored Fe^{3+} ions and red-colored $FeSCN^{2+}$ ions in solution. Set the reference tube aside so you can compare other tube colors to it.

4. The remaining four samples in the test tubes will be used in the four tests described below. Perform one test on each test tube. In each test, make sure the test tube contents are mixed well during heating or cooling and after test reagents are added. Then, note and record in Table 11.7 any differences in the color of the solution compared to the reference solution. Focus on the intensity (lighter or darker) of the orange color.

a. Add 2 drops of 0.1 M iron (III) nitrate solution ($Fe(NO_3)_3$) to one test tube.

b. Add 2 drops of 0.1 M potassium thiocyanate solution (KSCN) to one test tube.

c. Add 1 drop of 0.1 M silver nitrate solution ($AgNO_3$) to one test tube.

d. Cool one test tube in a small beaker of ice and water for 10 minutes.

$$Ca(C_2H_3O_2)_2(s) \; \rightleftharpoons \; Ca^{2+}(aq) \; + \; 2C_2H_3O_2^-(aq)$$
$$\text{white solid} \quad\quad \text{colorless} \quad\quad \text{colorless}$$

Eq. 11.9

5. Put 20 drops of a saturated solution of calcium acetate $(Ca(C_2H_3O_2)_2)$ into a clean, dry 10-cm test tube. Note that this solution contains the equilibrium concentration of both Ca^{2+} and $C_2H_3O_2^-$ ions that are involved in the equilibrium represented by Equation 11.9.

6. Record the appearance of the solution in Table 11.9. Then, put the test tube into a boiling water bath for 3 minutes. Agitate the test tube to mix the contents after heating for 3 minutes, then note and record the appearance of the sample in the test tube.

7. Cool the sample in a small beaker of ice and water for 3 minutes. Agitate the test tube two or three times during the cooling. When the cooling is completed, note and record the appearance of the sample.

DISPOSAL  Test tube contents and unused reagents in container labeled "Exp. 11, Used Chemicals Part D"

CALCULATIONS AND REPORT

A. Influence of Concentration on Reaction Rates

1. Notice that according to the data in Table 11.1, the amount of KIO_3 was the same in each reaction, but the amount of $NaHSO_3$ was changed, while the total volume was kept constant.

2. Represent the concentration of $NaHSO_3$ in the first trial by C and write the concentrations in the other trials in terms of C. For example, if 15 drops of $NaHSO_3$ had been used, the volume and therefore the concentration is 3 times that of Trial 1 in which 5 drops were used. Therefore, the concentration of $NaHSO_3$ is 3C. Record these concentrations in Table 11.2.

3. Reaction rates are inversely proportional to reaction times—longer reaction times mean the reactions are going slower. Represent each reaction rate by the reciprocal of the reaction time, $1/t$. Divide these fractions out and record them in Table 11.2. For example, a reaction time of 42 seconds would represent a rate of $1/42 = 0.024$ and should be recorded as 0.024.

4. Compare the concentration changes with the reaction-rate changes recorded in Table 11.2. Do this by taking the ratio of two concentrations and comparing the ratio to that of the two corresponding rates. For example, suppose a concentration of C corresponded to a reaction rate of 0.019 and a concentration of 2C corresponded to a reaction rate of 0.076. The concentration ratio is $2C/C = 2$. Therefore, the second concentration is double the first. The corresponding reaction rate ratio is $.076/.019 = 4$. Therefore, the second reaction rate is four times the first. Thus, a doubling of concentration (from C to 2C) caused a fourfold increase in reaction rate (from 0.019 to 0.076).

5. Record these results in Table 11.2 by writing the concentration changes in Trials 2 and 3 compared to the first trial (double, four times, and so on) and the corresponding reaction rate changes (three times, four times, and so on).

B. Influence of Temperature on Reaction Rates

1. Notice that according to the data in Table 11.3, the reactant concentrations were the same for each trial. Only the temperature was changed.

2. Calculate the reaction rates for each trial as $1/t$, as you did in Part A. Record these rates and the corresponding reaction temperatures in Table 11.4.
3. Use data for the reaction at room temperature as your reference and determine how the rate changed (doubled, one-fourth as great, and so forth) when the temperature was decreased by about 10°C and when it was increased by about 10°C. Record your results in Table 11.4.

C. Influence of Catalysts on Reaction Rates

1. On the basis of the data in Table 11.5, classify each of the solids tested according to their catalytic activity—high, low, or no activity. Record your conclusions in Table 11.6.
2. Notice that each solution tested provided two ions. For example, KCl provides both K^+ and Cl^-. On this basis and the data in Table 11.5, classify each ion according to its catalytic activity—high, low, or no activity. For example, suppose a $CaCl_2$ solution caused the peroxide to give off gas slowly, while a KCl solution had no effect. Since KCl had no effect, neither of its ions (K^+ or Cl^-) had any catalytic activity. The $CaCl_2$ had low catalytic activity, which must be caused by the Ca^{2+} ion since the Cl^- was already shown to have no activity by the KCl results. Thus, K^+ and Cl^- would be classified as having no activity, while Ca^{2+} would be classified as having low activity. Record your conclusions in Table 11.6.

D. Equilibrium and Le Chatelier's Principle

1. Use the test results recorded in Table 11.7 and decide in which direction the equilibrium of the reaction in Equation 11.8 shifted in each case. Remember to focus on the orange color of the solution. The more intense the orange color, the more red it contains. In Test C where $AgNO_3$ was added, ignore the cloudiness that resulted and, again, focus on the color. Record your conclusions in Table 11.8.
2. On the basis of the conclusion for the test that involved cooling, classify the reaction in Equation 11.8 as exothermic (heat is on the right) or endothermic (heat is on the left). Record your conclusion.
3. Use the test results in Table 11.9 and decide in which direction the equilibrium of the reaction in Equation 11.9 shifted when heat was applied. Use the shift resulting from heating to classify the reaction in Equation 11.9 as exothermic or endothermic. Record your conclusions in Table 11.10.

Experiment 11 ▪ Pre-Lab Review

Reaction Rates and Equilibrium

1. Are any specific safety alerts given in the experiment? List any that are given.

   _____

2. Are any specific disposal directions given in the experiment? List any that are given.

   _____

   _____

3. What experimental observation will you use to determine the end of the reaction in Parts A and B?

   _____

4. What reaction takes place in Parts A and B to produce the observation you gave as a response to Question 3?

   _____

   _____

5. What are the sources of the IO_3^- reactant and the HSO_3^- reactant in Parts A and B?

   _____

   _____

6. How do you measure the amounts of reactant solutions used in Parts A and B?

   _____

   _____

7. When do you begin timing the reactions of Parts A and B? When do you stop timing the reactions?

   _____

   _____

8. What are water baths used for in Part B? _____

   _____

9. In Part C, what observation allows you to detect the decomposition of hydrogen peroxide?

   _____

   _____

10. In Part D, what property will you observe that will allow you to determine the direction of equilibrium shift for the reaction $Fe^{3+}(aq) + SCN^-(aq) \rightleftharpoons FeSCN^{2+}(aq)$?

_____

_____

Experiment 11 ▪ Data & Report Sheet

Reaction Rates and Equilibrium

A. Influence of Concentration on Reaction Rates

TABLE 11.1 (data)

Trial Number	Amount of KIO₃ Soln.	Amount of Water	Amount of NaHSO₃ Soln.	Reaction Time
	10-cm Test Tube		15-cm Test Tube	
1	20 drops	35 drops	5 drops	_____
2	20 drops	30 drops	10 drops	_____
3	20 drops	20 drops	20 drops	_____

TABLE 11.2 (report)

Trial Number	NaHSO₃ Concentration	Reaction Rate		Concentration Change	Reaction-Rate Change
1	_____C_____	_____			
2	_____	_____	between Trials 1 & 2	_____	_____
3	_____	_____	between Trials 1 & 3	_____	_____

B. Influence of Temperature on Reaction Rates

TABLE 11.3 (data)

Trial Number	Amount of KIO₃ Soln.	Amount of Water	Amount of NaHSO₃ Soln.	Reaction Temp.	Reaction Time
	10-cm Test Tube		15-cm Test Tube		
1	20 drops	30 drops	10 drops	_____	_____
2	20 drops	30 drops	10 drops	_____	_____
3	20 drops	30 drops	10 drops	_____	_____

TABLE 11.4 (report)

Trial Number	Temperature	Reaction Rate		Reaction Rate Change
1	_____	_____		
2	_____	_____	for 10°C increase	_____
3	_____	_____	for 10°C decrease	_____

C. Influence of Catalysts on Reaction Rates

TABLE 11.5 (data)

Substance Tested	Rate of O_2 Evolution
Solids	
Iron (III) oxide (Fe_2O_3)	_____
Manganese dioxide (MnO_2)	_____
Solutions	
0.1 *M* iron (III) chloride ($FeCl_3$)	_____
0.1 *M* manganese (II) chloride ($MnCl_2$)	_____
0.1 *M* sodium chloride (NaCl)	_____
0.1 *M* potassium chloride (KCl)	_____
6 *M* sodium hydroxide (NaOH)	_____
6 *M* hydrochloric acid (HCl)	_____

TABLE 11.6 (report)

Substance Tested	Catalytic Activity
Solid Fe_2O_3	_____
Solid MnO_2	_____
Fe^{3+}	_____
Mn^{2+}	_____
K^+	_____
Na^+	_____
H^+	_____
Cl^-	_____
OH^-	_____

D. Equilibrium and Le Chatelier's Principle

TABLE 11.7 (data)

Test Performed	Color Change Relative to Reference
Add Fe(NO₃)₃	_____
Add KSCN	_____
Add AgNO₃	_____
Cool solution	_____

TABLE 11.8 (report)

Test Performed	Shift Direction of Equilibrium
Add Fe(NO₃)₃	_____
Add KSCN	_____
Add AgNO₃	_____
Cool solution	_____
Is the reaction in Equation 11.8 exothermic or endothermic?	_____

TABLE 11.9 (data)

Test Performed	Observations
Heat solution	_____
Cool the hot solution	_____

TABLE 11.10 (report)

Test Performed	Shift Direction of Equilibrium
Heat solution	_____
Cool the hot solution	_____
Is the reaction in Equation 11.9 exothermic or endothermic?	_____

QUESTIONS

1. Suppose you want to study the influence of IO_3^- concentration on the rate of the reaction used in Part A. Which of the following procedures would be appropriate?

   a. Use different amounts of KIO_3 solution with a constant amount of water and a constant amount of $NaHSO_3$ solution.
   b. Use different amounts of KIO_3 solution, constant amounts of $NaHSO_3$ solution, and different water amounts to give a constant total amount.
   c. Use different amounts of KIO_3 solution, different amounts of $NaHSO_3$ solution, and constant amounts of water.

   Explain your answer: _____

   _____

   _____

2. When food is cooked (a chemical reaction), which of the following influences on reaction rates is utilized?

   a. The influence of reactant concentrations   c. The influence of catalysts
   b. The influence of temperature

   Explain your answer: _____

   _____

   _____

3. Consider the following equilibrium:

   $$\text{Heat} \quad + \quad 2HI(g) \quad \rightleftharpoons \quad H_2(g) \quad + \quad I_2(g)$$
   $$\text{colorless} \qquad \text{colorless} \qquad \text{violet}$$

   If an equilibrium mixture is cooled, what will be the effect on the color of the mixture?

   a. It will become darker violet.   c. No change in color will occur.
   b. It will become lighter violet.

   Explain your answer: _____

   _____

   _____

4. Are the reaction-rate changes recorded in Table 11.4 consistent with the rule of thumb concerning the effect of a 10°C temperature increase on the rate of reactions?

   a. Yes   **b.** No

   Explain your answer: _____

   _____

   _____

5. Silver ions react with thiocyanate ions as follows:

$$Ag^+(aq) + SCN^-(aq) \rightarrow AgSCN(s)$$

   This reaction

   a. Explains the color change noted in Table 11.7 when $AgNO_3$ was added.
   b. Explains the shift direction of equilibrium recorded in Table 11.8 when $AgNO_3$ was added.
   c. Both a and b are correct.
   d. Neither a nor b is correct.

   Explain your answer: _____

   _____

   _____

   _____

6. What would happen to the equilibrium expressed by the reaction in Equation 11.8 if a reagent were added that lowered the Fe^{3+} ion concentration?

   a. Equilibrium would shift left.   **c.** Equilibrium would not change.
   b. Equilibrium would shift right.

   Explain your answer: _____

   _____

Acids, Bases, Salts, and Buffers

In this experiment, you will

- Neutralize an acid with a base.
- Demonstrate the effects of acids and bases on various substances.
- Demonstrate the color changes of various acid–base indicators.
- Measure the pH of a variety of common substances.

- Demonstrate hydrolysis reactions of salts.
- Prepare several buffer solutions and measure their buffer capacities.

INTRODUCTION

In 1884 Svante Arrhenius proposed the first theoretical model for acids and bases. Prior to that time, these chemically opposite substances were described in terms of properties such as their taste; their effects on metals, carbonates, and dyes (called **indicators**); their feel; and their ability to react with each other. According to Arrhenius, pure water dissociates to some extent to produce hydrogen ions (H^+) and hydroxide ions (OH^-). When this occurs, equal amounts of H^+ and OH^- ions are produced:

$$H_2O(l) \rightleftharpoons H^+(aq) + OH^-(aq) \qquad \text{Eq. 12.1}$$

An **acid,** according to Arrhenius, is any substance that liberates H^+ ions when placed in water. Such a contribution causes the concentration of H^+ ions to be higher than that of OH^- ions, and the solution is said to be **acidic.** Similarly, a **base** is defined as any substance that liberates OH^- ions when placed in water. The resulting solution has a higher concentration of OH^- ions than H^+ ions and is said to be **basic,** or **alkaline.**

In a more general and useful theory proposed by Brønsted and Lowry, acids are defined as substances that donate H^+ ions (protons) and therefore are called **proton donators,** while bases are substances that accept H^+ ions and are defined as **proton acceptors.** On this basis, the autoionization of water given in Equation 12.1 is written as follows:

$$H_2O(l) + H_2O(l) \rightleftharpoons H_3O^+(aq) + OH^-(aq) \qquad \text{Eq. 12.2}$$

In this reaction, one water molecule donates a proton, H^+, and behaves as an acid, while the other water molecule accepts the proton and behaves as a base. Since the reaction is written as reversible, we see that when it proceeds from right to left, the hydronium ion (H_3O^+) behaves as an acid (proton donor) while the hydroxide ion (OH^-) behaves as a base (proton acceptor).

Nitric acid (HNO_3) behaves as a typical Brønsted acid in water solution:

$$HNO_3(aq) + H_2O(l) \rightleftharpoons H_3O^+(aq) + NO_3^-(aq) \qquad \text{Eq. 12.3}$$

We see that according to this reaction, HNO_3 added to water generates H_3O^+ ions in solution. This causes the solution to contain more H_3O^+ than

159

OH^- and thus the solution is acidic. Similarly, a base such as methyl amine (CH_3-NH_2) accepts a proton from water, thus generating OH^- ions and the solution is basic:

$$CH_3-NH_2(aq) + H_2O(l) \rightleftharpoons CH_3-NH_3{}^+(aq) + OH^-(aq) \quad \text{Eq. 12.4}$$

The strength of an acid or base is a measurement of the extent to which reactions such as Equations 12.3 and 12.4 take place. The reaction of HNO_3 takes place almost completely—1 mol HNO_3 produces 1 mol H_3O^+ and 1 mol $NO_3{}^-$ ions when dissolved in water. Thus, nitric acid is classified as a strong acid. On the other hand, the reaction of methyl amine to form the $CH_3-NH_3{}^+$ ion (Equation 12.4) does not take place completely. Part of any dissolved CH_3-NH_2 remains in the form of undissociated molecules. Thus, 1 mol CH_3-NH_2 dissolved in water produces less than 1 mol $CH_3-NH_3{}^+$ and less than 1 mol OH^- ions. Because of this, methyl amine is called a weak base.

The characteristic negative ions (anions) from acids (such as $NO_3{}^-$ in Equation 12.3) and the characteristic positive ions (cations) from bases (such as $CH_3-NH_3{}^+$ in Equation 12.4) also behave as Brønsted bases and acids, respectively. Thus, the nitrate ion can accept protons and behave as a base:

$$NO_3{}^-(aq) + H_2O(l) \rightleftharpoons HNO_3(aq) + OH^-(aq) \quad \text{Eq. 12.5}$$

However, since HNO_3 gives up protons readily (the reaction in Equation 12.3 goes completely to the right), $NO_3{}^-$ is not going to accept protons readily. Thus, $NO_3{}^-$ does not react significantly as a Brønsted base.

The $CH_3-NH_3{}^+$ ion can act as a Brønsted acid:

$$CH_3-NH_3{}^+(aq) + H_2O(l) \rightleftharpoons CH_3-NH_2(aq) + H_3O^+(aq) \quad \text{Eq. 12.6}$$

Since CH_3-NH_2 does not readily accept protons, $CH_3-NH_3{}^+$ has a tendency to give up H^+. Thus, $CH_3-NH_3{}^+$ behaves as a significant Brønsted acid.

In general, the anions of weak acids behave as significant bases while the anions of strong acids do not. Similarly, the cations of weak bases behave as significant acids, while the cations of strong bases do not.

EXPERIMENTAL PROCEDURE

NOTE: You will need boiled distilled water for portions of the experiment. Put about 250 mL of distilled water into a clean 500-mL beaker and boil for about 5 minutes. Set this aside to cool and use as needed later.

A. Acid–Base Neutralization

The term **neutralization** is often used to describe a reaction in which equal amounts of acid and base react with each other. The products of the reaction are water and a salt in dissociated form. Thus, the neutralization of nitric acid with potassium hydroxide is represented by the reaction

$$HNO_3(aq) + KOH(aq) \rightarrow H_2O(l) + K^+(aq) + NO_3{}^-(aq) \quad \text{Eq. 12.7}$$

A solution identical to that resulting from the neutralization reaction could be made by dissolving the salt, KNO_3, in water.

PROCEDURE

1. Use a clean, dry 10-mL graduated cylinder to put 2.0 mL of 1.0 *M* sodium hydroxide solution (NaOH) into a 10-cm test tube.
2. Rinse and dry your 10-mL graduated cylinder, then use it to put 2.0 mL of 1.0 *M* hydrochloric acid solution (HCl) into another 10-cm test tube.

SAFETY ALERT

1.0 *M* solutions of NaOH and HCl can both attack and damage tissue. If you contact either of the solutions, wash the contacted area with cool tap water immediately.

3. Pour the contents of one of the test tubes into the other test tube and mix the contents well by repeatedly blowing air bubbles through it using one of your plastic droppers. Test the mixed solution with a strip of blue litmus paper. The paper should turn pink.
4. Pour the resulting neutralized solution into a clean evaporating dish and evaporate to dryness over a boiling water bath (see Figure 12.1).

FIGURE 12.1
Evaporation with a steam bath

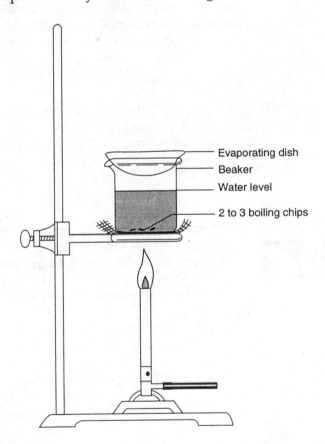

Evaporating dish
Beaker
Water level
2 to 3 boiling chips

5. While evaporation proceeds, go on to other parts of the experiment.
6. When evaporation is complete, collect the solid residue on a piece of filter paper, note and record its appearance in Table 12.1, and divide it into two equal portions.
7. Dissolve one portion in 20 drops of distilled water in a clean 10-cm test tube.
8. Test the dissolved sample for chloride ions (Cl⁻) by adding 2 drops of 0.1 *M* silver nitrate solution (AgNO₃). A distinct clouding of the

solution or the formation of a white precipitate indicates the presence of Cl⁻. Record the test results in Table 12.1.

9. Moisten the end of a clean glass stirring rod with distilled water.
10. Dip the moist end of the rod into the second portion of collected solid so that solid particles adhere to the rod.
11. Place the end of the rod with adhering solid into the flame of your burner. The appearance of a bright orange-yellow flame confirms the presence of sodium ions (Na⁺). Record the results of this flame test in Table 12.1.

DISPOSAL   Test tube contents in sink
Unused solid and filter paper in wastebasket

B. Effect of Acids and Bases on Materials

Acids and bases can react with many different substances. The extent and rate of such reactions often depends on the strength of the acids or bases involved. In some instances, the reactions are considered to be useful, as in the use of hydrochloric acid to clean newly laid brick. In other cases, the reactions are undesirable, as in the deterioration of stone statues and buildings by acid rain formed from air pollutants.

PROCEDURE

1. Place small samples of the following materials into separate, clean 15-cm test tubes: iron (Fe wire or coarse filings), zinc (Zn), filter paper, marble (CaCO₃), wool (cloth or yarn), cotton (cloth or yarn), nylon (cloth or yarn), and milk (use 2 to 3 mL).
2. Add enough 6 M hydrochloric acid (HCl) to just cover each sample. Use 20 to 30 drops of acid for the milk sample.

SAFETY ALERT

> 6 M hydrochloric acid can severely attack and damage tissue. If you contact it, immediately wash the contacted area with cool tap water, then inform your instructor.

3. Allow the samples to stand for 10 minutes or until an obvious change occurs, whichever comes first. Gentle stirring with a glass rod will help you detect the breakup of some samples. Note and record in Table 12.3 any changes that take place. Be especially aware of color changes, the evolution of gases (fizzing or bubbling), and the dissolving or breaking up of solids.

DISPOSAL   Test tube contents in container labeled "Exp. 12, Used Chemicals, Part B"

4. Repeat Steps 1 to 3 using 6 M sodium hydroxide (NaOH) in place of HCl. Record your data in Table 12.4.

SAFETY ALERT

> 6 M sodium hydroxide can severely attack and damage tissue. If you contact it, immediately wash the contacted area with cool tap water, then inform your instructor.

C. pH Measurement

The concentration of hydronium ions (H_3O^+) in solution is often expressed in terms of the pH scale devised by Søren Sørensen. On this scale, **pH** is defined as the negative logarithm of the H_3O^+ molar concentration:

$$pH = -\log[H_3O^+] \qquad \text{Eq. 12.8}$$

where $[H_3O^+]$ represents the molar concentration of H_3O^+.

In pure water, $[H_3O^+] = 10^{-7}$, and the pH = 7. This pH value is characteristic of neutral solutions. A pH value lower than 7 corresponds to an *acidic* condition, while a pH higher than 7 is *basic*.

PROCEDURE

1. Use pH paper with a range of 1 to 11 pH units and measure the pH of the substances listed in Table 12.6. Use a strip of paper about 5 cm (2 inches) long. Dip a clean glass stirring rod into the liquid being tested, then touch the rod to the pH paper. Compare the color of the spot on the paper to the chart that is located on the paper dispenser. Clean the glass stirring rod between each test. One 5-cm strip of paper should be enough for all the pH tests done in this part of the experiment.
2. If the samples in Table 12.6 are solids, dissolve about 0.1 g in 2 mL of boiled distilled water and test the resulting solution.
3. If the samples are liquids, use 5-drop samples in your smallest test tubes.
4. Record the measured pH values.

DISPOSAL Tested samples in sink
Used pH paper in wastebasket

D. Acid–Base Indicators

Dyes that change color as the pH changes are referred to as **indicators**. They are often used to locate the point at which exact amounts of acid and base have been reacted in neutralization reactions.

PROCEDURE

1. Place 20-drop samples of 0.05 *M* hydrochloric acid into each of five clean 10-cm test tubes.
2. Prepare five similar samples of 0.05 *M* sodium hydroxide.
3. Test each indicator listed in Table 12.8 by adding 1 drop of indicator to one of the HCl samples and 1 drop to one of the NaOH samples prepared in Steps 1 and 2. Mix well and record the resulting solution color in Table 12.8.

DISPOSAL Test tube contents in sink

E. Behavior of Salts in Solution

The neutralization of an acid by a base produces a **salt**. If the acid and base that reacted are both strong, the resulting cation and anion of the salt will not be significant Brønsted bases and acids, and the solution will have a pH near 7. However, if the salt produced during neutralization contains the cation of a strong base and the anion of a weak acid, the pH of the solution will not be 7. The cation of a strong base is not a significant Brønsted acid and will not influence the solution pH. However, the anion of a weak acid behaves as a significant Brønsted base that will accept protons from water to produce OH^- ions. Thus, the solution will contain more OH^- than H_3O^+, and the solution will be alkaline. For example, potassium nitrite (KNO_2) is the salt produced when potassium hydroxide, a strong base, reacts with nitrous acid, a weak acid:

$$KOH(aq) + HNO_2(aq) \rightarrow KNO_2(aq) + H_2O(l) \qquad \text{Eq. 12.9}$$

The salt dissociates to give K^+ and NO_2^- ions. The K^+ (from a strong base) is not a significant Brønsted acid. The NO_2^- (from a weak acid) is a significant Brønsted base and reacts with water in what is called a **hydrolysis reaction**:

$$NO_2^-(aq) + H_2O(l) \rightleftharpoons HNO_2(aq) + OH^-(aq) \qquad \text{Eq. 12.10}$$

PROCEDURE

1. Place 0.1-g samples of the following salts into separate, clean 10-cm test tubes. The tubes do not have to be dry. Identify each tube in some way.

   Salts: sodium acetate ($NaC_2H_3O_2$)   ammonium chloride (NH_4Cl)
                sodium carbonate (Na_2CO_3)   sodium chloride ($NaCl$)

2. Add 5 mL of boiled distilled water to each sample and agitate until the solid dissolves.

3. Use pH paper (range 1 to 11) to measure the pH of each solution and the pH of a separate sample of boiled distilled water. Record the measured pH values in Table 12.10.

DISPOSAL  Tested samples in sink
Used pH paper in wastebasket

F. Buffers

Solutions containing substances with the ability to donate protons and substances with the ability to accept protons have the capacity to maintain a fairly constant pH despite the addition of acids or bases. Solutions with this capability are called **buffers,** and the amount of acid or base they can accept without significant pH change is called the **buffer capacity.** A common type of buffer consists of a solution containing about equal amounts of a weak acid and the salt (anion) of the weak acid. Formic acid ($HCHO_2$) and its salt, sodium formate ($NaCHO_2$) could be used. Formic acid reacts with water as follows:

$$HCHO_2(aq) + H_2O(l) \rightleftharpoons H_3O^+(aq) + CHO_2^-(aq) \qquad \text{Eq. 12.11}$$

Thus, $HCHO_2$ behaves as a Brønsted acid, while CHO_2^- would behave as a Brønsted base. In a solution containing both species, the addition of acid (H^+) would not change the pH much because the CHO_2^- ions would react as follows to minimize the formation of H_3O^+:

$$CHO_2^-(aq) + H^+(aq) \text{ (from added acid)} \rightarrow HCHO_2(aq) \qquad \text{Eq. 12.12}$$

Similarly, any added base will deplete the H_3O^+ concentration, but the $HCHO_2$ molecules present will replenish them by undergoing the reaction from left to right in Equation 12.11.

PROCEDURE

1. Prepare each of the following solutions in separate, clean 10-cm test tubes. Mark the tubes so you can remember their contents.

    Solution A: 20 drops of 0.1 M sodium acetate solution ($NaC_2H_3O_2$) plus 20 drops of 0.1 M acetic acid solution ($HC_2H_3O_2$)
    Solution B: 20 drops of 0.1 M aqueous ammonia solution (NH_3) plus 20 drops of 0.1 M ammonium chloride solution (NH_4CL)
    Solution C: 20 drops of 0.1 M sodium acetate solution ($NaC_2H_3O_2$) plus 20 drops of boiled distilled water
    Solution D: 20 drops of 0.1 M sodium chloride solution (NaCl) plus 20 drops of boiled distilled water

2. Mix each solution well by using a clean dropper to bubble air through it, then pour half of it into another clean 10-cm test tube so you have two equal samples.
3. Use pH paper (range 1 to 11) to measure the pH of one sample of each solution. Record this initial pH value in Table 12.12.
4. Also measure and record the pH of a sample of boiled distilled water.
5. Add 8 drops of 0.05 M hydrochloric acid (HCl) to one sample of each solution and to a 20-drop sample of boiled distilled water. Mix well, measure, and record the resulting pH of each solution.
6. Repeat Step 5, using the other solution samples and a fresh sample of boiled distilled water, but add 0.05 M sodium hydroxide solution (NaOH) instead of HCl.

DISPOSAL Tested samples in sink
Used pH paper in wastebasket

REPORT

A. Acid–Base Neutralization

1. On the basis of the test results recorded in Table 12.1, identify the salt produced by the neutralization reaction. Record your response in Table 12.2
2. Write a reaction in Table 12.2 to represent the neutralization process that took place.

B. Effect of Acids and Bases on Materials

1. In Table 12.5, list materials that reacted with HCl and gave off gases as products.
2. Also list those that reacted but gave off no gases.
3. List those that appeared not to have reacted.
4. Repeat Steps 1 to 3 for the samples placed in sodium hydroxide.

C. pH Measurement

1. Refer to the data in Table 12.6. Arrange in Table 12.7 the substances tested in order of decreasing pH.

D. Acid–Base Indicators

1. Refer to Table 12.8. Identify an indicator that gave a very distinct color change between the acidic and basic solutions. List your response in Table 12.9.
2. Identify an indicator that gave only a slight color change.

E. Behavior of Salts in Solution

1. Compare the pH of the salt solutions listed in Table 12.10 to the pH of pure (boiled, distilled) water.
2. List in Table 12.11 the salts that produced solutions with a pH higher than that of pure water.
3. List the salts that produced solutions with a lower pH than that of pure water.
4. List the salts that had little effect on the pH of water.
5. Use this information and decide which cations behaved as significant Brønsted acids, which anions behaved as bases, and which ions did not show significant acid or base characteristics. Record your conclusions in Table 12.11.

F. Buffers

1. Refer to Table 12.12. List in Table 12.13 the solutions that fit the following categories: (a) able to buffer well only against added acid, (b) able to buffer well only against added base, (c) able to buffer well against added acid or base, or (d) had little or no ability to buffer against added acid or base.

Experiment 12 ▪ Pre-Lab Review

Acids, Bases, Salts, and Buffers

1. Are any specific safety alerts given in the experiment? List any that are given.

   _____

   _____

   _____

2. Are any specific disposal directions given in the experiment? List any that are given.

   _____

   _____

3. What material should you prepare at the beginning of the laboratory period for use in several parts of the experiment?

   _____

4. In Part A, what test results verify the presence of Cl^- and Na^+ ions? _____

   _____

   _____

5. In Part B, what observations indicate that an acid or base has caused changes to occur in materials?

   _____

   _____

   _____

6. Describe how pH paper is to be used in Part C to make pH measurements.

   _____

   _____

   _____

7. Describe how you are told to mix solutions together in 10-cm test tubes in Parts A and F.

   _____

   _____

8. In Part E, what observation is used to evaluate the behavior of salts in solution?

_____

_____

9. What observation is used in Part F to compare the buffering ability of various buffers?

_____

_____

_____

Experiment 12 ▪ Data & Report Sheet

Acids, Bases, Salts, and Buffers

A. Acid–Base Neutralization

TABLE 12.1 (data)

Appearance of solid residue _____

Result of adding AgNO$_3$ _____

Result of flame test _____

TABLE 12.2 (report)

Salt produced _____

Neutralization reaction _____

_____

B. Effects of Acids and Bases on Materials

TABLE 12.3 (data)

6 M HCl

Material	Changes Observed
Iron	_____
Zinc	_____
Paper	_____
Marble	_____
Wool	_____
Cotton	_____
Nylon	_____
Milk	_____

TABLE 12.4 (data)

6 M NaOH

Material	Changes Observed
Iron	_____
Zinc	_____
Paper	_____
Marble	_____
Wool	_____
Cotton	_____
Nylon	_____
Milk	_____

TABLE 12.5 (report)

Reactant	Substances Reacted and Gave Gases	Substances Reacted but Gave No Gases	Substances That Did Not React
Hydrochloric acid (HCl)	_____	_____	_____
Sodium hydroxide (NaOH)	_____	_____	_____

C. pH Measurement

TABLE 12.6 (data)

Substance	pH	Substance	pH	Substance	pH
Lemon juice	_____	Household ammonia	_____	Baking soda	_____
Orange juice	_____	Detergent (laundry)	_____	Aspirin	_____
Milk	_____	Detergent (dishwashing)	_____	Buffered aspirin	_____
Saliva (your own)	_____				

TABLE 12.7 (report)

Substance	pH	Substance	pH
_____	_____	_____	_____
_____	_____	_____	_____
_____	_____	_____	_____
_____	_____	_____	_____
_____	_____	_____	_____

D. Acid–Base Indicators

TABLE 12.8 (data)

Indicator	Color in 0.05 M HCl	Color in 0.05 M NaOH
Methyl red	_____	_____
Bromcresol green	_____	_____
Phenolpthalein	_____	_____
Methyl orange	_____	_____
Methyl violet	_____	_____

TABLE 12.9 (report)

Indicator showing very distinct color difference in acid and base	_____
Indicator showing only slight color difference in acid and base	_____

E. Behavior of Salts in Solution

TABLE 12.10 (data)

Salt	Solution pH
Sodium acetate	_____
Ammonium chloride	_____
Sodium carbonate	_____
Sodium chloride	_____
Boiled distilled water	_____

TABLE 12.11 (report)

Salts that produced pH higher than water	_____
Salts that produced pH lower than water	_____
Salts that had little effect on water pH	_____
Ions exhibiting significant Brønsted acid characteristics	_____
Ions exhibiting significant Brønsted base characteristics	_____
Ions exhibiting no significant Brønsted acid or base characteristics	_____

F. Buffers

TABLE 12.12 (data)

Buffer	Initial pH	pH After Adding HCl	pH After Adding NaOH
Solution A	_____	_____	_____
Solution B	_____	_____	_____
Solution C	_____	_____	_____
Solution D	_____	_____	_____
Boiled distilled water	_____	_____	_____

TABLE 12.13 (report)

Solutions able to buffer well against acid   _____

Solutions able to buffer well against base   _____

Solutions able to buffer well against acid or base   _____

Solutions with little or no buffering ability against acid or base   _____

QUESTIONS

1. What salt would result if sulfuric acid (H_2SO_4) was neutralized with lithium hydroxide (LiOH)?

   a. $LiSO_4$ b. $Li(SO_4)_2$ c. Li_2SO_4

   Explain your answer: _____

   _____

2. Which of the following would be most effective in removing hard-water deposits (containing calcium carbonate, $CaCO_3$, and other carbonates) from a sink?

   a. Vinegar (a dilute solution of acetic acid)
   b. Household ammonia (aqueous ammonia)

   Explain your answer: _____

   _____

3. Most cleaning agents (cleansers, soaps, and the like) tested in Part C are best classified as
   a. neutral.   b. acidic.   c. basic.

   Explain your answer: _____

   _____

4. Which of the following reactions best represents the hydrolysis reaction of a salt that gives an acidic solution when dissolved in water? Let C^+ and A^- represent the cation and anion of the salt.

   a. $CA + H_2O \rightarrow C^+ + A^- + H_2O$ c. $CA + H_2O \rightarrow COH + A^- + H^+$
   b. $CA + H_2O \rightarrow HA + C^+ + OH^-$ d. $CA + H_2O \rightarrow COH + HA$

   Explain your answer: _____

   _____

5. How could you tell which of the solutions that were able to buffer well against added acid has the greatest buffering capacity against acid?

   a. Add equal amounts of acid to each solution until the pH of one solution changes significantly.

   b. Add equal amounts of base to each solution until the pH of one solution changes significantly.

   c. The solution with the highest pH after adding the 2 ml of acid has the greatest buffering capacity.

   Explain your answer: _____

   _____

   _____

6. Amino acids, with the general formula $H_2N-CH-CO_2H$, can behave as buffers. Circle the

   $\quad\quad\quad\quad\quad\quad\quad\quad\quad\quad\quad\quad\quad\quad\quad\quad | $

   $\quad\quad\quad\quad\quad\quad\quad\quad\quad\quad\quad\quad\quad\quad\quad\quad R$

   response that represents buffering action against added base and draw a square around the response that represents buffering action against added acid.

   a. $H_2N-CH-CO_2H + H^+ \rightarrow H_3^+N-CH-CO_2H$
      $\quad\quad\quad | \quad\quad\quad\quad\quad\quad\quad\quad\quad\quad\quad\quad | $
      $\quad\quad\quad R \quad\quad\quad\quad\quad\quad\quad\quad\quad\quad\quad\quad R$

   b. $H_2N-CH-CO_2H \rightarrow H_3^+N-CH-CO_2^-$
      $\quad\quad\quad | \quad\quad\quad\quad\quad\quad\quad\quad\quad | $
      $\quad\quad\quad R \quad\quad\quad\quad\quad\quad\quad\quad\quad R$

   c. $H_2N-CH-CO_2H \rightarrow H_2N-CH-CO_2^- + H^+$
      $\quad\quad\quad | \quad\quad\quad\quad\quad\quad\quad\quad\quad | $
      $\quad\quad\quad R \quad\quad\quad\quad\quad\quad\quad\quad\quad R$

   Explain your answer: _____

   _____

   _____

7. On the basis of data in Table 12.10, which of the following anions is the strongest Brønsted base?

   a. Acetate $(C_2H_3O_2^-)$ b. Carbonate (CO_3^{2-}) c. Chloride (Cl^-)

   Explain your answer: _____

   _____

8. In Part D of the experiment, you are told to dispose of the used chemicals by pouring them into the sink. Consider the reaction that is possible between the used chemicals. What compound different from those in the used chemicals will be in the drain water as a result of this disposal?

   a. HCl   b. NaOH   c. NaCl

   Explain your answer: _____

   _____

   _____

Experiment 13

Analysis of Vinegar

In this experiment, you will

- Prepare a solution of sodium hydroxide.
- Standardize the sodium hydroxide solution by titration against potassium acid phthalate.

- Titrate vinegar with the standardized sodium hydroxide solution.

INTRODUCTION

The neutralization reaction that occurs between an acid and a base provides a convenient method for analyzing either acids or bases. In a general way, the **neutralization reaction** is written as

$$\text{Acid} + \text{base} \rightarrow \text{salt} + \text{water} \qquad \text{Eq. 13.1}$$

If the substance to be analyzed is an acid, a specific amount of the acid is reacted with a base solution of known concentration. The point at which all of the acid in the sample has completely reacted with the base is known as the **equivalence point** in titrations; this point is detected by observing a color change in an indicator added to the system (the end point of the titration) or by reading a pH meter. If a base is to be analyzed, the roles of acid and base described above are reversed.

EXPERIMENTAL PROCEDURE

A. Standardization of Sodium Hydroxide Solution

Samples to be analyzed can be in the form of pure solids, pure liquids, or solutions. When the sample is a solid or liquid, it is usually dissolved in water to form a solution before the analysis is carried out. The addition of the acid or base solution to the sample solution is called **titration** and is accomplished by using a buret. Before the analysis begins, the concentration of the acid or base used to titrate the sample must be established through a procedure known as **standardization**. The solutions of known concentration are called **standard solutions**. In this part of the experiment, you will prepare and standardize a solution of sodium hydroxide by titrating it into a solution containing a known mass of potassium acid phthalate ($KHC_8H_4O_4$), commonly referred to as KHP.

PROCEDURE

1. Use your 10-ml graduated cylinder to put 6 mL of 6 M sodium hydroxide solution (NaOH) into a clean plastic storage bottle with a capacity of at least 500 mL.

2. Use a 50- or 100-mL graduated cylinder to add 345 mL of distilled water to the plastic bottle.

3. Place a lid on the bottle and thoroughly mix the contents by repeated inverting and shaking. Set this solution aside for later use (Step 16).

4. Accurately weigh out three samples of potassium acid phthalate (KHP) with masses in the range of 0.3 to 0.4 g. Either an electronic balance or a centigram balance can be used; your instructor will tell you which to use. Each sample should be weighed by difference and placed in a clean 125-mL flask (the flasks do not need to be dry). The weighing procedures are reviewed in Steps 5 to 9 for centigram balances and in Steps 10 to 13 for electronic balances.

5. *If a centigram balance is used,* weigh an empty container (or piece of paper) accurately and record the mass in Table 13.1.

6. Adjust the weights on the balance to add an additional 0.3 g.

7. Add solid KHP to the container until the balance just trips.

8. Readjust the weights until balance is achieved. Read and record the mass of the container plus sample.

9. Carefully pour the sample from the container into a clean 125-mL flask identified as Sample 1. If a plastic weighing dish was used as a container, the last traces of KHP can be rinsed from the dish into the flask with a little distilled water from a wash bottle.

10. *If an electronic balance is used,* zero the balance, then place an empty container (or piece of paper) on the balance.

11. Record the mass of the empty container in Table 13.1 or tare the balance so it reads zero with the empty container on the pan. Record a mass of zero for the empty container if you tare the balance. Use the correct number of significant figures.

12. Carefully add KHP to the container until the steady balance reading is between 0.3 and 0.4 g.

13. Record the balance reading in Table 13.1, and transfer the sample to a 125-mL flask as described in Step 9.

14. Repeat Steps 5 to 9 or 10 to 13 to obtain a second and a third sample. Identify the 125-mL flasks used as Samples 2 and 3. The samples prepared this way do not have to be of exactly the same mass, but all should be in the range of 0.3 to 0.4 g.

15. Add about 20 mL of distilled water and 3 drops of phenolphthalein indicator to each sample. Allow the solid to dissolve, and the samples are ready for titration.

16. Prepare a 25- or 50-mL buret for use by washing it, rinsing it with tap water, and rinsing it twice with distilled water. Finally, rinse it twice with 5-mL portions of your sodium hydroxide solution prepared earlier in Steps 1 to 3.

17. Mount the buret on a ringstand (see Figure 3.5) and fill it above the zero mark with the prepared sodium hydroxide solution.

18. Open the stopcock and allow the buret to drain until the upper liquid level is at zero or below zero. Using this procedure, make sure all air bubbles are removed from the buret tip.

19. Read and record the initial buret reading in Table 13.1. Remember to estimate your reading to .01 mL.

20. Place the 125-mL flask containing the first KHP sample under the buret. Place a piece of white paper under the flask.

21. Open the buret stopcock and carefully add sodium hydroxide to the sample while gently swirling the flask contents. A pink color will be seen where the added NaOH mixes with the sample.

22. As the end point is approached, the pink color will persist for a longer duration. When this happens, slow the output of the buret and add the NaOH dropwise until a *single drop* causes the colorless flask contents to change to a pink color that persists for at least 15 seconds while the flask contents are swirled.

23. When this end point is reached, stop adding base from the buret and read and record the final buret reading. Approximately 15 to 20 mL of sodium hydroxide will be required for the titration.

24. Refill the buret with prepared NaOH solution and repeat Steps 18 to 23, using the second KHP sample. Record the initial and final buret readings.

25. Refill the buret and titrate the third KHP sample. Record the initial and final buret readings.

DISPOSAL  Titrated samples in sink

B. Titration of Vinegar

Acetic acid, the active ingredient in vinegar, reacts with sodium hydroxide to give the salt sodium acetate and water:

$$HC_2H_3O_2(aq) + NaOH(aq) \rightarrow NaC_2H_3O_2(aq) + H_2O(l) \qquad \text{Eq. 13.2}$$
$$\text{acetic acid} \qquad\qquad\qquad \text{sodium acetate}$$

Thus, the standardized NaOH prepared in Part A can be used to determine the amount of acetic acid in a vinegar sample.

PROCEDURE

1. Take a small, clean, dry beaker to the reagent area and get no more than 40 mL of commercial white vinegar.

2. Use a clean, dry 10-mL pipet twice and put 20.00 mL of commercial white vinegar into a 100-mL volumetric flask. Record the vinegar volume in Table 13.3.

3. Carefully fill the volumetric flask to the calibration mark using distilled water. Empty the contents of the flask into a clean, dry 250-mL beaker and stir the liquid well with a glass stirring rod.

4. Rinse the 10-mL pipet with distilled water once, then with two 5-mL portions of diluted vinegar solution from the beaker.

5. Use the rinsed 10-mL pipet to put 10.00 mL of diluted vinegar solution into each of three clean 125-mL flasks (they do not have to be dry). Record the volume of diluted vinegar used.

6. Add 3 drops of phenolphthalein indicator and about 20 mL of distilled water to the sample in each 125-mL flask.

7. Titrate each sample with the standard NaOH solution prepared and standardized in Part A. Record the appropriate buret readings in Table 13.3.

DISPOSAL Titrated samples and unused vinegar in sink. Do not dispose of your standard NaOH solution until you first check with your instructor. The standard NaOH can be used in Experiments 14 and 15. If your instructor tells you to dispose of the standard NaOH, pour it down the sink and flush it down the drain with cool tap water.

A. Standardization of Sodium Hydroxide Solution ▭▭▭

The reaction between KHP and sodium hydroxide is

$$KHC_8H_4O_4(aq) + NaOH(aq) \rightarrow NaKC_8H_4O_4(aq) + H_2O(l) \qquad \text{Eq. 13.3}$$

We see that 1 mol KHP reacts with 1 mol NaOH. This fact will be used in some of the calculations that follow.

1. Calculate and record in Table 13.2 the molecular weight of KHP ($KHC_8H_4O_4$).
2. Use data from Table 13.1 and calculate the mass of each KHP sample titrated. Record the masses.
3. Use the molecular weight calculated in Step 1 and the sample masses from Step 2 to calculate the number of moles of KHP in each sample. Remember

$$\text{Number of moles} = \frac{\text{sample mass}}{\text{sample molecular weight}} \qquad \text{Eq. 13.4}$$

   Record the number of moles in each sample in Table 13.2.
4. According to Equation 13.3, 1 mol NaOH reacts with 1 mol KHP. Use this fact and the result of Step 3 to calculate the number of moles of NaOH used to titrate (react with) each sample of KHP. Record this result.
5. Use the buret readings in Table 13.1 and determine the volume of NaOH solution used to titrate each KHP sample. Record these volumes.
6. Use the NaOH volumes of Step 5 and the moles of NaOH from Step 4 to calculate the molarity of the NaOH solution. Convert the volumes to liters and remember

$$\text{Molarity} = \frac{\text{moles of solute}}{\text{liters of solution}} \qquad \text{Eq. 13.5}$$

   and specifically,

$$\text{NaOH molarity} = \frac{\text{moles of NaOH}}{\text{liters of NaOH}} \qquad \text{Eq. 13.6}$$

7. Record the calculated molarities in Table 13.2. Then average the three results and record the average.

B. Titration of Vinegar ▭▭▭

The equivalence point of a titration represents a condition in which the acid or base in a sample has completely and exactly reacted with the standard base or acid added during the titration. The required volume of standard solution, obtained from appropriate buret readings, together with the standard solution concentration can be used to calculate the amount of acid or base added. Suppose, for example, that 18.22 mL of 0.107 M sodium hydroxide is required to titrate an acid sample to the proper end point. The number of moles of NaOH required is given by the following calculation:

$$\text{Moles of NaOH} = (\text{mL NaOH})\left(\frac{1\ L}{1000\ \text{mL}}\right)(\text{NaOH molarity}) \qquad \text{Eq. 13.7}$$

in which the conversion factor

$$\frac{1\ L}{1000\ mL}$$

is used to change milliliters to liters. When this equation is used with the data given above, the result is

$$\text{Moles of NaOH} = (18.22\ mL)\left(\frac{1\ L}{1000\ mL}\right)\left(\frac{0.107\ mol\ NaOH}{1\ L}\right) \quad \text{Eq. 13.8}$$

$$= 1.95 \times 10^{-3}\ mol$$

Assume the acid in the titrated sample is acetic acid ($HC_2H_3O_2$) and the volume titrated was 10.00 mL. According to Equation 13.2, 1 mol of acid reacts with 1 mol of base. Therefore, the number of moles of acid in the sample is identical to the number of moles of NaOH used in the titration and calculated in Equation 13.8: 1.95×10^{-3}. The molar concentration of the acid can then be calculated using Equation 13.5 since both the sample volume (10.00 mL) and moles (1.95×10^{-3}) are known:

$$HC_2H_3O_2\ \text{molarity} = \frac{1.95 \times 10^{-3}\ mol}{0.01000\ L} = 0.195\ \frac{mol}{L} \quad \text{Eq. 13.9}$$

Note in this calculation that the 10.00-mL sample volume was changed to 0.01000 liters (L).

1. Calculate the molecular weight of acetic acid ($HC_2H_3O_2$) and record it in Table 13.4.
2. Use the buret readings from Table 13.3 and determine the volume of standard NaOH used in titrating each diluted vinegar sample. Record these results.
3. Use the NaOH volume from Step 2, the average NaOH molarity from Table 13.2 and Equation 13.7 to calculate the number of moles of NaOH used to titrate each sample of vinegar. Record the results.
4. Use the results of Step 3 and the reaction relationship given in Equation 13.2 to determine the number of moles of acetic acid titrated (reacted) in each sample. Record the results.
5. Remember that each titrated vinegar sample contained 10.00 mL taken from the 100-mL diluted sample. This is ⅒ of the diluted sample and thus ⅒ of the commercial vinegar used to make the diluted sample. Use this fact and the volume of commercial vinegar used to make the 100 mL of diluted sample and calculate the volume of commercial vinegar present in each 10.00-mL titrated sample of diluted vinegar. Record this volume in Table 13.4.
6. Use the number of moles of acetic acid in each sample determined in Step 4 and the volume of commercial vinegar in each sample from Step 5 to calculate the molarity of commercial vinegar:

$$\text{Commercial vinegar molarity} = \frac{\text{moles of acetic acid in sample}}{\text{liters of commercial vinegar}} \quad \text{Eq. 13.10}$$

Record the three calculated values and the average of the three in Table 13.4.
7. Assume commercial vinegar has a density of 1.00 g/mL and use the volume from Step 5 to calculate the mass of commercial vinegar titrated in each sample. The mass in grams will equal the volume in milliliters. Record this result.
8. Use the molecular weight of acetic acid from Step 1, the number of moles of acetic acid from Step 4, and Equation 13.4 (rearranged) to

calculate the mass of acetic acid titrated in each sample. Record the result.

9. Use the masses calculated in Steps 7 and 8 to calculate the weight percent (w/w%) of acetic acid in commercial vinegar:

$$w/w\% \text{ Acetic acid} = \frac{\text{mass of acetic acid}}{\text{mass of vinegar}} \times 100 \qquad \text{Eq. 13.11}$$

Record the values obtained for each sample, and the average of the three.

Experiment 13 ▪ Pre-Lab Review

Analysis of Vinegar

1. Are any specific safety alerts given in the experiment? List any that are given.

   _____

2. Are any specific disposal directions given in the experiment? List any that are given.

   _____

   _____

3. What is a standard solution? _____

   _____

4. What observation lets you know when the end point of the titration is reached in Parts A and B?

   _____

   _____

5. What is the name of the indicator used in the titrations of this experiment?

   _____

6. What is the name and the formula of the solid acid used to standardize your prepared NaOH solution in Part A?

   _____

7. Briefly describe the proper way to obtain buret readings. _____

   _____

   _____

8. Briefly describe how you are to accurately weigh a solid acid sample, using a centigram balance in Part A.

   _____

   _____

   _____

9. Briefly describe how you are to accurately weigh a solid acid sample, using an electronic balance in Part A.

_____

_____

_____

10. List the steps used to prepare the NaOH solution that is to be standardized in Part A and used in Part B.

_____

_____

_____

11. Briefly describe how you prepare the diluted vinegar in Part B. _____

_____

_____

Experiment 13 ▪ Data & Report Sheet

Analysis of Vinegar

A. Standardization of Sodium Hydroxide Solution

TABLE 13.1 (data)

	Sample 1	Sample 2	Sample 3
Mass of container plus sample	_____	_____	_____
Mass of empty container	_____	_____	_____
Final buret reading	_____	_____	_____
Initial buret reading	_____	_____	_____

TABLE 13.2 (report)

Molecular weight of KHP _____

	Sample 1	Sample 2	Sample 3
Mass of KHP titrated	_____	_____	_____
Moles of KHP titrated	_____	_____	_____
Moles of NaOH required in titration	_____	_____	_____
Volume of NaOH used in titration	_____	_____	_____
Calculated NaOH molarity	_____	_____	_____
Average NaOH molarity		_____	

B. Titration of Vinegar

TABLE 13.3 (data)

Volume of commercial vinegar used to make diluted solution _____

	Sample 1	Sample 2	Sample 3
Volume of diluted vinegar	_____	_____	_____
Final buret reading	_____	_____	_____
Initial buret reading	_____	_____	_____

TABLE 13.4 (report)

Molecular weight of acetic acid _____

	Sample 1	Sample 2	Sample 3
Volume of NaOH used in titration	_____	_____	_____
Moles of NaOH used in titration	_____	_____	_____
Moles of acetic acid titrated	_____	_____	_____
Volume of commercial vinegar present	_____	_____	_____
Calculated commercial vinegar molarity	_____	_____	_____
Average commercial vinegar molarity		_____	
Mass of commercial vinegar present	_____	_____	_____
Mass of acetic acid titrated	_____	_____	_____
Percent acetic acid in commercial vinegar	_____	_____	_____
Average % acetic acid in commercial vinegar		_____	

QUESTIONS

1. Suppose that during the NaOH standardization procedure in Part A you misread the balance and recorded a KHP sample mass that was higher than the actual mass. How will this error influence the calculated NaOH concentration?

   a. Calculated concentration will be higher than the correct concentration
   b. Calculated concentration will be lower than the correct concentration.
   c. Calculated concentration will not be influenced.

   Explain your answer: _____

   _____

   _____

2. Suppose that during the NaOH standardization procedure in Part A you spill some of the weighed KHP sample before it can be dissolved in water. You continue on with the procedure as if nothing had happened. How will this error influence the calculated NaOH concentration.

   a. Calculated concentration will be higher than the correct concentration.
   b. Calculated concentration will be lower than the correct concentration.
   c. Calculated concentration will not be influenced.

   Explain your answer: _____

   _____

   _____

3. Suppose that while titrating a KHP sample in Part A you run past the end point and add more NaOH than you should. How will this error influence the calculated NaOH concentration?

   a. Calculated concentration will be higher than the correct concentration.
   b. Calculated concentration will be lower than the correct concentration.
   c. Calculated concentration will not be influenced.

   Explain your answer: _____

   _____

   _____

4. Suppose that while titrating a diluted vinegar sample in Part B you suddenly realize that you forgot to add the indicator. You add the proper amount of indicator to the flask, and the contents remain colorless. What should you do?

   a. Continue the titration as if no mistake had been made.
   b. Discard the sample and start over with a fresh diluted vinegar sample.
   c. Complete the titration but do not use the data in your calculations.

   Explain your answer: _____

   _____

   _____

5. Suppose that while titrating a vinegar sample with standard NaOH you misread the buret and the recorded NaOH volume is higher than the actual volume used. How will this error influence the final calculated percentage of acetic acid in the vinegar?

   a. Calculated percent will be higher than the correct percentage.
   b. Calculated percent will be lower than the correct percentage.
   c. Calculated percentage will not be influenced.

   Explain your answer: _____

   _____

   _____

6. Suppose you were confused about which of the two containers held standard NaOH and which held diluted vinegar. Which of the following would be the safest way to decide the contents of each container?

   a. Feel each solution.
   b. Taste each solution.
   c. Add one or two drops of phenolphthalein indicator to small samples of each solution.

   Explain your answer: _____

   _____

   _____

Determination of K_a for Weak Acids

In this experiment, you will

- Prepare a solution of NaOH if one is not available.
- Titrate a solution of unknown acid to the half-reacted point and use the pH of the solution at that point to determine K_a for the unknown acid.

- Titrate a solution containing a weighed amount of solid unknown acid to the half-reacted point and use the pH of the solution at that point to determine K_a for the unknown acid.

INTRODUCTION

The dissociation of a weak acid that is dissolved in water can be represented by the following general equation, where HA represents that part of the acid that is dissolved but not dissociated and A^- is the anion produced by that part of the dissolved acid that has dissociated:

$$HA(aq) + H_2O(l) \rightleftharpoons H_3O^+(aq) + A^-(aq)$$
Eq. 14.1

The equilibrium constant for this reaction is

$$K = \frac{[H_3O^+] [A^-]}{[HA] [H_2O]}$$
Eq. 14.2

In Equation 14.2, the brackets, [ ], indicate molar concentrations. The molar concentration of water, the solvent is essentially constant for dilute solutions, so Equation 14.2 may be written as follows:

$$K[H_2O] = K_a = \frac{[H_3O^+] [A^-]}{[HA]}$$
Eq. 14.3

K_a is called the **acid dissociation constant,** and the brackets again denote molar concentrations.

The strength of weak acids is indicated by the extent to which the process represented by Equation 14.1 occurs when an acid is dissolved in water. The stronger an acid, the greater is its tendency to donate H^+ ions to water to form H_3O^+ ions. Thus, solutions of stronger acids have higher H_3O^+ and A^- concentrations and lower HA concentrations at equilibrium than do solutions of weaker acids. Thus, according to Equation 14.3, stronger acids will have larger K_a values.

The value of K_a for a weak acid can be determined by a straightforward experimental procedure. If a solution is prepared that contains HA, H_3O^+, and A^- in equilibrium and if the concentrations are adjusted so that the concentrations of undissociated acid (HA) and acid anion (A^-) are made equal to each other, then the value of K_a will be equal to the concentration of H_3O^+ as shown in Equation 14.4:

$$K_a = \frac{[H_3O^+] [A^-]}{[HA]} = [H_3O^+]$$
Eq. 14.4

In Equation 14.4, the $[A^-]$ and $[HA]$ cancel because they have the same value.

The neutralization reaction that occurs between an acid and a base can be used to establish the condition in which $[A^-] = [HA]$. The neutralization reaction is written as

$$HA(aq) + NaOH(aq) \rightarrow H_2O(l) + Na^+(aq) + A^-(aq) \qquad Eq.\ 14.5$$

In Equation 14.5, the salt (NaA) is written in the dissociated form it takes when it is dissolved in water. If a solution is made by dissolving some weak acid (HA) in water, very little of the acid will dissociate, and so very little A^- will be present at equilibrium. If half the dissolved acid is reacted with NaOH to form the salt, the concentration of A^- from the salt that is produced will be the same as the concentration of unreacted acid that remains in solution (remember, dissolved salts dissociate completely). Thus, in a weak acid solution that is half-reacted, $[A^-] = [HA]$. Of course, this is only true if the amount of weak acid undergoing dissociation according to Equation 14.1 is negligible and provides only a negligible amount of A^-. This assumption is valid for most weak acids.

EXPERIMENTAL PROCEDURE

A. Preparation of a Solution of NaOH

In this experiment, you will titrate your acid samples with a solution of NaOH that is about 0.1 M. You can use the standardized solution you prepared and standardized in Experiment 13, a solution that is provided in the laboratory, or you can prepare a solution by following the steps given below. Check with your lab instructor for directions.

PROCEDURE

1. Use your 10-mL graduated cylinder to put 3.5 mL of 6 M sodium hydroxide solution (NaOH) into a clean 250-mL beaker.
2. Use a 50- or 100-mL graduated cylinder to add 200 mL of distilled water to the beaker. Stir the resulting solution well with a glass stirring rod and mark the beaker in some way so you remember that it contains your NaOH solution.

B. Determination of K_a for an Unknown Acid in Solution

In this part of the experiment, you will titrate a specific volume of a solution that contains a weak acid in solution. You will completely titrate two samples to the phenolphthalein end point in order to determine the volume of base that will be required to titrate a sample to the half-reacted point. You will then titrate a sample to the half-reacted point and measure the pH of the half-reacted sample. The unknown will either be available in the reagent area of the lab, or you will get a sample of unknown from the stockroom.

PROCEDURE

1. Obtain about 50 mL of unknown acid solution in a clean 100-mL beaker. Obtain the unknown from either the lab reagent area or from

the stockroom. If you are given an identification (ID) number for your sample, record it in the appropriate blank of Table 14.2.

2. Obtain a 25- or 50-mL buret from the stockroom. Wash it and rinse it with no more than 5 mL of the NaOH solution you will use for the titration.

DISPOSAL 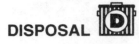 Rinse solution in sink

Mount the buret on a ringstand, fill it with the NaOH solution you will use, and prepare it for use in a titration. See Experiment 13, Part A, Steps 17 to 20 for a review if necessary.

3. Obtain a 10-mL pipet from your locker equipment or the stockroom. Rinse the pipet with about 5 mL of the unknown acid solution.

DISPOSAL 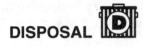 Rinse solution in sink

4. Pipet a 10.00-mL sample of unknown acid solution into a clean 125-mL flask, add 25.0 mL of distilled water (using a graduated cylinder), then add 3 drops of phenolphthalein indicator

5. Put the flask under the buret you set up in Step 2 and titrate the sample of acid to the phenolphthalein end point. Remember, this end point is reached when a single drop of NaOH solution from the buret causes the solution in the flask to have a pink color that remains for at least 15 seconds while the solution is swirled. Note and record in Table 14.1 the initial and final buret readings.

6. Repeat Steps 4 and 5 for a second 10.00-mL sample of the unknown acid solution. Record the initial and final buret readings in Table 14.1.

DISPOSAL 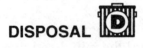 Titrated samples in sink

7. Use the initial and final buret readings recorded in Table 14.1 to determine the volume of NaOH solution required to completely titrate each sample of unknown acid solution. Record these volumes in Table 14.1. The two values should agree within 0.2 mL. If they do not, repeat Steps 4 and 5 for a third sample. Average the two NaOH volumes that agree most closely with each other (do not titrate more than three samples). Record the average in Table 14.2.

8. Divide the average value recorded in Step 7 by 2, and record this value in Table 14.2. This is the volume of NaOH solution that would be required to titrate a 10.00-mL sample of unknown acid solution to the half-reacted point.

9. Pipet a 10.00-mL sample of unknown acid solution into a clean 125-mL flask, add 25.0 mL of distilled water, and use your buret to carefully add the volume of NaOH solution required to titrate the sample to the half-reacted point. Record the initial and final buret readings in Table 14.2.

10. Mix the half-reacted sample well, then use long-range (1 to 11) pH paper to determine the approximate pH of the sample. Record the measured pH in Table 14.2.

11. Obtain a short-range pH paper that includes the pH value you measured in Step 10. Measure the sample pH with the short-range paper and record the value in Table 14.2.

12. If a pH meter is available, measure the pH of the half-reacted sample to the precision available on the meter. Obtain directions from your

lab instructor before you attempt to use the pH meter. Record the pH value in Table 14.2.

DISPOSAL 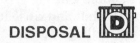 Half-reacted sample in sink

C. Determination of K_a for an Unknown Solid Acid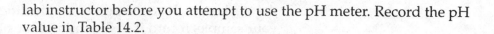

In this part of the experiment, you will weigh samples of an unknown solid acid, dissolve the samples in water, and completely titrate the dissolved samples to the phenolphthalein end point. You will use the mass of the samples and the volume of NaOH solution required to titrate them to calculate the amount of NaOH solution needed to titrate another weighed sample to the half-reacted point. You will then measure the pH of the half-reacted sample and use the data to calculate K_a for the unknown solid acid.

PROCEDURE

1. Obtain about 1 g of unknown solid acid from the reagent area of the lab or from the stockroom. If you are given an ID number for your sample, record it in the appropriate blank of Table 14.4.
2. Use a centigram or electronic balance to weigh out two samples of the unknown solid acid with masses in the range of 0.20 to 0.25 g. Record the masses in Table 14.3, then put each sample into a clean 125-mL flask. Identify the flasks in some way so you can tell which sample each one contains.
3. Add 25.0 mL of distilled water to each flask and 3 drops of phenolphthalein indicator. Swirl the flasks to speed the dissolving of the solid samples.
4. After the solid samples have dissolved, use your buret to titrate each sample to the phenolphthalein end point. Record the initial and final buret readings in Table 14.3.

DISPOSAL 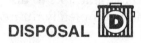 Titrated samples in sink

5. Use the buret readings recorded in Table 14.3 to calculate the volume of NaOH solution required to completely titrate each sample. Record the volumes in Table 14.3.
6. Divide the volume of NaOH required for each sample by the mass of the sample expressed in grams. This gives the volume of NaOH required to titrate each gram of sample (the vol-per-gram value). Record this value in Table 14.3. The two values should agree to within 1 mL of each other. If they do not agree, weigh out and titrate a third sample of unknown solid acid. Do the appropriate calculations and enter the values in Table 14.3.
7. Average the two vol-per-gram values calculated in Step 6 that agree most closely (do not titrate more than three solid acid samples). Record the average value in Table 14.3.
8. Weigh out and dissolve in 25.0 mL of water another sample of solid unknown acid in the range of 0.20 to 0.25 g. Record the mass of the sample in Table 14.4. Multiply the mass of the sample by the vol-per-gram value you calculated in Step 7. This will give you the volume of NaOH solution required to completely titrate the sample you just weighed out. Record the volume in Table 14.4. Divide this volume

by 2 and enter the value in Table 14.4 in the blank corresponding to the volume required to half-react your sample.

9. Use your buret and add to the dissolved sample the volume of NaOH solution required to half-react the sample. Record the initial and final buret readings in Table 14.4.

10. Mix the half-reacted sample well, then use long-range pH paper, short-range pH paper, and a pH meter (if available) as you did for your half-reacted sample in Part B, Steps 10 to 12. Record the appropriate pH values in Table 14.4.

DISPOSAL  Titrated sample, unused solid unknown, and excess NaOH solution in sink

---

CALCULATIONS AND REPORT

B. Determination of K_a for an Unknown Acid in Solution

As described earlier, the value of K_a is equal to $[H_3O^+]$ in a half-reacted sample of weak acid. In the experiment, you measured the pH of the half-reacted samples. Thus, it is necessary to convert pH to $[H_3O^+]$. The relationship between pH and $[H_3O^+]$ is

$$pH = -\log[H_3O^+] \qquad \text{Eq. 14.6}$$

Thus, the value of $[H_3O^+]$ is obtained by taking the antilog of $-pH$. On most calculators, this is done by the following steps:

1. Enter the pH value.
2. Change the sign of the pH value to negative.
3. Push the second function or inverse function key.
4. Push the log key.

Remember when you do this calculation, the calculated $[H_3O^+]$ value will have one less significant figure than the total number of figures used to express the pH. Thus, a pH value of 4.75 will give a value of 1.8×10^{-5} for $[H_3O^+]$.

1. Note in Table 14.2 the pH values obtained for the half-reacted sample using short-range pH paper and the pH meter. Convert each of these two pH values into a value for $[H_3O^+]$. Remember that the $[H_3O^+]$ value is the same as the value of K_a for the weak acid and record the K_a values in Table 14.2.

C. Determination of K_a for an Unknown Solid Acid

1. Note in Table 14.4 the pH values obtained for the half-reacted sample using short-range pH paper and the pH meter. Convert each of these two pH values into a value for $[H_3O^+]$ and record the K_a values in Table 14.4.

Experiment 14 ▪ Pre-Lab Review

Determination of K_a for Weak Acids

1. Are any specific safety alerts given in the experiment? List any that are given.

   _____

2. Are any specific disposal directions given in the experiment? List any that are given.

   _____

   _____

3. What are the possible sources of the NaOH solution needed in the experiment?

   _____

4. In Part B, why do you titrate some samples completely but only half-react one sample?

   _____

   _____

   _____

5. What indicator is used in all titrations that are done to completion? _____

   _____

6. How do you know when you have titrated to the proper end point in the complete titrations done in the experiment?

   _____

7. What two different methods are used to measure samples to be titrated in Parts B and C?

   _____

   _____

8. What size samples are titrated in Parts B and C? _____

   _____

9. Describe how the pH of half-reacted samples is measured using pH paper.

   _____

   _____

Experiment 14 ▪ Data & Report Sheet

Determination of K_a for Weak Acids

B. Determination of K_a for an Unknown Acid in Solution

TABLE 14.1 (data)

	Sample 1	Sample 2	Sample 3
Final buret reading	_____	_____	_____
Initial buret reading	_____	_____	_____
Vol. of NaOH required for complete titration	_____	_____	_____

TABLE 14.2 (report)

Unknown ID number	_____
Average volume for complete titration	_____
Volume required to half-react sample	_____
Final buret reading	_____
Initial buret reading	_____
Half-reacted sample pH (long range)	_____
Half-reacted sample pH (short range)	_____ K_a _____
Half-reacted sample pH (pH meter)	_____ K_a _____

C. Determination of K_a for an Unknown Solid Acid

TABLE 14.3 (data)

	Sample 1	Sample 2	Sample 3
Sample mass			
Final buret reading			
Initial buret reading			
Vol. of NaOH required for complete titration			
Vol-per-gram value			
Average vol-per-gram value			

TABLE 14.4 (report)

Unknown ID number	
Sample mass	
Vol. of NaOH required for complete titration	
Vol. of NaOH required to half-react sample	
Final buret reading	
Initial buret reading	
Half-reacted sample pH (long range)	
Half-reacted sample pH (short range)	K_a
Half-reacted sample pH (pH meter)	K_a

QUESTIONS

1. Suppose you had to prepare some NaOH solution according to the directions given in Part A. However, while using a 50-mL graduated cylinder to add the distilled water, you miscounted and only added 150 mL instead of 200 mL. How will this mistake influence the K_a value determined in Part B?

    a. Make it larger than the correct value.
    b. Make it smaller than the correct value.
    c. Will not influence it.

    Explain your answer: _____

    _____

    _____

    _____

2. Suppose in Part B a stockroom employee gave you a 15-mL pipet to use rather than the 10-mL type called for in the experiment. Without noticing the mistake, you use the pipet and do everything else in the experiment according to the written directions. How will this mistake influence the volume of NaOH solution required to titrate the two samples?

    a. The volume of NaOH required for the first sample would be greater than the volume required for the second sample.
    b. The volume of NaOH required for the first sample would be less than the volume required for the second sample.
    c. The volume of NaOH required should be the same for the two samples.

    Explain your answer: _____

    _____

    _____

3. How would the mistake described in Question 2 influence the K_a value determined in Part B?

    a. Make it larger than the correct value.
    b. Make it smaller than the correct value.
    c. Will not influence it.

    Explain your answer: _____

    _____

    _____

4. Suppose in Part C one of your samples that was titrated completely had a mass of 0.221 g (larger than the maximum mass given in the directions). How would this mistake influence the vol-per-gram value calculated for that sample compared to a sample that had a mass of 0.208 g (within the prescribed range)?

   a. The larger sample would have a larger vol-per-gram value than the smaller sample.
   b. The larger sample would have a smaller vol-per-gram value than the smaller sample.
   c. The vol-per-gram values of the two samples should be the same.

   Explain your answer: _____

   _____

   _____

5. How would the mistake described in Question 4 influence the K_a value determined in Part C?

   a. Make it larger than the correct value.
   b. Make it smaller then the correct value.
   c. Would not influence it.

   Explain your answer: _____

   _____

   _____

6. Suppose in Part B that a student pipets a 10.00-mL sample in Step 9 and adds 25 mL of distilled water and 3 drops of phenolphthalein indicator. Then, the student realizes that the directions do not call for the addition of any indicator. What should the student do to be most efficient in using lab time?

   a. Discard the sample and start over with a new one.
   b. Proceed to add enough NaOH solution to half-react with the sample.
   c. Titrate the sample to the phenolphthalein end point.

   Explain your answer: _____

   _____

   _____

   _____

Experiment 15

The Acidic Hydrogens of Acids

In this experiment, you will

- Prepare a solution of NaOH if one is not available.
- Titrate solutions of unknown acids and use the data to determine the formula of the acid.

- Titrate a solution of phosphoric acid and use the data to determine the number of hydrogens reacted per molecule of acid.

INTRODUCTION

Acids are often classified on the basis of the number of hydrogen atoms contained in each molecule of acid. Thus, hydrochloric acid (HCl) is a monoprotic acid because it contains one (*mono*) hydrogen per molecule. The *protic* part of the designation is based on the Brønsted–Lowry definition of acids as proton (H^+) donors. Similarly, oxalic acid ($H_2C_2O_4$) is classified as a diprotic acid. Sometimes, the formula of an acid can be misleading when this type of classification is attempted. For example, acetic acid has the formula $C_2H_4O_2$. On the basis of this formula, acetic acid might be classified as a tetraprotic acid. However, it is found experimentally that only one of the four hydrogen atoms is an acidic hydrogen that becomes involved in the acidic reactions characteristic of acetic acid. This discrepancy between the number of hydrogen atoms in the formula and the number that are acidic is a common characteristic of organic acids such as acetic. When it is important to represent the number of acidic hydrogens in an acid, it is the practice to separate the acidic hydrogens from the others and put the acidic hydrogens first in the formula. Thus, acetic acid would be represented by the formula $HC_2H_3O_2$.

The number of acidic hydrogens in a molecule of acid can be determined by using one of the acid properties such as the reaction of an acid with a base. This reaction is used when acids are titrated with bases. In this experiment, acid solutions of the same molarity (the same number of moles of acid per liter of solution) will be titrated with a solution of NaOH to the phenolphthalein end point. The acid samples titrated will all have the same volume. Since the acid concentrations are the same, each sample will contain the same number of moles of acid. Thus, the volume of NaOH required to titrate the samples will be proportional to the number of acidic hydrogens per molecule. For example, a 10-mL sample of a 0.1 *M* solution of a diprotic acid would require twice the volume of NaOH solution as would a 10-mL sample of a 0.1 *M* solution of a monoprotic acid because the diprotic acid solution would have twice as many acidic hydrogens to react.

A. Preparation of a Solution of NaOH

In this experiment, you will titrate your acid samples with a solution of NaOH that is about 0.1 *M*. You can use the standardized solution you prepared and standardized in Experiment 13, a solution that is provided in the laboratory, or you can prepare a solution by following the steps given below. Check with your lab instructor for directions.

PROCEDURE

1. Use your 10-mL graduated cylinder to put 2.5 mL of 6 *M* sodium hydroxide solution (NaOH) into a clean 250-mL beaker.
2. Use a 50- or 100-mL graduated cylinder to add 150 mL of distilled water to the beaker. Stir the resulting solution well with a glass stirring rod and mark the beaker in some way so you can remember that it contains your NaOH solution.

B. The Number of Acidic Hydrogens in Acids

In this part of the experiment, you will titrate specific volumes of three solutions that contain different acids. The molar concentration of acid in each solution is the same. At least one of the acids is monoprotic. You will use the volume of NaOH solution required to titrate each acid sample to determine which of the acids is/are monoprotic and the number of acidic hydrogens in the acids that are not monoprotic.

PROCEDURE

1. Obtain about 35 mL of unknown acid solution A, B, or C (you may start with any one of them) in a clean, dry 100-mL beaker.
2. Obtain a 25- or 50-mL buret from the stockroom. Wash it and rinse it with no more than 5 mL of the NaOH solution you will use for the titration.

DISPOSAL Rinse solution in sink

Mount the buret on a ringstand, fill it with the NaOH solution you will use, and prepare it for use in a titration. See Experiment 13, Part A, Steps 17 to 20 for a review if necessary.

3. Obtain a 10-mL pipet from your locker equipment or the stockroom. Rinse the pipet with about 5 mL of the unknown acid solution.

DISPOSAL Rinse solution in sink

4. Pipet a 10.00-mL sample of the unknown acid solution into a clean 125-mL flask, add 25.0 mL of distilled water (use a graduated cylinder), then add 3 drops of phenolphthalein indicator.
5. Put the flask under the buret you set up in Step 2 and titrate the sample of acid to the phenolphthalein end point. Remember, this end point is reached when a single drop of NaOH solution from the buret causes the solution in the flask to have a pink color that remains for at least 15 seconds while the solution is swirled. Note and record in Table 15.1 the initial and final buret readings under the proper unknown acid heading (A, B, or C).

6. Repeat Steps 4 and 5 for a second 10.00-mL sample of the unknown acid solution. Record the initial and final buret readings in Table 15.1.
7. Repeat Steps 1, 3 (only the pipet rinsing), 4, 5, and 6 for a second unknown acid. Be sure you get your acid sample in a clean beaker that is dry.
8. Repeat Step 7 for the third unknown acid solution.

 **DISPOSAL** Titrated samples and unused unknown acid solutions in sink

C. Acidic Hydrogens of Phosphoric Acid

The term *polyprotic* is used to describe in a general way acids that have more than one acidic hydrogen. In polyprotic acids, the acid hydrogens do not all behave in exactly the same way. One difference is the ability to leave the parent acid in a Brønsted–Lowry acid reaction.

Phosphoric acid (H_3PO_4) is a polyprotic acid that is an important constituent of living organisms. It is found in the nucleic acids DNA and RNA, which are involved in the transfer of genetic information from generation to generation. It is also found in adenosine triphosphate (ATP), an energy-storage compound of living cells.

The three acidic hydrogens of phosphoric acid vary greatly in their ability to leave the molecule in a Brønsted–Lowry acid reaction. This is represented by the K_a values given below for the acid dissociation reactions in water.

$$H_3PO_4(aq) + H_2O(l) \rightleftharpoons H_3O^+(aq) + H_2PO_4^-(aq) \quad K_a = 7.1 \times 10^{-3} \quad \text{Eq. 15.1}$$
$$H_2PO_4^-(aq) + H_2O(l) \rightleftharpoons H_3O^+(aq) + HPO_4^{2-}(aq) \quad K_a = 6.3 \times 10^{-8} \quad \text{Eq. 15.2}$$
$$HPO_4^{2-}(aq) + H_2O(l) \rightleftharpoons H_3O^+(aq) + PO_4^{3-}(aq) \quad K_a = 1.2 \times 10^{-12} \quad \text{Eq. 15.3}$$

The HPO_4^{2-} ion is such a weak Brønsted acid that it is difficult to react it completely with a base in a titration. The other two acids, H_3PO_4 and $H_2PO_4^-$, are strong enough acids to be titrated completely by NaOH. In this part of the experiment, you will titrate a sample of H_3PO_4 solution to the phenolphthalein end point and use the resulting data to determine how many of the three hydrogens reacted per molecule. The molar concentration of H_3PO_4 in the solution is the same as the concentration of the acid solutions titrated in Part B.

PROCEDURE

1. Obtain about 35 mL of phosphoric acid solution in a clean, dry 100-mL beaker.
2. Rinse a 10-mL pipet with distilled water, then with about 5 mL of the phosphoric acid solution.

 **DISPOSAL** Rinse solution in sink

3. Pipet a 10.00-mL sample of phosphoric acid solution into a clean 125-mL flask, then add 25.0 mL of distilled water and 3 drops of phenolphthalein indicator.
4. Titrate the sample to the phenolphthalein end point with the same NaOH solution used in Part B. Record the initial and final buret readings in Table 15.3.
5. Repeat Steps 3 and 4 for a second sample of phosphoric acid.

DISPOSAL 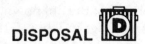 Titrated samples, unused NaOH solution, and unused phosphoric acid solution in sink

CALCULATIONS AND REPORT

B. The Number of Acidic Hydrogens in Acids

At least one of the acids titrated in this part of the experiment was a monoprotic acid. Since all of the acid solutions were the same concentration and identical-sized samples of the acids were titrated with the same NaOH solution, the volumes of NaOH required should reflect the number of hydrogens that reacted. Thus, a monoprotic acid can be identified as the one that required the smallest volume of NaOH solution for titration. A diprotic acid will require twice the volume of a monoprotic acid.

1. Refer to the data recorded in Table 15.1, and determine the volume of NaOH solution required to titrate each sample. Record that volume in Table 15.1, then average the two values for each unknown acid, and record the average in Table 15.2.
2. Based on the average volumes for reaction that are recorded in Table 15.2, select the acid that corresponds to a monoprotic acid. If more than one acid qualifies, use either one for Step 3.
3. Divide the average volume of NaOH required to titrate the monoprotic acid you selected in Step 2 into itself and into the average volume required for reaction by the other two acids. Record the results of this calculation in Table 15.2 as the volume ratio. If the results are not whole numbers, round each one to the nearest whole number before you record them.
4. Use the results of Step 3 and classify each of the three acids you titrated as monoprotic, diprotic, triprotic, and so on. Record the classification in Table 15.2 and write a formula for the acid in the form H_xA, where the x will reflect the number of acidic hydrogen atoms in one molecule of the acid. Record the formulas in Table 15.2.

C. Acidic Hydrogens of Phosphoric Acid

The same NaOH solution used in Part B to titrate the acid samples was used to titrate your samples of phosphoric acid. Thus, the number of phosphoric acid hydrogens that reacted can be determined by dividing the volume of NaOH required to titrate the phosphoric acid samples by the average volume of NaOH required to titrate the monoprotic acid identified in Step 2 of Part B.

1. Use the data recorded in Table 15.3 to determine the volume of NaOH required to titrate each sample of phosphoric acid. Record the volume in Table 15.3.
2. Average the two volumes calculated in Step 1 and record the average in Table 15.3.
3. Divide this average volume by the average volume of NaOH required to titrate the monoprotic acid identified in Step 2 of Part B. Do *not* round this result but record it to the nearest 0.1 unit in Table 15.3 as the volume ratio. This result indicates the average number of hydrogens that were reacted per molecule when phosphoric acid was titrated to the phenolphthalein end point using NaOH.

Experiment 15 ▪ Pre-Lab Review

The Acidic Hydrogens of Acids

1. Are any specific safety alerts given in the experiment? List any that are given.

   _____

2. Are any specific disposal directions given in the experiment? List any that are given.

   _____

   _____

3. What are the possible sources of the NaOH solution used in the experiment?

   _____

4. In Part B, which unknown acid sample should you begin with? _____

   _____

5. In Part B, what volume of unknown acid solution do you get in your 100-mL beaker?

   _____

6. In Parts B and C, what size of acid solution samples are titrated? _____

   _____

7. What measuring device is used to measure the samples of acid solution to be titrated in Parts B and C?

   _____

8. What indicator is used in the titrations of the experiment? _____

   _____

9. How do you know when you have titrated to the proper end point in Parts B and C?

   _____

   _____

   _____

Experiment 15 ▪ Data & Report Sheet

The Acidic Hydrogens of Acids

B. The Number of Acidic Hydrogens in Acids

TABLE 15.1 (data)

	Acid A Samples		Acid B Samples		Acid C Samples	
	1	2	1	2	1	2
Final buret reading	_____	_____	_____	_____	_____	_____
Initial buret reading	_____	_____	_____	_____	_____	_____
Vol. of NaOH required	_____	_____	_____	_____	_____	_____
Average vol. of NaOH required	_____		_____		_____	

TABLE 15.2 (report)

	Acid A	Acid B	Acid C
Average volume of NaOH required	_____	_____	_____
Volume ratio	_____	_____	_____
Acid classification	_____	_____	_____
Acid formula	_____	_____	_____

C. Acidic Hydrogens of Phosphoric Acid

TABLE 15.3 (data and report)

	Sample 1	Sample 2
Final buret reading	_____	_____
Initial buret reading	_____	_____
Vol. of NaOH required	_____	_____
Average vol. of NaOH required	_____	
Volume ratio	_____	

QUESTIONS

1. Suppose you had to prepare some NaOH solution according to the directions given in Part A. However, while using a 50-mL graduated cylinder to add the distilled water, you miscounted and added 200 mL instead of 150 mL. How will this mistake influence the values of the volume ratios calculated in Parts B and C, compared to the values that would have been obtained if the mistake had not been made?

   a. Make them larger.   b. Make them smaller.   c. Would not influence them.

   Explain your answer: _____

   _____

   _____

   _____

2. Suppose in Part B a stockroom employee gave you a 15-mL pipet to use rather than the 10-mL type called for in the experiment. Without noticing the mistake, you use the pipet and do everything else in the experiment according to the written directions. How will this mistake influence the volume of NaOH solution required to titrate the acid samples in Parts A and B, compared to the volume that would have been required if the mistake had not been made?

   a. Make it larger.   b. Make it smaller.   c. Would not influence it.

   Explain your answer: _____

   _____

   _____

   _____

3. How would the mistake of Question 2 influence the calculated values of the volume ratios compared to the values obtained without the mistake?

   a. Make them larger.   b. Make them smaller.   c. Would not influence them.

   Explain your answer: _____

   _____

   _____

   _____

4. Suppose the procedures described in Part B were used to determine the number of acidic hydrogens in a sample of a solution of phosphorus acid (H_3PO_3) that was made up to the same molar concentration as the other acid solutions in the experiment. The average volume of NaOH required to titrate the H_3PO_3 samples was 17.92 mL, and the average volume required for the identified monoprotic acid was 8.93 mL. How would you classify H_3PO_3?

   a. Monoprotic   b. Diprotic   c. Triprotic

   Explain your answer: _____

   _____

   _____

   _____

The Use of Melting Points in the Identification of Organic Compounds

In this experiment, you will

- Gain experience in the technique of melting-point determination.
- Observe the effect of impurities on the melting points of compounds.

- Identify an unknown solid on the basis of melting points.

INTRODUCTION

One of the most common and useful techniques used to characterize organic compounds is the determination of melting points. The **melting point** is the temperature at which a solid becomes liquid, or more correctly, the temperature at which solid and liquid exist together in equilibrium. The melting point of a pure substance is very definite. Liquid water, for example, is in equilibrium with ice at exactly 0.00°C. Substances ordinarily thought of as liquids are simply substances with melting points lower than room temperature.

SAFETY ALERT

> The organic compounds used in this experiment are all considered to be toxic and hazardous, but to varying degrees. All should be used with care. If any come in contact with the skin, the contacted area should be washed thoroughly with soap and water. Hands should be thoroughly washed after the lab is completed before food (or anything else) is handled. Care should be taken during the lab period to avoid spilling the compounds in the reagent-dispensing area and at your lab bench work space. Any spilled material should be collected and put into the disposal container that is indicated later in the experiment.

EXPERIMENTAL PROCEDURE

A. Thermometer Correction

The temperatures obtained from the readings of most laboratory thermometers contain errors. For example, ice and water in equilibrium represent a correct temperature of 0.00°C, but a thermometer placed in such a mixture might read 0.5°C. For this reason, it is necessary to correct thermometers if temperatures are to be measured accurately. Correction is done by measuring the melting point of several substances and comparing their known values against the thermometer readings. In this part of the experiment you will measure and record the melting points of three reference compounds: benzophenone, biphenyl, and vanillin (Table 16.1).

PROCEDURE

1. Carry out the following procedure using separate samples of the three reference compounds: benzophenone, biphenyl, and vanillin.
2. Place a small amount of finely powdered solid on a filter paper.
3. Scoop part of the solid into the open end of a melting-point capillary and gently tap the closed end on the bench top until the solid falls to the bottom. Repeat the process until about 3 to 4 mm of solid is collected in the bottom of the capillary as illustrated in Fig. 16.1(a).
4. Use a small rubber band to attach the sample-containing capillary to a thermometer. Adjust the capillary so that the sample is next to the bulb of the thermometer (Fig. 16.1(b)).
5. Place the thermometer and capillary into a 100-mL beaker about three-quarters full of cold water. Position the thermometer and capillary about ½ inch above the bottom of the beaker (Fig. 16.1(c)). Some types of clamps will not require the use of a split stopper to hold the thermometer. If a split stopper is required, adjust the split so the entire scale of the thermometer is visible.
6. Put a boiling chip in the water bath, then heat the bath with a Bunsen burner. At first it can be heated fairly rapidly. But when the temperature is about 10°C below the expected melting point, heat the bath slowly so the temperature is not increasing more than 3 degrees per minute in the neighborhood of the expected melting point.
7. If a compound fails to melt sharply, record its melting range; that is, record both the temperature at which melting starts and the temperature at which the entire sample is melted. Record the observed melting point (M.P.) or melting-point range in Table 16.1.

FIGURE 16.1
Determination of melting points

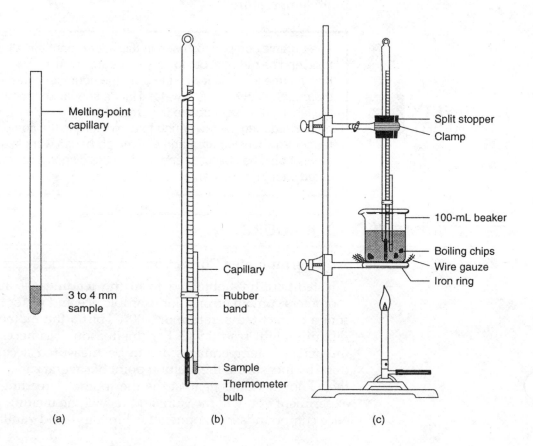

(a) (b) (c)

8. *Be sure the water bath is cooled to a temperature lower than the melting point of the next sample before you begin that sample. The simplest and quickest way to do this is to empty the water bath and refill it with cool water. Follow this practice in all melting-point determinations you carry out.*

DISPOSAL 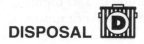 Used melting-point capillaries and unused solids in container labeled "Exp. 16, Used Chemicals"

B. Effect of Impurities on Melting Points

The presence of impurities in a substance influences both the melting point and melting range. Therefore, the melting point can be used as a criterion of purity for a substance. Ordinarily, a substance is assumed to be reasonably pure if it melts while the temperature changes by no more than 1 or 2 degrees (a melting range of 1 or 2 degrees).

PROCEDURE

1. Record in Table 16.2 the melting point or melting-point range of biphenyl that you measured in Part A.
2. Thoroughly mix together on a filter paper small but approximately equal amounts of finely powdered biphenyl and coumarin.
3. Measure and record the melting point or melting-point range of the mixture. The melting point of such a mixture is called a **mixed melting point.**
4. Place small but equal portions of solid *p*-toluidine and *o*-nitrophenol together in a mortar. Use the pestle and grind the mixture.
5. Record your observations in Table 16.2.

DISPOSAL  Used melting-point capillaries, mixed solids, and surplus solids in container labeled "Exp. 16, Used Chemicals"

C. Unknown Identification

The marked influence of impurities on the melting point of organic compounds provides a method for confirming the identity of a compound. If the unknown identity is suspected but still in doubt, a small amount can be mixed with a sample of known identity and the melting point measured. If no difference is found between the melting point of the mixture of known and unknown compounds, and the melting point of the pure known compound, the unknown and known are identical compounds.

PROCEDURE

1. Obtain an unknown sample and record its identification number in Table 16.4. The unknown is one of the nine compounds listed below.

Melting-Point Range of Possible Unknown Compounds

48 to 52°C	68 to 70°C	80 to 85°C
Benzophenone	Biphenyl	Acetamide
Ethyl Carbamate	Coumarin	Acetoacetanilide
Indole	Stearic acid	Vanillin

2. Measure and record the melting point of your unknown.
3. Obtain small samples of the three possible unknown compounds that form a vertical column in the table with a melting point range nearest that of your unknown.
4. Use small amounts of your unknown to determine mixed melting-point ranges with each of the three possible unknown compounds. On a dry filter paper, thoroughly mix together about 3 parts of your unknown and 1 part of the appropriate possible unknown compound for each determination. Record the observed mixed melting-point range and the identity of the possible unknown compound used in each case.

   NOTE: The melting points of mixtures formed from the possible unknown compounds in the 48 to 52°C range might be lower than room temperature (the mixtures will liquefy or appear "wet" as soon as the compounds are mixed together). If this happens, record "lower than room temperature" in the measured melting-point blank of Table 16.4.

DISPOSAL Used melting-point capillaries and unused solids in container labeled "Exp. 16, Used Chemicals"

REPORT

A. Thermometer Correction

1. Use the graph included on the Data and Report Sheet and plot the correct melting points listed in Table 16.2 versus the melting points measured with your thermometer. If melting-point ranges were recorded, use the range midpoint as the measured melting point. See Appendix A for a review of graphing techniques.
2. Draw a smooth line through the points. The line might be straight or curved. The result is called a **thermometer calibration plot.**

B. Effect of Impurities on Melting Points

1. Use the thermometer calibration plot prepared in Part A and correct the measured melting points listed in Table 16.2. Use the midpoint of any recorded melting-point ranges as the melting point. Record the corrected values in Table 16.3.

C. Unknown Identification

1. Use the thermometer calibration plot prepared in Part A and correct the measured melting points listed in Table 16.4. Record the corrected value in Table 16.5. Once again, use the midpoint of any range as the melting point.
2. Use the corrected melting point and the mixed melting-point concept to identify your unknown as one of the nine compounds listed earlier in the table of possible unknowns. Record the name of the unknown and the identification (ID) number in Table 16.5.

Experiment 16 ▪ Pre-Lab Review

The Use of Melting Points in the
Identification of Organic Compounds

1. Are any specific safety alerts given in the experiment? List any that are given.

   _____

   _____

   _____

   _____

2. Are any specific disposal directions given in the experiment? List any that are given.

   _____

   _____

3. How is a thermometer corrected to give accurate temperature readings?

   _____

   _____

   _____

4. How much solid should be placed in a melting-point capillary tube? _____

   Shade the capillary tube drawing below to show the correct amount:

   ⊂_____⊃

5. Where should a melting-point capillary be positioned in relation to the thermometer?

   _____

6. How slowly should the water bath be heated when the temperature is near the expected melting point of the sample?

   _____

7. Describe the steps used to determine the melting point of a solid. _____

   _____

   _____

   _____

8. What is the difference between a melting point and a melting-point range?

_____

_____

9. In Part A, what special precaution should be taken before a solid sample is placed in a water bath that has already been used in a melting-point determination?

_____ _____

_____

10. What is meant by the term *mixed melting point?* _____

_____

_____

11. Explain how the suspected identity of a compound is confirmed using mixed melting points in Part C.

_____

_____

12. A mixture of two substances was heated in a melting-point capillary that was properly mounted in a water bath. The solid sample first appeared to be wet at 64°C and was completely melted into a clear liquid at 78°C. What temperature would be used as the melting point?

_____

Experiment 16 ■ Data & Report Sheet

The Use of Melting Points in the Identification of Organic Compounds

A. Thermometer Correction

TABLE 16.1 (data)

Reference Compound	Correct M.P.	Measured M.P. or Range
Benzophenone	48°C	_____
Biphenyl	70°C	_____
Vanillin	80°C	_____

Thermometer
calibration plot

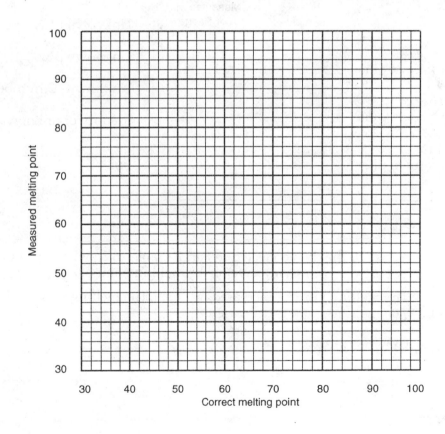

B. Effect of Impurities on Melting Points

TABLE 16.2 (data)

Measured melting point
   (or range) of pure biphenyl _____

Measured melting point
   (or range) of mixture of
   biphenyl and coumarin _____

Results of mixing together
   p-toluidine and *o*-nitrophenol _____

TABLE 16.3 (report)

Sample	Corrected M.P.
Biphenyl	_____
Biphenyl + coumarin	_____

C. Unknown Identification

TABLE 16.4 (data)

Unknown ID number _____

Melting point of unknown _____

Sample	Measured M.P. or Range
Mixture of unknown plus _____	_____
Mixture of unknown plus _____	_____
Mixture of unknown plus _____	_____

TABLE 16.5 (report)

Sample	Corrected M.P.
Unknown	_____
Unknown plus _____	_____
Unknown plus _____	_____
Unknown plus _____	_____
Unknown ID number	_____
Identity of unknown	_____

QUESTIONS

1. According to the data collected in Part B (Table 16.2), what effect does the presence of an impurity (such as coumarin) have on the melting point of a pure substance (such as biphenyl)?

   a. The melting point is increased.   **c.** The melting point is not changed.
   b. The melting point is decreased.

   Explain your answer: _____

   _____

2. What accounts for the behavior observed in Part B when *p*-toluidine and *o*-nitrophenol were ground and mixed together in a mortar?

    a. The friction of grinding raised the temperature enough to melt both pure substances.
    b. Water absorbed from the air dissolved the mixture.
    c. The melting point of the resulting mixture was higher than the melting point of either pure substance.
    d. The melting point of the resulting mixture was lower than room temperature.

    Explain your answer: _____

    _____

3. Suppose in Part C you mixed your unknown with one of the three possible unknown compounds and the mixture immediately melted. What would you conclude about the identity of the unknown?

    a. The unknown and possible unknown compounds are identical.
    b. The unknown and possible unknown compounds are different.
    c. No conclusion can be made.

    Explain your answer: _____

    _____

4. Suppose you carry out the procedure described in Question 3, but the resulting mixture remains a solid. Without further testing, what conclusion can be made about the unknown identity?

    a. The unknown and possible unknown compounds are identical.
    b. The unknown and possible unknown compounds are different.
    c. No conclusion can be made.

    Explain your answer: _____

    _____

Isolation and Purification of an Organic Compound

In this experiment, you will

- Identify the best solvent to use in the isolation of a compound.
- Study the effect of solvent quantity on the isolation of a compound.
- Gain experience in purifying an organic compound by the technique of recrystallization.

INTRODUCTION

Organic compounds isolated from their natural sources or prepared synthetically in the laboratory are usually impure. The impurities might be compounds with properties similar to those of the desired compound, reaction-starting materials, products of side reactions, or simple dust or soil. Pure substances are not only important to chemists in their studies of matter but also are vital in other areas of study such as medicine where those substances might be used as medications. **Recrystallization,** a frequently used method for purifying organic solids, is based on the fact that the solubility of an organic compound in a specific solvent will increase significantly as the solvent is heated. When an impure organic solid is dissolved in such a heated solvent and then the solution is cooled to room temperature or below, the solid will usually recrystallize from solution in a pure form. This allows the solid to be isolated in a pure form if the impurities in the solid either fail to dissolve at the elevated temperature or remain dissolved when the solution is cooled. A separation-and-purification procedure based on these ideas is given in flowchart form in Figure 17.1 for the isolation and purification of acetylsalicylic acid (commonly called aspirin) from a mixture of acetylsalicylic acid, sand, and sugar. You will use this flowchart in Part C of this experiment.

EXPERIMENTAL PROCEDURE

A. Choosing a Suitable Solvent

The choice of solvent is crucial in a separation-and-purification scheme such as the one shown in Figure 17.1. One essential characteristic of a suitable solvent is that the desired compound must be considerably more soluble in hot solvent than in cool solvent. A second desirable characteristic is that impurities should be either insoluble in the solvent at all temperatures or very soluble in the cool solvent. In Part C of this experiment, acetylsalicylic acid (we will call it aspirin from that point on) will be separated from the impurities sugar and sand. The sand is insoluble in all solvents that might be used to purify aspirin by recrystallization, so we will focus our

FIGURE 17.1
Flowchart for the isolation
and purification of
acetylsalicylic acid

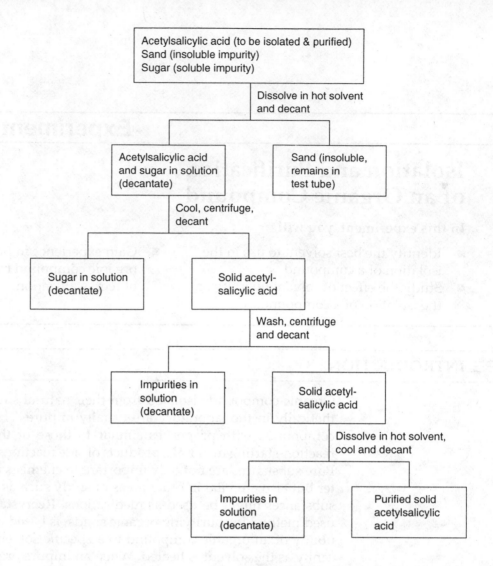

attention on the solubility characteristics of acetylsalicylic acid (aspirin) and sugar in various solvents in this part of the experiment.

PROCEDURE

1. Label four clean, dry 10-cm test tubes with the numbers 1, 2, 3, and 4.
2. Put a 0.10-g sample of acetylsalicylic acid into each test tube. It is important that the samples be weighed carefully even though the sample masses are not recorded in any data table. Use plastic weighing dishes and either a centigram or electronic balance to weigh each sample to ±.01 g.
3. Add 2.0 mL of ethanol to test tube 1 (measure all solvent volumes with a 10-mL graduated cylinder). Rinse the graduated cylinder with water and dry it with a paper or cloth towel. Add 2.0 mL of 2-propanol (isopropyl alcohol) to tube 2. Rinse and dry the cylinder and add 2.00-mL of distilled water to tube 3. Rinse and dry the cylinder and add 2.0 mL of toluene to tube 4.

SAFETY ALERT

Toluene, ethanol, and 2-propanol are flammable. Be certain no open flames are nearby when you do Step 4.

4. Grasp each test tube between your thumb and forefinger and agitate each tube vigorously for 2 minutes or until the solid dissolves, whichever comes first. Observe and record in Table 17.1 the results of this solubility test. Use the symbol + to represent those solvents in which the solute totally dissolved and – to represent those solvents in which some solid remained undissolved after the agitation was completed.

5. Solid should have remained in two of the solvents used in Step 4. Put only the two test tubes containing those solvents and undissolved solid into a boiling-water bath. The bath is made by half-filling a 250-mL beaker with water and adding one or two boiling chips (see Figure 9.1). Allow each of the two test tubes to remain in the bath for 5 minutes. Agitate each tube several times while the heating proceeds. After 5 minutes of heating, observe the test tubes and use the + and – symbols to represent the solubility behavior in Table 17.1.

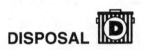

DISPOSAL Test tube that contains toluene (test tube 4) in container labeled "Exp. 17, Used Toluene Parts A and B"
Test tubes that contain ethanol, 2-propanol, and water in sink

6. On the basis of the results obtained so far, you should be able to identify one or two solvents that have suitable solubility behavior toward the acetylsalicylic acid that you want to separate and purify. Identify any suitable solvents by putting an x in the blanks under their names in Table 17.1.

7. Determine the solubility behavior of sugar in cool samples of the solvent or solvents identified in Step 6 as suitable. Do this by putting 0.10-g samples of sugar into clean, dry 10-cm test tubes and adding 2.0 mL of the appropriate solvent (remember, do not heat the solvents). Agitate each test tube as you did in Step 4. Record the behavior in Table 17.1, using the + and – symbols as before in the blanks under the recorded names.

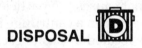

DISPOSAL Any test tube that contains toluene in container labeled "Exp. 17, Used Toluene Parts A and B." Other test tubes in sink.

Based on your observations, identify a single solvent that has solubility characteristics toward sugar that make it suitable for use in the separation and purification of acetylsalicylic acid you will do in Part C. Remember, a suitable solvent will be one in which the sugar will dissolve at a low temperature. Write the name of the solvent in Table 17.1.

B. The Effect of the Amount of Solvent Used

After selecting a good solvent for a recrystallization, care must be taken to use the proper amount of solvent. Enough solvent must be used so that the sample dissolves when heated, but if too much solvent is used the solution does not become saturated on cooling and nothing is recovered. The most common error in recrystallizations is using too much solvent. To demonstrate this principle, you will observe the effect of water quantity on the recrystallization of aspirin.

PROCEDURE

1. Label three 10-cm test tubes with the numbers 1, 2, and 3.
2. Put 0.10-g samples of acetylsalicylic acid into each test tube.
3. Add 2.0 mL of distilled water to test tube 1, 4.0 mL of distilled water to test tube 2, and 6.0 mL of distilled water to test tube 3.
4. Put all three test tubes into a boiling-water bath and allow them to heat until all the acetylsalicylic acid has dissolved in each test tube. Agitate each test tube occasionally during the heating.
5. After the acetylsalicylic acid has completely dissolved in each test tube, remove all test tubes from the bath and allow them to cool for 10 minutes. After 10 minutes, agitate each tube vigorously, allow the solid to settle and record in Table 17.2 the amount of solid that has come back out of solution. Use relative words like *none, some,* and *a lot* to describe the amounts.

DISPOSAL 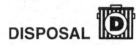   Test tube contents in sink

C. Isolation and Recrystallization of Aspirin

In this part of the experiment, it will be necessary to separate solids from liquids. A common laboratory method for doing this is filtration in which the mixture of solid and liquid is poured through a filter paper held in a funnel. The solid is retained by the filter paper, and the liquid passes through. Filtration is not effective when the quantities of material being separated are small. Small quantities of materials such as the quantities used in this part of the experiment can be separated effectively by using the technique known as centrifugation followed by decantation.

Suspended solids can be forced to settle quickly from solution by subjecting the solids to an increased settling force. One way to do this is to use a centrifuge (Figure 17.2). In a centrifuge, a mixture of solid and liquid contained in a test tube is spun rapidly in a circle. The spinning subjects the solids in the sample to an increased settling force that causes the particles of solid to settle rapidly to the bottom of the test tube. When a centrifuge is used, it is necessary that test tubes of equal mass be located opposite each other in the rotating head to provide balance. Failure to do this results in

FIGURE 17.2
The laboratory centrifuge

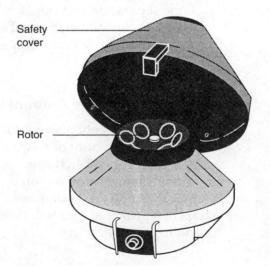

Safety cover

Rotor

excessive wear and damage to the centrifuge. Thus, when you use the centrifuge for a single-sample test tube, put a similar tube containing a volume of water equal to the sample volume in the rotor opposite the position of the sample tube. In the interest of efficient use of the centrifuges, it is better to cooperate with other students in the lab and put sample tubes of equal mass opposite one another in the rotor rather than tubes of water. If you do this, mark the tubes so that you can identify yours.

After a sample has been spun in a centrifuge for an appropriate time (usually about 1 minute), the solid will have settled to the bottom as shown in Figure 17.3(a). The solid will not collect evenly on the bottom of the test tube but will be located toward one side as shown because of the way the test tube is oriented in the centrifuge. Decantation, the separation of the liquid (the **decantate**) from the solid (the **precipitate**), can be accomplished in two ways. When the precipitate is well compacted and stuck to the bottom of the test tube, the liquid decantate can be poured from the test tube (the word *decant* means "to pour"). If pouring is not possible, decantation is easily done by inserting the tip of a previously squeezed dropper into the liquid as shown in Figure 17.3(b) and slowly and carefully drawing the liquid into the dropper. Care must be taken to make certain the precipitate is not disturbed and no solid is drawn off with the liquid. Notice the orientation of the dropper in Figure 17.3(b).

FIGURE 17.3
Decantation using
a dropper

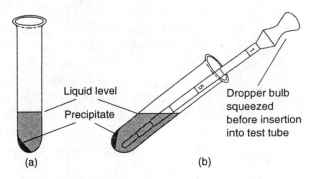

Liquid level

Precipitate

Dropper bulb
squeezed
before insertion
into test tube

(a)   (b)

PROCEDURE

1. Obtain a sample of impure aspirin from the stockroom.
2. Use a centigram or electronic balance to accurately weigh a sample of impure aspirin with a mass in the range of 0.25 to 0.30 g.
3. Record the sample mass in Table 17.3, then put the sample into a clean 10-cm test tube.
4. Add to the test tube 2.5 mL of the single solvent identified in Table 17.1 as suitable. Put the test tube into a boiling-water bath and allow it to remain in the bath for a minimum of 5 minutes. Agitate the tube several times while it is heating.
5. After the tube has been in the boiling-water bath for 5 minutes, agitate the tube, then remove it from the bath and, as quickly as you can, use a dropper to decant the clear liquid from the sand that will remain undissolved in the bottom of the tube. Be careful not to remove any of the sand with the liquid. Put the clear liquid into another clean, dry 10-cm test tube and discard the sand.

DISPOSAL 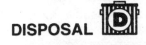   Sand in wastebasket

6. Allow the tube to cool for 5 minutes. Agitate the tube several times while it is cooling.

7. After 5 minutes of cooling, the test tube should contain a significant amount of solid aspirin. Put the test tube into a centrifuge and allow it to spin for about 1 minute (remember to balance the centrifuge with a tube containing water or another student's sample).

8. After the centrifuging is completed, decant the liquid from the test tube using a dropper (see Figure 17.3). Discard the decanted liquid.

DISPOSAL  Decanted liquid in sink

9. Wash the solid aspirin in the test tube by adding 30 drops of distilled water to the test tube and agitating the contents to break up the solid. Return the test tube to a centrifuge and allow it to spin for 1 minute.

10. Remove the test tube from the centrifuge, decant and discard the liquid, and then add 30 drops of distilled water to the solid and put the test tube into a boiling-water bath. Allow the test tube to remain in the bath until all the solid has dissolved.

11. While the solid is dissolving, weigh a clean, dry watch glass on the same balance you used in Steps 2 and 3. The watch glass should have a diameter large enough to allow it to cover the mouth of the beaker used to make your boiling-water bath. Record the mass of the watch glass in Table 17.3. After the solid has all dissolved, pour the resulting clear liquid onto the watch glass. Allow the liquid to cool to room temperature (about 5 minutes). Purified aspirin will crystallize from the liquid during the cooling.

12. If any liquid remains on the watch glass after cooling, remove it by touching the edge of the liquid with the tip of a piece of filter paper or paper towel. The liquid will absorb into the paper. Discard the paper and liquid (**wastebasket**), then put the watch glass with the solid onto the top of your boiling-water bath. Allow the watch glass to remain on the boiling-water bath for 10 to 15 minutes to dry the purified aspirin. The time required for drying will vary depending on the room temperature and the humidity.

13. After the aspirin is dry, remove the watch glass from the boiling-water bath, dry the bottom of the watch glass with a towel, and allow it to cool for 5 minutes. Weigh the cool watch glass and purified aspirin on the same balance used in Step 11. Record the combined mass of the watch glass and purified aspirin in Table 17.3

14. Scrape the aspirin from the watch glass, put an appropriate amount into a melting-point capillary, and determine the melting point of the purified aspirin. If necessary, see Experiment 16, Part A, for a review of the technique to use. You will have to use a glycerol-filled bath to heat the sample. If glycerol baths are not provided, construct one from a small, dry beaker. The bath will look like the one shown in Figure 16.1(c), except you will use glycerol instead of water in the beaker. Stop heating the bath when your thermometer reads 120° C. The temperature will continue to increase after the heating is stopped.

SAFETY ALERT
The glycerol in the bath gets very hot, and care must be exercised to avoid serious burns.

15. Record the melting point of your purified aspirin in Table 17.3. Pure aspirin has a melting point of 135°C. If you put together your own glycerol bath, dispose of the glycerol according to the directions below. If a glycerol bath was provided for your use, leave it as you found it.

DISPOSAL

Used glycerol from your glycerol bath in container labeled "Exp. 17, Used Glycerol"
Used capillary tube in broken-glass container
Unused purified aspirin in wastebasket

CALCULATIONS AND REPORT

The calculations and report for Parts A and B were completed when Tables 17.1 and 17.2 were completed.

C. Isolation and Recrystallization of Aspirin

1. Refer to the data recorded in Table 17.3 and calculate the mass of purified aspirin you recovered. Record this value and the mass of impure aspirin you used in Table 17.4.
2. Calculate the percentage of purified aspirin recovered from the impure sample by using the following formula:

$$\% \text{ Recovered} = \frac{\text{mass of purified aspirin}}{\text{mass of impure aspirin}} \times 100$$

3. Record in Table 17.4 the melting-point range of the purified aspirin from Table 17.3.

Experiment 17 ▪ Pre-Lab Review

Isolation and Purification
of an Organic Compound

1. Are any specific safety alerts given in the experiment? List any that are given.

    _____

    _____

2. Are any specific disposal directions given in the experiment? List any that are given.

    _____

    _____

    _____

3. What device is used to measure the size of the solid samples used in the experiment?

    _____

    _____

4. What device is used to measure the volume of solvent used in the experiment?

    _____

    _____

5. What characteristics make a solvent suitable to use in the isolation and purification of an organic compound?

    _____

    _____

6. What figure should you look at to get information about putting together a boiling-water bath?

    _____

    _____

7. Which solvents do you test with both aspirin and sugar in Part A? _____

    _____

    _____

8. In which parts of the experiment do you use a boiling-water bath? _____

    _____

9. How are solids separated from liquids in Part C of the experiment? _____

_____

_____

_____

10. What procedure is followed to balance centrifuges before they are used?

_____

_____

11. In Part C, what mass should your sample of impure aspirin have? _____

_____

12. In Part C, what object should you weigh while you are waiting for the last test tube to be heated in a boiling-water bath?

_____

Experiment 17 ▪ Data & Report Sheet

Isolation and Purification
of an Organic Compound

A. Choosing a Suitable Solvent

TABLE 17.1 (data and report)

Solid tested: Acetylsalicylic acid

Solvents Tested

	Ethanol (1)	2-Propanol (2)	Water (3)	Toluene (4)
Solubility in cool solvent	_____	_____	_____	_____
Solubility in heated solvent			_____	_____
Solvent(s) with suitable solubility behavior toward acetylsalicylic acid	_____	_____	_____	_____

Solid tested: Sugar

Solubility in cool solvent			_____	_____

Solvent suitable for use in the separation
and purification of acetylsalicylic acid _____

B. The Effect of the Amount of Solvent Used

TABLE 17.2 (data and report)

Volume of Water Used	Amount of Solid Present After Cooling
2.0 mL	_____
4.0 mL	_____
6.0 mL	_____

C. Isolation and Recrystallization of Aspirin

TABLE 17.3 (data)

Mass of impure aspirin sample _____

Mass of empty watch glass _____

Mass of watch glass with
   recovered aspirin _____

Observed melting-point range of
   purified aspirin _____

TABLE 17.4 (report)

Mass of impure aspirin sample _____

Mass of purified aspirin recovered _____

Percentage of aspirin recoverd _____

Melting-point range of purified
   aspirin _____

QUESTIONS

1. What would be the result if too much solvent were used in Step 10 of Part C?

   a. The solid obtained would not be of high purity.
   b. Less solid would be obtained.
   c. Impurities would be less likely to dissolve.

   Explain your answer: _____

   _____

   _____

2. In Part C, what would be the result if warm water rather than room temperature water were used to wash the collected purified aspirin?

   a. The mass of recovered solid would be reduced.
   b. The solid would become more impure.
   c. The melting point of the recovered solid would be lowered.

   Explain your answer: _____

   _____

   _____

3. Which of the following melting-point ranges indicate the purest aspirin (literature melting point 135°C)?

    a. 120 to 130°C   **b.** 130 to 134°C   **c.** 133 to 134°C

    Explain your answer: _____

    _____

    _____

    _____

4. Suppose in Part C that a student fails to completely dry the purified aspirin before weighing it. How would this mistake influence the calculated percentage of purified aspirin that was isolated?

    a. Make it greater than it should be.
    b. Make it less than it should be.
    c. Would not influence it.

    Explain your answer: _____

    _____

    _____

    _____

5. Suppose in Part C that a student got busy cleaning up the work area and allowed the watch glass with purified aspirin to heat on the boiling-water bath for 20 minutes. How would this influence the calculated percentage of purified aspirin that was isolated?

    a. Make it greater than it should be.
    b. Make it less than it should be.
    c. Would not influence it.

    Explain your answer: _____

    _____

    _____

6. In Step 8 of Part C, what dissolved materials are likely to be in the discarded decantate?

    a. Only sand   **b.** Only aspirin   **c.** Only sugar   **d.** Two of the three solids listed

    Explain your answer: _____

    _____

    _____

7. Consider the procedure followed in Part C to obtain purified aspirin and decide which of the following statements about the percentage of purified aspirin actually recovered is most likely to be true.

   a. It is equal to the percentage really in the impure sample.
   b. It is less than the percentage really in the impure sample.
   c. It is greater than the percentage really in the impure sample.

   Explain your answer: _____

   _____

   _____

8. Refer to the melting-point range of purified aspirin recorded in Table 17.4. Remember, pure aspirin melts at 135°C. Which of the following best describes the purity of the aspirin you isolated?

   a. Very pure
   b. More pure than the impure sample but still contains impurities
   c. Very impure

   Explain your answer: _____

   _____

   _____

Hydrocarbons

In this experiment, you will

■ Compare the chemical properties of three classes of hydrocarbons.

■ Gain experience in the technique of boiling-point determination.

■ Identify an unknown hydrocarbon on the basis of its chemical properties and boiling point.

INTRODUCTION

Hydrocarbons, organic compounds containing only carbon and hydrogen, can be conveniently grouped into three classes: saturated, unsaturated, and aromatic. Compounds belonging to different classes react differently with certain reagents. These differences can be used in some instances to distinguish between the three classes.

Saturated hydrocarbons, or **alkanes,** contain no multiple bonds between carbon atoms. Hydrocarbons classified as **unsaturated** belong to one of two groups, the alkenes or the alkynes. **Alkenes** contain one or more double bonds between the carbon atoms of the molecule, while **alkynes** contain one or more triple bonds.

Aromatic hydrocarbons are characterized by the presence in the molecule of one or more of the six-carbon rings that comprise benzene, the simplest member of the class. This ring is often represented as ⬡, where each corner of the hexagon represents a carbon atom to which a hydrogen atom is bonded. The circle represents three pairs of electrons that move freely around the ring. Complete molecular structures for benzene are:

$$
\begin{array}{c}
\text{H} \\
\text{H---C} \quad \overset{\text{C}}{} \quad \text{C---H} \\
\text{H---C} \quad \overset{\text{C}}{} \quad \text{C---H} \\
\text{H} \quad \text{C} \quad \text{H} \\
\text{H}
\end{array}
\quad \text{or} \quad
\begin{array}{c}
\text{H} \\
\text{H---C} \quad \overset{\text{C}}{} \quad \text{C---H} \\
\text{H---C} \quad \overset{\text{C}}{} \quad \text{C---H} \\
\text{H} \quad \text{C} \quad \text{H} \\
\text{H}
\end{array}
$$

SAFETY ALERT

All liquid hydrocarbons are very flammable. Do no experiments with an open flame nearby, except as specifically directed. Do not use samples any larger than specified in the directions.

A. Reaction with Bromine

Alkanes are unreactive with most common laboratory reagents, but they will react with halogens such as chlorine (Cl_2) and bromine (Br_2) in the presence of sunlight to produce substituted halide compounds.

$$CH_3-CH_3 + Br_2 \xrightarrow{\text{light}} CH_3-CH_2-Br + HBr \qquad \text{Eq. 18.1}$$

ethane   bromine   ethyl bromide
(red color)   (colorless)

Unsaturated compounds readily undergo addition reactions with a number of substances including halogens. The reaction with bromine is illustrated in Equation 18.2 for 2-butene, an alkene:

$$CH_3-\underset{\underset{\text{H}}{|}}{C}=\underset{\underset{\text{H}}{|}}{C}-CH_3 + Br_2 \rightarrow CH_3-\overset{\overset{\text{Br}}{|}}{\underset{\underset{\text{H}}{|}}{C}}-\overset{\overset{\text{Br}}{|}}{\underset{\underset{\text{H}}{|}}{C}}-CH_3 \qquad \text{Eq. 18.2}$$

2-butene   bromine   2,3-dibromobutane
(red color)   (colorless)

Bromine can also react with benzene. However, a catalyst such as iron is needed to make the reaction rate reasonably high.

$$\text{benzene} + Br_2 \xrightarrow{\text{Fe}} \text{bromobenzene} + HBr \qquad \text{Eq. 18.3}$$

Other aromatic hydrocarbons that contain attached side chains react with bromine in the presence of light just as alkanes do.

$$\text{ethylbenzene}-CH_2-CH_3 + Br_2 \xrightarrow{\text{light}} -CH_2-CH_2-Br + HBr$$

$$\text{Eq. 18.4}$$

ethylbenzene   1-bromo-2-phenylethane
(colorless)

In general, alkenes can be differentiated from alkanes and aromatics by noting the solution color immediately after adding the red-colored bromine solution. Alkanes and aromatics with the exception of benzene remain red but become colorless after a few minutes because of the reaction that occurs in light. (Equations 18.1 and 18.4). Alkenes turn colorless immediately (Equation 18.2).

PROCEDURE

1. Obtain a sample of unknown hydrocarbon from the stockroom and record its identification (ID) number in Table 18.1.
2. Perform the following test on separate samples of hexane, hexene, toluene, and your unknown.

3. Put 10 drops of the hydrocarbon sample into a clean, *dry* 10-cm test tube and add 3 drops of a 1% solution of Br_2.

SAFETY ALERT

Bromine must be handled and used with care. It will attack tissue vigorously. Use the solution only in the hood and avoid getting any on your skin. If you do contact it, immediately wash the contacted area with soap and water, then inform your instructor.

4. Mix each tube and record in Table 18.1 the color immediately after mixing.

DISPOSAL   Test tube contents in container labeled "Exp. 18, Used Chemicals"

B. Reaction with Potassium Permanganate

Alkenes are readily oxidized by a dilute solution of potassium permanganate ($KMnO_4$). As the reaction takes place, the characteristic color of the permanganate ion (MnO_4^-) is replaced by a brown to black color that is characteristic of solid manganese dioxide (MnO_2), one of the products of the reaction. The color change of a reaction mixture from deep purple to dark brown or black indicates the presence of an alkene or an alkyne. Solutions of alkanes or aromatic compounds remain unreacted and retain the purple color. The reaction for 2-butene is

$$CH_3 - \underset{\underset{H}{|}}{C} = \underset{\underset{H}{|}}{C} - CH_3 + KMnO_4(aq) \rightarrow CH_3 - \underset{\underset{H}{|}}{\overset{\overset{OH}{|}}{C}} - \underset{\underset{H}{|}}{\overset{\overset{OH}{|}}{C}} - CH_3 + MnO_2(s) \quad \text{Eq. 18.5}$$

2-butene (colorless)   potassium permanganate (purple)   2,3-butanediol (colorless)   manganese dioxide (brown-black solid)

PROCEDURE

1. Perform the following test on separate samples of the hydrocarbons hexane, hexene, toluene, and your unknown.
2. Put 10 drops of acetone (a solvent for the hydrocarbon) and 2 drops of the hydrocarbon sample into a clean, dry 10-cm test tube. Add 2 drops of 1% potassium permanganate solution ($KMnO_4$) and mix the tube by agitation.
3. Note the color of the contents of each tube and record the color in Table 18.2.

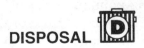

DISPOSAL   Test tube contents in container labeled "Exp. 18, Used Chemcials"

C. Flammability of Hydrocarbons

An extremely important reaction of hydrocarbons is combustion, in which carbon dioxide and water are formed as products:

$$C_3H_8 + 5O_2 \rightarrow 3CO_2 + 4H_2O \qquad \text{Eq. 18.6}$$

propane   carbon dioxide

The greater the extent of unsaturation in a compound, the more luminous and smoky are the resulting flames when it is burned in air. Generally speaking, alkanes burn cleanly with a blue to blue-yellow flame. Unsaturated compounds burn with a slightly smoky flame, and aromatic compounds burn with a very smoky flame.

PROCEDURE

1. Perform the following test on separate samples of hexane (an alkane), hexene (an alkene), toluene (an aromatic), and your unknown.
2. Place 6 drops of the hydrocarbons into a clean, dry evaporating dish.
3. Ignite the sample and note especially the smokiness of the flame. If a flame is very sooty, blow it out after making your observation.
4. Record your observations in Table 18.3.

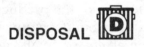

DISPOSAL   Unburned hydrocarbons in container labeled "Exp. 18, Used Chemicals"

D. Boiling-Point Determination

Hydrocarbons, like other organic compounds, can be characterized to some extent on the basis of physical properties. One of these properties, the melting point, was investigated in Experiment 16. In this experiment, we will work only with liquid compounds, for which the boiling point is an important property. The **boiling point** of a liquid is the temperature at which the liquid vapor pressure is equal to the pressure exerted on the liquid by the atmosphere. Pure compounds have constant boiling points, but mixtures usually boil over a range of temperatures. Since the boiling point is related to the prevailing atmospheric pressure, measured boiling points may vary with weather conditions and elevation. Tabulated boiling points are based on an atmospheric pressure of 760 torr and are called **normal boiling points.** Thus, variations in barometric pressure (weather and elevation), plus differences and variations in the equipment used to measure boiling points, may cause observed values to be different than tabulated normal values by several degrees. The effects of such variations will be eliminated by the thermometer-correction procedure that is used.

Boiling points will be determined using a micro boiling-point apparatus (Figure 18.1) consisting of a thermometer with an attached boiling tube containing the liquid sample and a capillary melting-point tube. The air in the capillary tube is displaced by sample vapor as the temperature of the sample increases. Bubbles are seen to come from the capillary tube while this takes place. Above the boiling point of the liquid, the vapor leaves the end of the capillary tube very rapidly in a continuous stream of bubbles. As the sample cools, this bubbling slows. At the boiling point of the liquid, the pressure of the sample vapor inside the capillary tube is equal to the at-

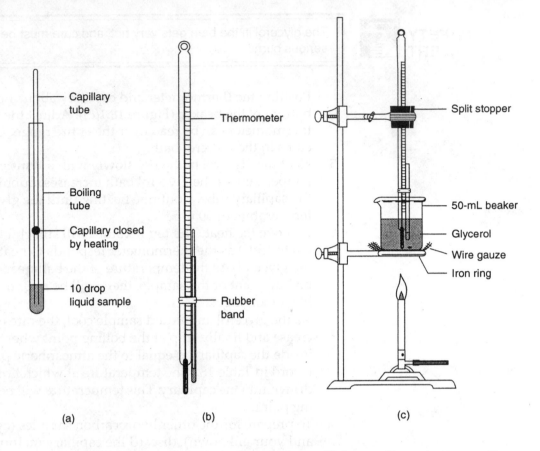

FIGURE 18.1
Micro boiling-point
apparatus

(a) — Capillary tube; Boiling tube; Capillary closed by heating; 10 drop liquid sample

(b) — Thermometer; Rubber band

(c) — Split stopper; 50-mL beaker; Glycerol; Wire gauze; Iron ring

mospheric pressure, and the bubbling stops. If the cooling continues, liquid sample is drawn into the capillary tube. The use of this technique requires only small samples of the liquids and minimizes the chances of an accidental fire.

PROCEDURE

1. Prepare four boiling-point capillary tubes by heating each tube carefully in the outer edge of a burner flame at a location about 2 cm from the open end of the tube. When the tube has softened sufficiently, twist it to close the tube and form a small chamber about 2 cm long on the end of the tube. Your instructor will demonstrate the technique to use.

SAFETY ALERT

The capillary tubes cool quickly but are quite hot for a short time. Use care to avoid burns.

2. Add 10 drops of the hydrocarbon sample (hexane) to a boiling tube consisting of a 5-cm length of 6-mm (outside diameter) glass tubing sealed at one end.
3. Put a capillary melting-point tube (open end down) into the boiling tube. Secure the assembly to a thermometer using a small rubber band (Figure 18.1(b)).
4. Place the thermometer and boiling tube into a 100-mL beaker about ¾ full of glycerol.

Position the thermometer and capillary about 1 cm above the bottom of the beaker (Figure 18.1(c)). Adjust the split stopper so the thermometer can be read over the entire range. Do not put boiling chips in the glycerol bath.

5. Heat the glycerol bath very slowly with a Bunsen burner. As the temperature of the glycerol bath increases, bubbles will emerge from the capillary tube. Continue heating until the glycerol bath reaches a temperature of 50 to 55°C.

6. Remove the heat. The temperature will still climb several degrees (10 to 15°C) as the thermometer responds to heat convected through the glycerol. As the temperature of the bath reaches and passes the boiling point of the sample, there will be a continuous stream of bubbles emerging.

7. As the glycerol and liquid sample cool, the rate of bubbling will decrease and finally stop at the boiling point where the vapor pressure inside the capillary is equal to the atmospheric pressure. Note and record in Table 18.5 the temperature at which liquid just begins to be drawn into the capillary. This temperature will be the recorded boiling point.

8. To prepare for the other hydrocarbon samples (cyclohexene, toluene, and your unknown), discard the capillary melting-point tube and any hydrocarbon sample remaining in the boiling tube. Rinse the boiling tube with 2 drops of the next sample and discard the rinse material.

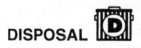

DISPOSAL

Used boiling-point samples in container labeled "Exp. 18, Used chemicals"
Used capillary tubes in broken-glass container
Surplus unknown return to stockroom

Add 10 drops of the next sample to the boiling tube and insert a clean capillary tube. Boiling points should be determined in order of increasing boiling point so that the oil bath does not need excessive cooling. In each determination, stop heating the glycerol bath when the temperature is 10 to 15°C lower than the expected boiling point. After determining the boiling point of toluene, set aside the hot-glycerol bath and prepare a second glycerol bath to be used in determining the boiling point of your unknown.

After you finish this part of the experiment, dispose of the used glycerol if you prepared your own bath. If a glycerol bath was provided for your use, leave it as you found it. Return the boiling tube to the stockroom.

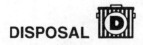

DISPOSAL

Used glycerol in container labeled "Exp. 18, Used Glycerol"

Possible Unknowns and their Normal Boiling Points (in °C)

Saturated Compounds	Unsaturated Compounds	Aromatic Compounds
Pentane (35°)	2-Pentene (36°)	
Cyclopentane (50°)	Cyclopentene (46°)	
Hexane (68°)	1-Hexene (66°)	
Cyclohexane (80°)	Cyclohexene (85°)	
Heptane (98°)	1-Heptene (94°)	
2-Methylheptane (118°)	Cycloheptene (115°)	Toluene (111°)
Octane (125°)	1-Octene (121°)	
Ethylcyclohexane (132°)	1-Octyne (132°)	Ethylbenzene (135°)

REPORT

A. Reaction with Bromine

1. Use test results recorded in Table 18.1 to decide what result is characteristic of alkanes, alkenes, and aromatics.
2. Record these characteristic results in Table 18.4.

B. Reaction with Potassium Permanganate

1. Use the test results recorded in Table 18.2 to decide what result is characteristic of alkanes, alkenes, and aromatics.
2. Record these characteristics in Table 18.4.

C. Flammability of Hydrocarbons

1. Use the test results recorded in Table 18.3 to decide what result is characteristic of alkanes, alkenes, and aromatics.
2. Record these characteristic results in Table 18.4.

Classification of Unknown

1. Use the unknown test result recorded in Table 18.1 and the summary of Table 18.4 to decide what classification (alkane, alkene, or aromatic) is indicated for your unknown on the basis of the bromine test. Record your conclusion in Table 18.6.
2. Use the unknown test result recorded in Table 18.2 and classify your unknown on the basis of the potassium permanganate test. Record your conclusion in Table 18.6.

3. Use the flammability test result for your unknown listed in Table 18.3 and the characteristic flammability results summarized in Table 18.4 to decide what classification is indicated by this test for your unknown.
4. Record this classification based on the flammability test in Table 18.6.
5. On the basis of the three conclusions listed in Table 18.6, decide what classification should be assigned to your unknown. Record this conclusion in Table 18.7.

D. Boiling-Point Determination

1. Use the graph paper provided on the Data and Report sheet to plot the measured boiling points of the reference compounds recorded in Table 18.5 versus the normal boiling points.
2. Draw a smooth line through the points to give a boiling-point calibration plot. The line may be straight or curved. See Appendix A if you need a review of graphing techniques.
3. Use the plot prepared in Step 2 to convert the measured boiling point of your unknown (from Table 18.5) to a normal boiling point. This procedure eliminates errors arising from thermometer or other variations. Record this result in Table 18.7.

Identification of Unknown

1. Record the unknown identification (ID) number (from Table 18.1) in Table 18.7.
2. Use the classification and normal boiling point recorded in Table 18.7 to determine which of the possible unknowns listed earlier in the experiment is your unknown. Record the name of your unknown in Table 18.7.

Experiment 18 ▪ Pre-Lab Review

Hydrocarbons

1. Are any specific safety alerts given in the experiment? List any that are given.

   _____

   _____

2. Are any specific disposal directions given in the experiment? List any that are given.

   _____

   _____

3. What results are expected when an aromatic hydrocarbon is burned?

   _____

   _____

4. What results are expected when a bromine solution is added to an alkene?

   _____

   _____

5. What results are expected when a potassium permanganate solution is added to an alkene?

   _____

6. How must capillary tubes be modified before they are used to measure boiling points in Part D?

   _____

   _____

7. What observation in Part D indicates that the glycerol bath temperature is above the boiling point of the sample?

   _____

   _____

8. At what point should the boiling temperature of a sample be observed and recorded?

   _____

   _____

Experiment 18 ▪ Data & Report Sheet

Hydrocarbons

A. Reaction with Bromine

TABLE 18.1 (data)

Unknown ID Number _____

Sample	Solution Color After Adding Br_2
Hexane	_____
Hexene	_____
Toluene	_____
Unknown	_____

B. Reaction with Potassium Permanganate

TABLE 18.2 (data)

Sample	Color After 2 minutes
Hexane	_____
Hexene	_____
Toluene	_____
Unknown	_____

C. Flammability of Hydrocarbons

TABLE 18.3 (data)

Sample	Amount of Smoke or Soot
Hexane	_____
Hexene	_____
Toluene	_____
Unknown	_____

TABLE 18.4 Summary of Characteristic Test Results (report)

Test	Characteristic Results		
	For Alkanes	For Alkenes	For Aromatics
Bromine	_____	_____	_____
Potassium permanganate	_____	_____	_____
Flammability	_____	_____	_____

D. Boiling-Point Determination

TABLE 18.5 (data)

Sample	Measured B.P.	Normal B.P.
Hexane	_____	68°C
Cyclohexene	_____	85° C
Toluene	_____	111°C
Unknown	_____	

TABLE 18.6 Classification of Unknown (report)

Test	Classification
Bromine	_____
Potassium permanganate	_____
Flammability	_____

TABLE 18.7 Identification of Unknown (report)

Unknown ID Number _____

Unknown normal B.P. _____

Unknown classification _____

Unknown identity _____

Boiling-point calibration
plot (report)

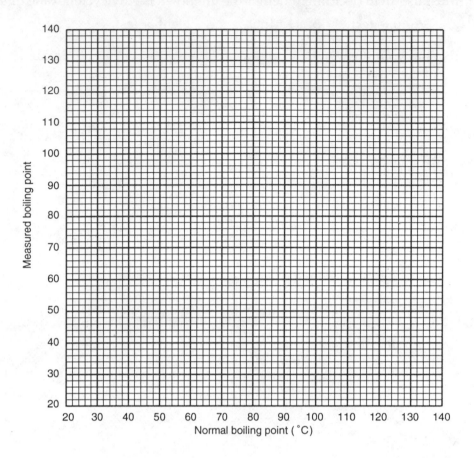

QUESTIONS

1. Suppose that in completing Table 18.6 it was hard to decide whether the flammability-test result was characteristic of an alkene or aromatic, but both the Br_2 and $KMnO_4$ tests were characteristic of aromatics. How would you classify your unknown?

   a. As an alkane   **b.** As an alkene   **c.** As an aromatic

   Explain your answer: _____

   _____

2. The following test results were obtained for an unknown:

Test	Result
Bromine	Solution was red immediately after mixing
$KMnO_4$	Solution was purple
Flammability	Very little smoke or soot
Normal B.P.	115.6°C

   Which of the following is the best choice for the identity of the unknown?

   a. 2-Methylheptane   **b.** Cycloheptene   **c.** Toluene

   Explain your answer: _____

   _____

3. An unknown has a normal boiling point of 133°C. Which of the following tests would be must useful in deciding whether the unknown is ethylcyclohexane or ethylbenzene?

a. Bromine test   b. KMnO$_4$ c. Neither test

Explain your answer: _____

_____

Reactions of Alcohols and Phenols

In this experiment, you will

- Examine some of the properties and characteristic reactions of alcohols and phenols.

- Use chemical properties to identify the functional group in two unknown compounds

INTRODUCTION

Alcohols are hydrocarbon derivatives in which a hydrogen atom of the hydrocarbon is replaced by a hydroxyl group (—OH). When the replaced hydrogen belongs to the aromatic ring, the resulting compound is known as a **phenol**.

$$R-OH$$

an alcohol   a phenol

Thus, the chemistry of alcohols and phenols is the chemistry of a hydroxyl functional group.

EXPERIMENTAL PROCEDURE

A. Solubility in Water

Low-molecular-weight alcohols and some phenols are completely miscible with water because of the formation of hydrogen bonds.

As the molecular weight of alcohols or phenols increases, the water solubility decreases, and the compounds become more soluble in nonpolar solvents. In general, one alcohol hydroxyl group can solubilize three to four carbon atoms.

PROCEDURE

1. Obtain one unknown from the stockroom and record its identification (ID) number in Table 19.1. This unknown may be either an alcohol or a phenol. The other unknown used in this experiment is an organic compound often used as an antiseptic. This unknown is located with the other reagents in a bottle labeled "antiseptic."

2. Put 5-drop samples of the following into separate, labeled or numbered 10-cm test tubes: ethanol, *t*-butanol, cyclohexanol, ethylene glycol, the unknown antiseptic, and your unknown.

3. To each tube, add 20 drops of distilled water, agitate to mix, and record your observations concerning solubility of the compounds in Table 19.1.

DISPOSAL Test tube contents in container labeled "Exp. 19, Used Chemicals Part A"

B. The Ceric Nitrate Test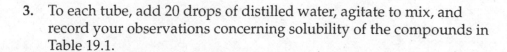

The hydroxyl functional group reacts with ceric nitrate to produce colored compounds. When a yellow-colored ceric nitrate solution is added to an alcohol, a red product forms. A dark brown precipitate forms when the ceric nitrate solution is added to a phenol. Thus, the ceric nitrate test is a general test for alcohols and phenols. It can also be used to differentiate an alcohol from a phenol because of the different color of the products. The general reaction with alcohols is

$$[Ce(NO_3)_6]^{2-} + ROH \rightarrow [CeOR(NO_3)_5]^{2-} + HNO_3 \qquad Eq. \ 19.1$$

PROCEDURE

1. Put 3-drop samples of the following into separate, clean 10-cm test tubes: ethanol, *t*-butanol, cyclohexanol, liquified phenol, 35% resorcinol, antiseptic, and your unknown. Phenol and resorcinol are solid compounds at room temperature, but we will use them in the form of very concentrated solutions for convenience.

SAFETY ALERT

> Phenol will vigorously attack tissue even when it is in solution. Contact with the phenol solution should be avoided. If you get any on your skin, wash it off immediately with soap and water and inform your instructor.

2. Add 8 drops of ceric nitrate solution to each of the seven test tubes and mix by agitation. Note the colors formed in each tube and record your observations in Table 19.2.

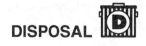

DISPOSAL Test tube contents in container labeled "Exp. 19, Used Chemicals Part B"

C. The Lucas Test

Alcohols are classified as *primary, secondary,* and *tertiary* on the basis of the number of carbon atoms attached to the carbon that is attached to the hydroxyl group. In primary alcohols, the number of carbon atoms is one, in secondary alcohols it is two, and in tertiary alcohols it is three, as illustrated by the following examples:

$$CH_3CH_2-OH \qquad \langle \quad \rangle -OH \qquad CH_3-\overset{\overset{\displaystyle CH_3}{|}}{\underset{\underset{\displaystyle CH_3}{|}}{C}}-OH$$

ethanol	cyclohexanol	*t*-butanol
(a primary alcohol)	(a secondary alcohol)	(a tertiary alcohol)

The hydroxyl group of alcohols can be replaced by a Cl from hydrochloric acid (HCl) if zinc chloride ($ZnCl_2$) is present as a catalyst. A mixture of hydrochloric acid and zinc chloride is called a Lucas reagent.

$$R-OH + HCl \xrightarrow{ZnCl_2} R-Cl + H-OH \qquad \text{Eq. 19.2}$$

The Lucas test is based on the different rates at which primary, secondary, and tertiary alcohols react with this reagent. Tertiary alcohols react almost immediately with the Lucas reagent to form a cloudy mixture that sometimes separates into layers. Secondary alcohols dissolve in Lucas reagent to form a clear solution that becomes cloudy within two minutes when the tube is put into a boiling water bath. Primary alcohols react only at elevated temperatures, and even under these conditions some remain unreacted. Phenols also remain unreacted.

PROCEDURE

1. Put 5-drop samples of ethanol, t-butanol, and cyclohexanol into separate, clean, dry 10-cm test tubes. Prepare similar samples of antiseptic and unknown.
2. Add 10 drops of Lucas reagent to each sample and mix the contents well by agitation.

 **SAFETY ALERT**
> Lucas reagent contains concentrated hydrochloric acid (HCl) that will vigorously damage tissue. Contact should be avoided. If you get any on your skin, wash it off immediately with cool water and inform your instructor.

3. Observe the tubes within a minute after mixing and note whether or not a cloudiness has developed. Put any tubes with clear solutions into a boiling water bath and observe them again after two minutes.
4. Record your observations in Table 19.3.

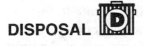 **DISPOSAL** Test tube contents in container labeled "Exp. 19, Used Chemicals Part C"

D. Oxidation with Chromic Acid

Primary and secondary alcohols can be oxidized by strong oxidizing agents such as chromic acid. In this process, two H atoms are removed from the alcohol—one from the hydroxyl group and one from the carbon atom to which the hydroxyl group is attached. Tertiary alcohols have no H atoms attached to the hydroxyl-bearing carbon and thus cannot be oxidized by this process. Most phenols are also resistant to oxidation.

$$3R-\overset{\overset{\displaystyle H}{|}}{\underset{\underset{\displaystyle R'}{|}}{C}}-O-H + 2H_2CrO_4 + 6H^+ \rightarrow 3R-\overset{\overset{\displaystyle O}{\|}}{C}-R' + 2Cr^{3+} + 8H_2O \qquad \text{Eq. 19.3}$$

chromic acid (orange) chromic ion (green)

The disappearance of the red-orange chromic acid color and the appearance of the blue-to-green colored Cr^{3+} ions indicate that the alcohol has been oxidized.

PROCEDURE

1. Place 1-drop samples of ethanol, t-butanol, cyclohexanol, antiseptic, and your unknown into separate 10-cm test tubes.

2. Add 20 drops of acetone to each test tube. The acetone acts as a solvent for the alcohols.
3. Agitate each test tube to mix the contents.
4. Add 1 drop of 10% chromic acid solution to each test tube and mix well by agitation.

SAFETY ALERT

Chromic acid contains chromium in a +6 oxidation state. Chromium in a +6 state is carcinogenic and should be handled with care. Try to avoid contact with it and immediately wash any contacted area with cool water.

5. Observe the test tubes after 3 minutes and record in Table 19.4 the color of the contents of each one.

DISPOSAL   Test tube contents in container labeled "Exp. 19, Used Chromic Acid Part D"

E. Iodoform Test

$$OH$$
$$|$$

Alcohols that have the structure $CH_3 - CH - R$, where a $CH_3 -$ group is attached to the hydroxyl-bearing carbon, produce iodoform (CHI_3) when treated with iodine in an alkaline solution. The CHI_3 forms as a yellow precipitate.

PROCEDURE

1. Put 10 drops of the following compounds into separate 10-cm tubes: ethanol, 1-propanol, 2-propanol, antiseptic, and your unknown.
2. Add 20 drops of $I_2 - KI$ solution to each tube and agitate to mix.
3. To each tube add 3 drops of 10% NaOH solution. Agitate to mix. In a positive test the brown color of the reagent disappears and a light-yellow cloudiness forms, which settles as a yellow precipitate.
4. After one minute, look for the formation of any yellow cloudiness or yellow precipitates and record your observations in Table 19.5.

DISPOSAL   Test tube contents in container labeled "Exp. 19, Used Chemicals Part E"

F. The Acidity of Phenols

Phenols are structurally similar to the alcohols; however, the fact that the —OH group is attached to a benzene ring causes their properties to be somewhat different than alcohols. For example, phenols are stronger acids than alcohols and produce acidic solutions with a pH lower than pure water when dissolved in water. Phenols that are water-insoluble dissolve readily in basic solutions.

PROCEDURE

1. Place 2-drop samples of 1-propanol, liquified phenol, the antiseptic, and your unknown into separate 10-cm test tubes.

Review the safety alert given earlier in Part B concerning the concentrated solution of phenol.

2. Add 20 drops of distilled water to each test tube and agitate to mix the contents.
3. Add 20 drops of distilled water to a clean 10-cm test tube.
4. Using wide-range pH paper, measure the pH of each solution and the pure water. Record each pH value in Table 19.6.

 Test tube contents in container labeled "Exp. 19, Used Chemicals Part F"

G. Reaction with Ferric Chloride

Phenols form colored complexes when mixed with ferric chloride. This test may be used to distinguish most phenols from alcohols that usually do not react.

PROCEDURE

1. Put 2-drop samples of the following into separate 10-cm test tubes: liquified phenol, 35% resorcinol solution, 2-propanol, antiseptic, and your unknown.
2. Add 20 drops of water to each test tube and agitate to mix the contents.
3. To each test tube, add 2 drops of 1% ferric chloride solution, swirl the contents, and record the results in Table 19.7.

 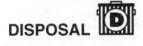Test tube contents and surplus unknown in container labeled "Exp. 19, Used Chemicals Part G"

REPORT

Characterization of Unknowns

1. Record the identification (ID) number of your unknown in Table 19.8.
2. Collect from Tables 19.2 through 19.7 the various test results obtained for your unknown and the antiseptic. Record these test results in Table 19.8.
3. Use the characteristic test results to decide what conclusions can be drawn relative to your unknown and the antiseptic. Note that when solution pH values are considered, be sure to compare them to the pH of pure water.
4. On the basis of your conclusions, decide what structural features are present in your unknown and the antiseptic and indicate these features in the answers to Questions I, II, and III that follow Table 19.8.

Experiment 19 ▪ Pre-Lab Review

Reactions of Alcohols and Phenols

1. Are any specific safety alerts given in the experiment? List any that are given.

    _____

    _____

    _____

2. Are any specific disposal directions given in the experiment? List any that are given.

    _____   _____

    _____   _____

    _____

    _____

3. What types of alcohols would you expect to be water-soluble? _____

    _____

    _____

4. What results are expected when the ceric nitrate test is done on

    a. an alcohol? _____

    b. a phenol? _____

5. What results are expected when the Lucas test is done on a

    a. primary alcohol? _____

    b. secondary alcohol? _____

    c. tertiary alcohol? _____

    d. phenol? _____

6. What results are expected when the chromic acid test is done on a

    a. primary alcohol? _____

    b. secondary alcohol? _____

    c. tertiary alcohol? _____

    d. phenol? _____

7. What structural feature is confirmed by a positive iodoform test? _____

_____

8. How is a positive test for phenols using ferric chloride recognized? _____

_____

Experiment 19 ▪ Data & Report Sheet

Reactions of Alcohols and Phenols

A. Solubility in Water

TABLE 19.1 (data)

Unknown ID number _____

Sample	Result of Adding H_2O
Ethanol	
t-Butanol	
Cyclohexanol	
Ethylene glycol	
Antiseptic	
Unknown	

B. The Ceric Nitrate Test

TABLE 19.2 (data)

Sample	Color After Adding $[Ce(NO_3)_6]^{2-}$
Ethanol	
t-Butanol	
Cyclohexanol	
Phenol	
Resorcinol	
Antiseptic	
Unknown	

C. The Lucas Test

TABLE 19.3 (data)

Sample	Appearance After 1 Minute	Appearance After Heating
Ethanol		
t-Butanol		
Cyclohexanol		
Antiseptic		
Unknown		

D. Oxidation with Chromic Acid

TABLE 19.4 (data)

Sample	Result of Adding H_2CrO_4
Ethanol	
t-Butanol	
Cyclohexanol	
Antiseptic	
Unknown	

E. Iodoform Test

TABLE 19.5 (data)

Sample	Result of Adding $I_2 - KI$
Ethanol	
1-Propanol	
2-Propanol	
Antiseptic	
Unknown	

F. The Acidity of Phenols

TABLE 19.6 (data)

Sample	Solution pH	Sample	Solution pH
Pure water	_____	Antiseptic	_____
1-Propanol	_____	Unknown	_____
Phenol	_____		

G. Reaction with Ferric Chloride

TABLE 19.7 (data)

Sample	Result of Adding FeCl₃
Phenol	_____
Resorcinol	_____
2-Propanol	_____
Antiseptic	_____
Unknown	_____

Characterization of Unknowns

TABLE 19.8 (report)

Unknown ID number _____

Test	Observation for Unknown	Conclusion for Unknown	Observation for Antiseptic	Conclusion for Antiseptic
Ceric nitrate	_____	_____	_____	_____
Lucas reagent	_____	_____	_____	_____
Chromic acid	_____	_____	_____	_____
Iodoform test	_____	_____	_____	_____
Solution pH	_____	_____	_____	_____
Ferric chloride	_____	_____	_____	_____

I. Circle the letter that best describes your unknown.

    a. Phenol   **c.** Secondary alcohol
    b. Primary alcohol   **d.** Tertiary alcohol

II. Which of the above possibilities best describes the antiseptic? _____

III. Check below if a $CH_3 - \overset{\displaystyle OH}{\overset{|}{CH}} -$ group is present in

    a. your unknown? _____   **b.** the antiseptic? _____

QUESTIONS

1. Which of the following reagents or tests would be most useful in quickly differentiating between

$$\text{Ph} - CH_2CH_2 - OH \quad \text{and} \quad \overset{CH_3}{\text{Ph}} - OH$$

    a. Lucas reagent   **b.** Iodoform test   **c.** Ferric chloride

Explain your answer: _____

_____

_____

2. Which of the following reagents or tests would be useful in differentiating between *n*-propyl alcohol ($CH_3CH_2CH_2 - OH$) and isopropyl alcohol ($CH_3\overset{\displaystyle OH}{\overset{|}{CH}}CH_3$)?

    a. Chromic acid   **b.** Lucas reagent   **c.** Aqueous bromine

Explain your answer: _____

_____

_____

3. Which would be less soluble in water?

    a. 1-Pentanol   **b.** 1-Butanol   **c.** 1-Heptanol

Explain your answer: _____

_____

_____

4. A liquid compound is subjected to a variety of tests with the following results:

   Lucas test: Clear solution formed, which became cloudy when heated for 2 minutes.

   Chromic acid test: Solution color changed from pale orange to pale green.

   Iodoform test: A yellow precipitate formed.

   Solution pH: Water solution of the compound had a pH near that of pure water.

   Which of the following formulas could best represent the liquid compound?

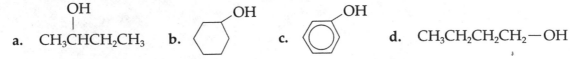

        OH

   a. $CH_3CHCH_2CH_3$ b.   c.   d. $CH_3CH_2CH_2CH_2-OH$

   Explain your answer: _____

   _____

   _____

5. A water-insoluble compound is discovered to be soluble in 10% NaOH. Which of the following substances is most likely to be the compound?

   a. CH_3CH_2-OH b.   c. $CH_3(CH_2)_8CH_2-OH$ d. $\bigcirc\!-OH$

       OH

       $(CH_2)_4CH_3$

   Explain your answer: _____

   _____

   _____

6. Which of the following diagrams correctly depicts hydrogen bonding in pure ethanol?

   a. CH_3CH_2-O $\overset{H\cdots H}{\diagup\quad\diagdown}$ $O-CH_2CH_3$ c. $CH_3CH_2-O\cdots H-\overset{\overset{H}{|}}{\underset{\underset{H}{|}}{C}}-CH_2-O\overset{H}{\diagup}\diagdown^H$

   b. $CH_3CH_2-O\cdots \overset{H}{\underset{\diagdown CH_2CH_3}{O}}\overset{\overset{H}{|}}{}$ d. $CH_3CH_2-O\cdots H-O\overset{H}{\diagup}\underset{\diagdown CH_2CH_3}{}$

   Explain your answer: _____

   _____

   _____

Reactions of Aldehydes and Ketones

In this experiment, you will

■ Investigate some of the properties and characteristic reactions of aldehydes and ketones.

■ Identify the functional group in two unknown compounds through their chemical reactivity.

INTRODUCTION

Aldehydes and ketones are both characterized by the carbonyl group

$$-\overset{\overset{\displaystyle O}{\|}}{C}-.$$ **Aldehydes** have at least one hydrogen bonded to the carbonyl carbon, whereas in **ketones,** both bonds are to carbon atoms. The carbonyl group is polar and capable of participating in hydrogen bonding in water solutions. Thus, low-molecular-weight aldehydes and ketones are water-soluble. As the length of the carbon chain increases, water solubility tends to decrease.

EXPERIMENTAL PROCEDURE

A. DNP Test for Aldehydes and Ketones

A characteristic property of the carbonyl group is the reaction with 2,4-dinitrophenylhydrazine (DNP) to form a yellow-orange precipitate. This readily detectable reaction is characteristic of both aldehydes and ketones.

PROCEDURE

1. Obtain one unknown from the stockroom and record its identification (ID) number in Table 20.5. The other unknown used in this experiment is an organic compound often used as a flavoring agent. This unknown is located with the other reagents in a bottle labeled "flavoring agent."
2. Number four clean 10-cm test tubes 1, 2, 3, and 4. Make sure that none of the tubes have been contaminated with acetone (which gives a positive DNP test) in the cleaning process.
3. Put 10 drops of DNP reagent into each of the four test tubes.

SAFETY ALERT

The DNP reagent should be handled carefully and kept off skin and clothing. If contact occurs, wash the contacted area immediately with cool water.

Add 2 drops of acetone to tube 1, 2 drops of butyraldehyde to tube 2, 2 drops of flavoring agent to tube 3, and 2 drops of your unknown to tube 4. Mix well by agitation and note the almost instantaneous formation of a precipitate in tubes 1 and 2. Record your observations for all samples in Table 20.1.

B. Tollens' Reagent

Aldehydes and ketones can be differentiated on the basis of the ease with which they are oxidized. Aldehydes are readily oxidized in basic solutions while ketones are more resistant. Thus, weak oxidizing agents react with aldehydes but not with ketones. Tollens' reagent, a weak oxidizing mixture containing an alkaline silver-ammonia complex, reacts with aldehydes to produce metallic silver. The silver will appear as a black precipitate or as a silver mirror on the walls of clean glass containers.

$$\underset{\text{aldehyde}}{\overset{\displaystyle O}{\overset{\displaystyle \|}{RC-H}}} + \underset{\substack{\text{Tollens'} \\ \text{reagent}}}{2Ag(NH_3)_2^+} + 2OH^- \rightarrow \underset{\substack{\text{carboxylate} \\ \text{salt}}}{\overset{\displaystyle O}{\overset{\displaystyle \|}{R-C-O^-NH_4^+}}} + \underset{\substack{\text{metallic} \\ \text{silver}}}{2Ag} + 3NH_3 + H_2O \quad \text{Eq. 20.1}$$

PROCEDURE

1. Number four clean 10-cm test tubes 1, 2, 3, and 4.
2. Put 15 drops of dilute (6 M) sodium hydroxide (NaOH) into each test tube and allow them to stand for at least 1 minute. This prepares the glass surface for the Tollens' test.

3. After 1 minute, empty the NaOH from the tubes and *without rinsing* the tubes, add 15 drops of 0.1 M silver nitrate (AgNO₃) solution.

4. Next, add 15 drops of 6 M ammonia (NH₃ or NH₄OH), shaking until the precipitate that forms initially is essentially dissolved.
5. Add 5 drops of acetone to tube 1, 5 drops of butyraldehyde to tube 2, 5 drops of flavoring agent to tube 3, and 5 drops of your unknown to tube 4.
6. Agitate the mixtures and set the tubes aside. Observe them after 3 minutes.
7. Record your observations in Table 20.2. If a tube with a positive test is allowed to stand for about 10 minutes, a silver mirror will develop.

DISPOSAL  Test tube contents in sink. Flush down the drain with running tap water.

C. Chromic Acid Test for Aldehydes

Aldehydes can be oxidized by chromic acid, the same reagent used in Experiment 19 for the oxidation of alcohols. Ketones are resistant to oxidation under these acidic conditions.

$$3R-\overset{\overset{\displaystyle O}{\|}}{C}-H + 2H_2CrO_4 + 6H^+ \rightarrow 3R-\overset{\overset{\displaystyle O}{\|}}{C}-OH + 2Cr^{3+} + 5H_2O \quad \text{Eq. 20.2}$$

| an aldehyde | chromic acid (orange) | a carboxylic acid | chromic ion (green) |

The disappearance of the red-orange chromic acid color and the appearance of the blue-to-green colored Cr^{3+} ions indicates that the aldehyde has been oxidized.

Aliphatic aldehydes react with chromic acid almost immediately (within 10 seconds), while aromatic aldehydes take up to several minutes to react. This difference in reactivity can be used to identify an aldehyde as aromatic or aliphatic.

PROCEDURE

1. Label five 10-cm test tubes 1, 2, 3, 4, and 5.
2. Place 20 drops of acetone in each tube.
3. Acetone, a ketone, is the solvent for this test. It will be tested alone (tube 1) to be certain it does not react.
4. Add 2 drops of butyraldehyde to tube 2, 2 drops of p-tolualdehyde (an aromatic aldehyde) to tube 3, 2 drops of flavoring agent to tube 4, and 2 drops of unknown to tube 5.
5. Agitate each test tube to mix the contents.
6. Add 1 drop of 10% chromic acid solution to each test tube and mix well by agitation.
7. Observe the test tubes after 10 seconds and again after 12 minutes. Record in Table 20.3 the color of the contents of each tube.

SAFETY ALERT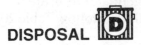

Chromic acid contains chromium in a +6 oxidation state. Chromium in a +6 state is carcinogenic and should be handled with care. Try to avoid contact with it and immediately wash any contacted area with cool water.

DISPOSAL Test tube contents in container labeled "Exp. 20, Used Chromic Acid Part C"

D. Iodoform Test

Acetaldehyde, $CH_3-\overset{\overset{\displaystyle O}{\|}}{C}-H$, and other methyl ketones react with iodine in basic solution to form iodoform, a light-yellow precipitate with a medicinal odor, and other products.

$$R-\overset{\overset{\displaystyle O}{\|}}{C}-CH_3 + 3I_2 + 4NaOH \rightarrow CHI_3 + R-\overset{\overset{\displaystyle O}{\|}}{C}-O^-Na^+ + 3NaI + 3H_2O \quad \text{Eq. 20.3}$$

a methyl   iodoform
ketone   (a yellow solid)

The test is positive for aldehydes and ketones with the $CH_3-\overset{\overset{\displaystyle O}{\|}}{C}-$ group

and, as noted in Experiment 19, for alcohols with $CH_3\overset{\overset{\displaystyle OH}{|}}{CH}-$ as a structural feature.

PROCEDURE

1. Label five 10-cm test tubes 1, 2, 3, 4, and 5.
2. Place 10 drops of acetone in tube 1, 10 drops of butyraldehyde in tube 2, 10 drops of 2-butanone in tube 3, 10 drops of flavoring agent in tube 4, and 10 drops of unknown in tube 5.
3. Add 20 drops of I_2 — KI solution to each tube.
4. Add 3 drops of 10% NaOH to each tube and agitate to mix. In a positive test, the brown color of the reagent disappears and a light-yellow cloudiness forms, which settles as a yellow precipitate. Some compounds giving a negative test form a brown layer. Record your observations in Table 20.4.

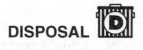 **DISPOSAL** Test tube contents and surplus unknown in container labeled "Exp. 20, Used Chemicals Part D"

REPORT

Characterization of Unknowns

1. Collect from Tables 20.1 through 20.4 the various test results obtained for your unknown and the flavoring agent. Record these test results in Table 20.5.
2. Use the characteristic test results to decide what conclusions can be drawn relative to your unknown and the flavoring agent.
3. On the basis of your conclusions, decide what structural features are present in your unknown and the flavoring agent and indicate these features in the answers to Questions I, II, and III that follow Table 20.5.

Experiment 20 ▪ Pre-Lab Review

Reactions of Aldehydes and Ketones

1. Are any specific safety alerts given in the experiment? List any that are given.

   _____

   _____

2. Are any specific disposal directions given in the experiment? List any that are given.

   _____

   _____

3. What types of aldehydes and ketones would you expect to be water-soluble?

   _____

4. What results are expected when the DNP test is done on

   a. an aldehyde? _____

   b. a ketone? _____

5. What size test tubes are used in Part B? _____

6. What results are expected when a Tollens' test is done on

   a. an aldehyde? _____

   b. a ketone? _____

7. What type of aldehyde reacts quite readily with chromic acid? _____

   _____

8. What structural features give positive results in the iodoform test? _____

   _____

Reactions of Alkynes and Ketones

Experiment 20 ▪ Data & Report Sheet

Reactions of Aldehydes and Ketones

A. DNP Test for Aldehydes and Ketones

TABLE 20.1 (data)

Sample	Tube No.	Result of Adding DNP
Acetone	1	
Butyraldehyde	2	
Flavoring agent	3	
Unknown	4	

B. Tollens' Reagent

TABLE 20.2 (data)

Sample	Tube No.	Observations
Acetone	1	
Butyraldehyde	2	
Flavoring agent	3	
Unknown	4	

C. Chromic Acid Test for Aldehydes

TABLE 20.3 (data)

Sample	Tube No.	Color After 10 Seconds	Color After 12 Minutes
Acetone	1		
Butyraldehyde	2		
p-Tolualdehyde	3		
Flavoring agent	4		
Unknown	5		

D. Iodoform Test

TABLE 20.4 (data)

	Tube No.	Observations
Acetone	1	
Butyraldehyde	2	
2-Butanone	3	
Flavoring agent	4	
Unknown	5	

Characterization of Unknowns

TABLE 20.5 (report)

Unknown ID number _____

Test	Observation for Unknown	Conclusion for Unknown	Observation for Flavoring Agent	Conclusion for Flavoring Agent
DNP				
Tollens'				
Chromic acid				
Iodoform				

I.   Circle the letter that best describes your unknown.

    a. Aliphatic aldehyde   **b.** Aromatic aldehyde   **c.** Ketone

II. Which of the above possibilities best describes the flavoring agent? _____

III. Check below if a $CH_3 - \overset{\displaystyle O}{\overset{\|}{C}} -$ group is present in

    a. your unknown? _____   **b.** the flavoring agent? _____

QUESTIONS

1. Which of the following reagents or tests would you use to differentiate between

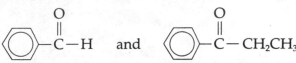

    a. Iodoform   **b.** Chromic acid   **c.** Tollens'   **d.** DNP

    Explain your answer: _____

    _____

    _____

2. Which of the following reagents or tests would you use to differentiate between

$$CH_3CH_2CH_2 - \overset{\overset{\displaystyle O}{\|}}{C} - H \quad \text{and} \quad \text{⬡} - \overset{\overset{\displaystyle O}{\|}}{C} - H$$

    a. Iodoform   **b.** Chromic acid   **c.** Tollens'   **d.** DNP

    Explain your answer: _____

    _____

    _____

3. An unknown compound is tested with several reagents, yielding the following results:

    DNP test: A thick, yellow precipitate formed.

    Tollens' test: No precipitate or mirror was detected.

    Iodoform test: A yellow precipitate formed.

    Which of the following formulas could represent the compound?

    a. 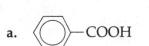—COOH   **b.** 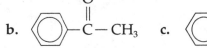   **c.** 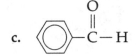   **d.** $CH_3CH_2CH_2 - \overset{\overset{\displaystyle O}{\|}}{C} - H$

    Explain your answer: _____

    _____

    _____

Reactions of Carboxylic Acids, Amines, and Amides

In this experiment, you will

- Study the chemical properties of carboxylic acids, amines, and amides.
- Identify the functional group in an unknown as carboxylic acid, amine, or amide.

- Isolate aspirin from Alka-Seltzer®.
- Saponify an amide.

INTRODUCTION

Carboxylic acids are organic compounds that contain the carboxyl functional group, $-\overset{\overset{\displaystyle O}{\|}}{C}-OH$. Carboxylic acids are widely found in nature and are responsible for the sour taste and some of the pungent odors of foods such as vinegar, cheese, and some fruits.

Amines are organic derivatives of ammonia (NH_3) in which one or more of the hydrogens are replaced by an aromatic or alkyl group (represented by R):

$$R - NH_2$$
an amine

The characteristic functional group of **amides** is a carbonyl group attached to nitrogen. The single bond linking the carbon and nitrogen atoms in the group is called an amide linkage.

$$-\overset{\overset{\displaystyle O}{\|}}{C}-\overset{\overset{\displaystyle }{\underset{\displaystyle |}{N}}}- \quad \text{amide linkage}$$

amide functional group

All three of these functional groups are capable of participating in hydrogen bonding in water solutions. As a result, low-molecular-weight carboxylic acids, amines, and amides are water-soluble. As the length of the carbon chain attached to the functional group increases, the water solubility tends to decrease.

EXPERIMENTAL PROCEDURE

A. Water-Soluble Carboxylic Acids

Carboxylic acids are weak acids that dissociate to some extent when dissolved in water to form hydronium cations and carboxylate anions:

$$R-\underset{\substack{\| \\ O}}{C}-OH + H_2O \rightleftarrows R-\underset{\substack{\| \\ O}}{C}-O^- + H_3O^+ \qquad \text{Eq. 21.1}$$

a carboxylic   a carboxylate   hydronium

acid   anion   ion

Carboxylic acids in solution can be detected by measuring the pH and by adding a solution of bicarbonate. A water solution of a carboxylic acid will have a pH that is lower than the pH of pure water. When bicarbonate is added to a carboxylic acid solution, gaseous carbon dioxide (CO_2) is liberated and can be detected as a fizzing or as bubbles (Equation 21.2):

$$R-\underset{\substack{\| \\ O}}{C}-OH + NaHCO_3 \rightleftarrows R-\underset{\substack{\| \\ O}}{C}-O^-Na^+ + H_2O + CO_2 \qquad \text{Eq. 21.2}$$

PROCEDURE

1. Obtain an unknown from the stockroom and record its identification (ID) number in Table 21.7 of the Data and Report Sheet.
2. Number four 10-cm test tubes 1, 2, 3, and 4.
3. Put 5 drops of distilled water into each tube.
4. Add 5 drops of glacial acetic acid to tube 1, 5 drops of butyric acid to tube 2, and 5 drops of liquid unknown or 0.1 g of solid unknown to tube 3.

SAFETY ALERT

> These acids will attack tissue. If contact is made, wash the contacted area with cool water. You do not know the identity of your unknown, so it should be treated as if it were toxic and dangerous to get on the skin. If contact is made, wash the contacted area with soap and water.

5. Agitate each tube to mix the contents and dissolve the samples.
6. Use wide-range (pH 1 to 11) pH paper to determine the pH of each solution and of a sample of distilled water (use test tube 4). Record the measured pH values in Table 21.1.
7. Add 5 drops of saturated sodium bicarbonate solution ($NaHCO_3$) to each test tube and agitate to mix. Note and record in Table 21.1 whether a gas (CO_2) is produced as indicated by fizzing or the appearance of bubbles.

DISPOSAL   Test tube contents in container labeled "Exp. 21, Used Chemicals Part A"

B. Water-Insoluble Carboxylic Acids

Like other acids, carboxylic acids react with bases such as sodium hydroxide to produce a salt and water:

$$\underset{\substack{\| \\ O}}{C}-OH \bigcirc \quad + \quad NaOH \quad \rightarrow \quad \underset{\substack{\| \\ O}}{C}-O^-Na^+ \bigcirc \quad + \quad H_2O \qquad \text{Eq. 21.3}$$

benzoic acid   sodium benzoate

(water-insoluble)   (water-soluble)

Because they are ionic, the salts of carboxylic acids are usually soluble in water. This includes the salts of water-insoluble carboxylic acids. Indeed, water-insoluble carboxylic acids can often be differentiated from other water-insoluble compounds by testing the solubility of the compounds in a solution of base; the acid will dissolve in the basic solution. When a strong mineral acid such as hydrochloric acid is added to a solution that contains the salt of a water-insoluble carboxylic acid, the salt is converted back to the acid, which will then precipitate from solution.

 + HCl →  + NaCl   Eq. 21.4

sodium benzoate   benzoic acid
(water-soluble)   (water-insoluble)

PROCEDURE

1. Number three 10-cm test tubes 1, 2, and 3 and put 20 drops of distilled water into each tube.
2. Add 4 drops of caproic acid to tube 1, a few crystals of benzoic acid to tube 2, and 4 drops of unknown to tube 3.
3. Agitate each tube and note in Table 21.2 which acids are insoluble.
4. Add 10 drops of 3 M sodium hydroxide solution (NaOH) to each tube, and agitate to mix the contents. Record your observations of the solubility behavior of each sample in Table 21.2 of the Data and Report Sheet.
5. Now add 10 drops of 6 M hydrochloric acid to each tube and agitate to mix the contents. Again, record your observations in Table 21.2.

SAFETY ALERT

Both 3 M NaOH and 6 M HCl are corrosive and will attack tissue. Wash any contacted area with cool water.

DISPOSAL

Test tube contents in container labeled "Exp. 21, Used Chemicals Part B"

C. Aspirin from Alka-Seltzer®

The original Alka-Seltzer® antacid and pain-reliever formulation consists of sodium bicarbonate (NaHCO₃, the antacid), aspirin (the pain reliever), and citric acid. The purpose of the citric acid is to supply hydronium ions when the solid tablet dissolves in water. The hydronium ions react with the sodium bicarbonate to provide the fizz for the final liquid that is ingested. In the process, the citric acid is converted into the salt, sodium citrate. Aspirin (acetylsalicylic acid) also contains a carboxyl group, and it reacts with some of the sodium bicarbonate to form a water-soluble sodium salt and carbon dioxide:

$$\text{aspirin} + NaHCO_3 \rightarrow \text{salt (water-soluble)} \qquad \text{Eq. 21.5}$$

aspirin — salt (water-soluble)

Thus, when an Alka-Seltzer® tablet is added to water, a solution is formed that contains any excess, unreacted sodium bicarbonate and the dissolved sodium salts of citric acid and aspirin. When a mineral acid such as HCl is added to the solution, three reactions take place, as shown below:

$$NaHCO_3 + HCl \rightarrow NaCl + H_2O + CO_2 \qquad \text{Eq. 21.6}$$

$$\text{sodium citrate} + HCl \rightarrow NaCl + \text{citric acid} \qquad \text{Eq. 21.7}$$
$$\text{(water-soluble)}$$

$$\text{sodium acetylsalicylate} + HCl \rightarrow NaCl + \text{acetyl salicylic acid} \qquad \text{Eq. 21.8}$$
$$\text{aspirin salt} \qquad \qquad \text{aspirin precipitate}$$

The water-insoluble (at room temperature) aspirin precipitates from solution and is thus separated from the water-soluble NaCl and citric acid, which remain in solution.

PROCEDURE

1. Weigh a sample of Alka-Seltzer® with a mass of about 0.8 g. This is equal to ¼ a tablet, so if large pieces or whole tablets are available, break them into appropriate sizes and skip the weighing step. Put the solid pieces of tablet into a 50-mL beaker. Record your observations for the following steps in Table 21.3.
2. Add 4.0 mL of distilled water to the beaker and swirl it occasionally while the tablet dissolves.
3. When the tablet is essentially all dissolved, add 4 drops of 6 *M* hydrochloric acid (HCl) to the beaker and swirl the contents.
4. Pour the contents of the beaker onto a clean watch glass and slowly add 2 mL of 6 *M* HCl dropwise to the liquid.
5. Note the formation of a white precipitate of aspirin. If no solid forms, stir the liquid on the watch glass to start the crystallization.

SAFETY ALERT

> Remember, the HCl is very corrosive. Wash any contacted area with cool water.

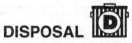

DISPOSAL   Mixture on the watch glass in sink

D. Amine Properties

Amines are organic compounds that behave as weak bases that react with water to form alkaline solutions:

$$R-NH_2 + H_2O \rightleftharpoons R-NH_3^+ + OH^- \qquad \text{Eq. 21.9}$$

Thus, water-soluble amines can be detected by measuring the pH of a water solution. Solutions of water-soluble amines will have pH values sigificantly higher than the pH of pure water. Amines that are not water-soluble are usually differentiated from other water-insoluble compounds by their solubility in dilute hydrochloric acid (HCl). The amines react with HCl to form ionic salts that are water-soluble:

$$R—NH_2 + HCl \rightarrow R—NH_3^+Cl^- \qquad \text{Eq. 21.10}$$
$$\text{amine salt}$$
$$\text{(water-soluble)}$$

Sometimes the amine salts also form in the gaseous state, appearing as a cloudy vapor in the test tube. Amines with low molecular weights have rather foul, fishy, or ammonia-like odors that disappear or significantly decrease as the amines react with acids to form salts.

PROCEDURE

1. Put 20 drops of distilled water and 3 drops of triethylamine into a clean 10-cm test tube and agitate to mix.

SAFETY ALERT

> Triethylamine will attack tissue. If contact is made, wash the contacted area with soap and water.

2. Cautiously note the odor of the solution and check its pH with wide-range (pH 1 to 11) pH paper. Record your observations in Table 21.4.
3. Add 5 drops of dilute (6 M) hydrochloric acid (HCl) to the solution and agitate to mix the contents of the tube.

SAFETY ALERT

> Remember, dilute HCl is corrosive. Wash any contacted area with cool water.

4. Once again, cautiously check the odor of the solution and record your observations in Table 21.4.
5. Repeat Steps 1 to 4, but use N,N-dimethylaniline in place of the triethylamine. Record your observations.
6. If your unknown did not show characteristics in Parts A and B that lead you to conclude it is an acid, repeat Steps 1 and 2 using your unknown in place of the triethylamine (if unknown is a solid, use 0.1 g sample). If the pH of the water solution is basic (significantly higher than that of pure water), also carry out Steps 3 and 4 (if unknown is a solid, use 0.1 g sample).

SAFETY ALERT

> Treat the N,N-dimethylaniline and your unknown as toxic and corrosive materials.

DISPOSAL   Tested materials in container labeled "Exp. 21, Used Chemicals Part D"

E. Amide Tests

Amides are neutral compounds that show none of the acidic properties of carboxylic acids nor the basic properties of amines. Unsubstituted amides (those that contain an —NH₂ group) can be detected by **saponification** with a strong base. The saponification reaction converts unsubstituted amides into carboxylate salts and ammonia gas. The ammonia can be detected by its ability to turn red litmus paper to a blue color.

$$\underset{\text{amide}}{R-\overset{\displaystyle O}{\overset{\|}{C}}-NH_2} + NaOH \rightarrow \underset{\underset{\text{salt}}{\text{carboxylate}}}{R-\overset{\displaystyle O}{\overset{\|}{C}}-O^-Na^+} + \underset{\text{ammonia}}{NH_3} \qquad \text{Eq. 21.11}$$

PROCEDURE

1. If your unknown has not been shown to be a carboxylic acid or an amine by the results of Parts A, B, and D, you will test it here to see if it is an amide. If you have already identified it, ignore the references to your unknown in the following steps to prevent getting a false positive test for an amide.
2. Number two 10-cm test tubes 1 and 2.
3. Put 0.1 g of acetamide into tube 1 and 4 drops (or 0.1 g) of your unknown into tube 2.
4. Add 15 drops of dilute (6 M) sodium hydroxide solution (NaOH) to each tube and agitate to mix well.

 **SAFETY ALERT**

> Dilute sodium hydroxide will vigorously attack tissue and is especially damaging to the eyes. Wash any contacted area with cool water and, if the eyes are involved, inform your instructor.

5. Put both tubes into a boiling-water bath.
6. After heating the tubes for 1 to 2 minutes, insert the end of a strip of red litmus paper that has been moistened with a drop of distilled water into the mouth of each tube where it can react with any liberated ammonia gas. Do not let the litmus paper touch the sides of the test tube or the liquid inside. Observe and record in Table 21.5 any color changes that the litmus undergoes.

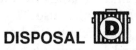 **DISPOSAL**

Test tube contents and surplus unknown in container labeled "Exp. 21, Used Chemicals Part E"
Used litmus paper in wastebasket

F. Saponification of Benzamide

The saponification of benzamide demonstrates much of the chemistry of carboxylic acids, amines, and amides studied in the other parts of this experiment. Saponification involves the treatment of benzamide with a strong base. The reaction produces ammonia and a water-soluble carboxylate salt of benzoic acid:

$$\underset{\text{benzamide}}{C_6H_5-\overset{\displaystyle O}{\overset{\|}{C}}-NH_2} + NaOH \rightarrow \underset{\underset{\text{(water-soluble)}}{\text{sodium benzoate}}}{C_6H_5-\overset{\displaystyle O}{\overset{\|}{C}}-O^-Na^+} + \underset{\text{ammonia}}{NH_3} \qquad \text{Eq. 21.12}$$

Acidification of the resulting solution of the benzoic acid salt (sodium benzoate) converts the salt to benzoic acid, which is water-insoluble, and the acid precipitates from solution.

sodium benzoate
(water-soluble)

$+$ HCl $\rightarrow$

benzoic acid
(water-insoluble)

$+$ NaCl

Eq. 21.13

PROCEDURE

1. Use a weighing dish and a centigram balance to weigh out a sample of benzamide with a mass of about 0.10 g.
2. Put the sample of benzamide into a 10-cm test tube and add 2.0 mL of distilled water.
3. Add 20 drops of 6 M sodium hydroxide solution (NaOH) to the tube and put the tube in a boiling-water bath.

SAFETY ALERT

Remember the hazardous nature of 6 M NaOH and follow the directions given earlier if contact occurs.

4. After the test tube has been heating in the boiling-water bath for 3 minutes, insert a strip of moistened red litmus paper into the mouth of the test tube. Do not let the litmus paper touch the sides of the test tube or the liquid inside. Note and record in Table 21.6 any changes in color of the litmus paper.
5. Allow the test tube to continue heating in the boiling-water bath for another 8 to 10 minutes, then remove it and allow it to cool for at least 5 minutes in a cool-water bath.
6. If any solid forms as the test tube cools, it is unreacted benzamide. After the cooling is completed, check to see if any formed solid has settled completely to the bottom of the test tube. If necessary, centrifuge the test tube to force any solid to collect in the bottom. Refer to Experiment 17, Part C, if you feel you need a review of the procedures to follow when using a centrifuge. If no solid forms, add HCl as directed in step 7.
7. Use a dropper to decant the clear liquid from the solid in the test tube without getting any of the solid with the liquid. Put the clear liquid into a clean 10-cm test tube and add to it 20 drops of dilute (6 M) hydrochloric acid (HCl).

SAFETY ALERT

Remember the corrosive nature of dilute hydrochloric acid. Wash any contacted area with cool water.

8. Note and record in Table 21.6 the result of adding HCl in Step 7.

DISPOSAL 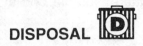 Test tube contents in container labeled "Exp. 21, Used Chemicals Part F"

REPORT

Most of the report for the experiment was completed when Tables 21.1 21.2, 21.4 and 21.5 were completed. Refer to the information recorded for your unknown in those Tables. Note that you might not have information concerning your unknown in all of those tables. On the basis of the information, classify the functional group in your unknown as a carboxylic acid, an amine, or an amide. Record the identity of the functional group in Table 21.7. Also write in Table 21.7 the evidence you used as the basis for coming to your conclusion about the functional group identity. Keep in mind that some tests are more definitive than others. For example, carboxylic acids will always give solutions a low pH value, but seeing bubbles when $NaHCO_3$ is added to a solution is sometimes more difficult. An unknown water-soluble amine will always produce solutions with a basic (high) pH, but detecting changes in odor intensity is more elusive.

Remember that if your unknown is an amine, it might also give a positive test for an amide. The false result is the reason you should not test for an amide if you have already concluded that your unknown is an amine.

Experiment 21 ▪ Pre-Lab Review

Reactions of Carboxylic Acids, Amines, and Amides

1. Are any specific safety alerts given in the experiment? List any that are given.

   _____

   _____

   _____

2. Are any specific disposal directions given in the experiment? List any that are given.

   _____

   _____

   _____

3. How are water-soluble carboxylic acids detected in solution? _____

   _____

   _____

4. How are water-insoluble carboxylic acids differentiated from other water-insoluble compounds?

   _____

5. What two methods could be used to get a sample of Alka-Seltzer® of appropriate size in Part C?

   _____

   _____

6. What reagent is added to the dissolved Alka-Seltzer® to cause the aspirin to precipitate?

   _____

7. How are water-soluble amines detected in solution? _____

   _____

8. How are water-insoluble amines differentiated from other water-insoluble compounds?

   _____

   _____

9. What compound produced as a result of the saponification of unsubstituted amides causes red litmus paper to turn blue?

10. Under what circumstances will you not have any information concerning your unknown recorded in Tables 21.4 or 21.5?

Experiment 21 ▪ Data & Report Sheet

Reactions of Carboxylic Acids, Amines, and Amides

A. Water-Soluble Carboxylic Acids

TABLE 21.1 (data and report)

Sample Tested	Solution pH	Result of Adding Bicarbonate
Acetic acid	_____	_____
Butyric acid	_____	_____
Unknown	_____	_____
Dist. water	_____	_____

B. Water-Insoluble Carboxylic Acids

TABLE 21.2 (data and report)

Sample Tested	Solubility in Water	Solubility in NaOH Solution	Result of Adding HCl
Caproic acid	_____	_____	_____
Benzoic acid	_____	_____	_____
Unknown	_____	_____	_____

C. Aspirin from Alka-Seltzer®

TABLE 21.3 (data and report)

Step	Observations
2	_____
3	_____
4	_____
5	_____

D. Amine Properties

TABLE 21.4 (data and report)

Sample Tested	Solution Odor	Solution pH	Solution Odor After Adding HCl
Triethylamine			
N,N-dimethylaniline			
Unknown			

E. Amide Tests

TABLE 21.5 (data and report)

Sample Tested	Result of Red Litmus Test
Acetamide	
Unknown	

F. Saponification of Benzamide

TABLE 21.6 (data and report)

Test Performed	Test Result
Red litmus paper	
Add HCl to liquid	

Identification of Unknown Functional Group

TABLE 21.7 (report)

Unknown ID number _____

Functional group in unknown _____

Basis for conclusion _____

_____

_____

QUESTIONS

1. Suppose an unknown dissolves in water to produce a solution with a pH of 2.4. Which of the following reagents would be useful in obtaining more information about the functional group in the unknown?

    a. 6 *M* HCl solution   **b.** Saturated sodium bicarbonate solution
    c. 3 *M* NaOH solution

    Explain your answer: _____

    _____

2. Suppose a solid unknown is insoluble in water. Which of the following reagents would be useful in obtaining more information about the functional group in the unknown?

    a. 6 *M* HCl solution   **b.** 3 *M* NaOH solution   **c.** Dilute NH_3 solution
    d. More than one answer is correct.

    Explain your answer: _____

    _____

    _____

3. Suppose a student is doing Part C of the experiment. The student gets to the point where all the 6 *M* HCl has been added, but no precipitate of aspirin forms. What should the student do next?

    a. Conclude that the original sample contained no aspirin.
    b. Add more HCl even though the directions do not call for it.
    c. Stir the liquid to which the HCl has been added.

    Explain your answer: _____

    _____

    _____

4. It is suspected that a water-insoluble solid unknown is an amine. Which of the following would be useful in determining the validity of the suspicion?

    a. 6 *M* HCl solution   **b.** 3 *M* NaOH solution   **c.** Red litmus paper

    Explain your answer: _____

    _____

    _____

5. A water-soluble unknown is found to form a solution with a pH near that of pure water. Which of the following functional groups is indicated by this information?

    a. Carboxylic acid   **b.** Amine   **c.** Amide   **d.** No answer is correct.

    Explain your answer: _____

    _____

    _____

6. Which of the following would be useful in helping you identify the functional group indicated in Question 5?

    a. Red litmus paper   **b.** 6 M NaOH solution   **c.** 6 M HCl solution
    d. More than one answer is correct.

    Explain your answer: _____

    _____

    _____

7. Which of the following would be formed if the amide shown below was saponified?

$$CH_3-(CH_2)_6-\overset{\displaystyle O}{\overset{\|}{C}}-NH_2$$

a. $CH_3-(CH_2)_6-\overset{\displaystyle O}{\overset{\|}{C}}-OH$ **b.** $CH_3-(CH_2)_6-\overset{\displaystyle O}{\overset{\|}{C}}-O^-Na^+$ **c.** NH_3 **d.** NaCl
e. More than one answer is correct.

    Explain your answer: _____

    _____

    _____

    _____

The Synthesis of Aspirin and Other Esters

In this experiment, you will

- Synthesize aspirin and oil of wintergreen.
- Use a qualitative test to judge the purity of your prepared aspirin.

- Calculate a percentage yield for your prepared aspirin.
- Prepare several fragrant esters.

INTRODUCTION

Esters are organic compounds that are often pleasant smelling and, in some cases, are responsible for the flavors and fragrances of fruits and flowers. Fats, oils, and waxes are naturally occurring esters of great biological importance.

Esters can be prepared by reacting an alcohol with a carboxylic acid. However, the process is reversible and equilibrium is reached slowly unless a catalyst, such as sulfuric acid, is used.

$$\underset{\substack{\text{carboxylic} \\ \text{acid}}}{R-\overset{\overset{\displaystyle O}{\|}}{C}-OH} + \underset{\text{alcohol}}{R'-OH} \underset{}{\overset{H_2SO_4}{\rightleftharpoons}} \underset{\text{ester}}{R-\overset{\overset{\displaystyle O}{\|}}{C}-O-R'} + H_2O \quad \text{Eq. 22.1}$$

EXPERIMENTAL PROCEDURE

A. Synthesis of Aspirin

Salicylic acid is a carboxylic acid that, along with its derivatives, has long been used as a mild analgesic (pain reliever) and antipyretic (fever reducer). Salicylic acid molecules contain both a phenol and a carboxylic acid group. Therefore, it can form esters by behaving as an alcohol and reacting with an acid or by acting as an acid in a reaction with an alcohol. Aspirin (acetyl salicylic acid) is an ester in which a reaction has taken place between the phenol group of salicylic acid and the carboxyl group of acetic acid, although it is usually prepared by using the more reactive acetic anhydride in place of acetic acid. The preparation of aspirin is illustrated by the following reaction:

salicylic acid   acetic anhydride   aspirin   acetic acid
                                                    (acetylsalicylic acid)

Aspirin is undoubtedly the most widely used of all drugs. Its daily consumption in the United States alone is estimated to be more than 20 tons.

PROCEDURE

1. Set up a hot-water bath made from a 250-mL beaker half full of water and containing 1 or 2 boiling chips (see Figure 9.1).
2. Use a weighing dish and a centigram or an electronic balance to accurately weigh out approximately 0.250 g of salicylic acid. Record the mass of the salicylic acid sample in Table 22.1 and put the sample into a clean, dry 10-cm test tube.
3. Heat and stir the water bath until it reaches a temperature of 85 to 90°C, then remove the heat source.
4. Carefully add 15 drops of acetic anhydride and 2 drops of concentrated sulfuric acid to the salicylic acid sample in the test tube and agitate the test tube to mix the contents. Put the test tube into the water bath.

SAFETY ALERT

Concentrated sulfuric acid and acetic anhydride both attack tissue vigorously. Both materials should be handled with care. If contact is made with either material, wash it from the skin with large amounts of cool water.

---

5. Allow the test tube to remain in the hot-water bath for 15 minutes. Agitate the tube several times during the first 5 minutes to make certain all the solid is dissolved. Do *not* heat the bath during the 15 minutes.
6. After 15 minutes, remove the test tube from the bath and add 40 drops of ice water, 5 drops at a time, with agitation between each 5-drop addition. After the water addition is completed, put the test tube into a mixture of ice and water contained in a small beaker. Allow the test tube to cool in the ice-and-water bath for 5 minutes.
7. After 5 minutes, remove the test tube from the cooling bath and observe its appearance. If you can see distinct crystals of solid forming, put the tube back into the ice-and-water bath for another 5 to 10 minutes. If the test tube contents appear uniform and milky, it means your aspirin has not crystallized. Try to get it to crystallize by stirring the mixture with a small spatula or glass stirring rod and, while stirring, scratching the inside surface of the test tube with the spatula or stirring rod. Continue this stirring and scratching for a

Eq. 22.2

minimum of 5 minutes or until the sample begins to crystallize, whichever comes first.

8. If your sample begins to crystallize, return it to the ice-and-water bath for 5 to 10 minutes. If it does not begin to crystallize after 5 minutes of stirring and scratching, inform your instructor who will give you directions for adding a few seed crystals of aspirin to start the crystallization. Once crystallization has been started, return the test tube to the ice-and-water bath for 5 to 10 more minutes of cooling.

9. After the cooling is completed, remove the tube from the cooling bath and centrifuge it to cause the solid to settle. Remember to balance the centrifuge (see Experiment 17, Part C). After centrifuging, decant the liquid from the solid, using a dropper (see Figure 17.3).

DISPOSAL  Decanted liquid in sink

10. Wash the solid in the test tube by adding 20 drops of ice water and agitating the tube or stirring the mixture until all the solid is suspended. Centrifuge the sample again; decant and dispose of the liquid as you did in Step 9.

11. Accurately weigh a clean, dry watch glass on the same balance used in Step 1 and record the mass in Table 22.1.

12. Add 20 drops of ice water to the solid in the tube and agitate or stir the contents to suspend the solid. Quickly pour the suspension out of the test tube and onto the watch glass you weighed in Step 11. Put the watch glass on top of a boiling-water bath and allow it to remain for a minimum of 10 minutes to dry the aspirin you prepared. A significant amount of water can be removed from the watch glass by touching the edge of the liquid with the corner of a torn piece of filter paper or kitchen towel. Do this carefully so no solid is removed. Discard the wet paper in the wastebasket.

13. After the sample has dried for 10 minutes, remove the watch glass and allow it to cool for 2 to 3 minutes, then dry off the bottom and weigh it on the same balance used in Step 11. Record the combined mass of the watch glass and dried aspirin in Table 22.1. Keep the aspirin for use in Part C.

B. Synthesis of Oil of Wintergreen

Methylsalicylate (oil of wintergreen) is an ester in which the carboxyl group of the salicylic acid has reacted with the alcohol group of methyl alcohol. Interestingly, oil of wintergreen is used both as a topical analgesic in rubs for sore muscles and as a flavoring agent.

$$
\underset{\text{salicylic acid}}{\begin{array}{c}\text{O}\\ \|\\ \text{C}-\text{OH}\\ \text{OH}\\ \bigcirc \end{array}} + \underset{\text{methyl alcohol}}{\text{CH}_3-\text{OH}} \xrightarrow{\text{H}^+} \underset{\substack{\text{oil of wintergreen}\\ \text{(methylsalicylate)}}}{\begin{array}{c}\text{O}\\ \|\\ \text{C}-\text{O}-\text{CH}_3\\ \text{OH}\\ \bigcirc \end{array}} + \text{H}_2\text{O} \qquad \text{Eq. 22.3}
$$

PROCEDURE

1. Place 0.1 g of salicylic acid into a clean, dry 10-cm test tube.
2. Add 6 drops of methyl alcohol and agitate the tube until the contents are well mixed.
3. Then add 1 drop of concentrated sulfuric acid.

4. Agitate the tube contents and place the tube in a beaker of boiling water for 3 minutes.
5. After the heating is completed remove the tube from the bath and add 15 drops of water to the tube contents.
6. Cautiously note the odor of the products in the test tube by passing the tube back and forth under your nose, slowly bringing it closer to your nose until you can detect the odor. Record the odor in Table 22.3.

DISPOSAL 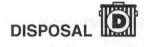 Test tube contents in container labeled "Exp. 22, Used Chemicals"

C. Qualitative Test for Phenols

Many phenolic compounds (aromatic ring with an attached —OH group) react with iron chloride to give highly colored products. In the presence of phenols, the yellow color of iron chloride turns to a violet color.

PROCEDURE

1. Into four separate, small, and clean test tubes place a few crystals of salicylic acid, commercial aspirin, your prepared aspirin from Part A, and 1 drop of commercial oil of wintergreen. For the three solid samples, use an amount that would fit inside this ○.
2. Add 20 drops of ethyl alcohol to each of the four test tubes. The ethyl alcohol acts as a solvent.
3. Add 1 drop of 1% iron (III) chloride solution ($FeCl_3$) to each test tube.
4. Shake each tube and record the color of the solutions in Table 22.4.

DISPOSAL  Test tube contents in container labeled "Exp. 22, Used Chemicals"
Prepared aspirin from Part A in wastebasket

D. Synthesis of Some Fragrant Esters

A number of other commercially important esters can also be synthesized using the procedure that yielded oil of wintergreen.

$$\text{alcohol} + \text{carboxylic acid} \xrightarrow{H_2SO_4} \text{ester} + H_2O \qquad \text{Eq. 22.4}$$

In this experiment, you will synthesize four esters used commercially as flavorings (see Table below).

Some Esters Used Commercially as Flavorings

Carboxylic Acid	Alcohol	Ester	Fragrance
Acetic acid	*n*-Propyl	*n*-Propyl acetate	Pear
Acetic acid	Isoamyl	Isoamyl acetate	Banana
Butyric acid	*n*-Butyl	*n*-Butyl butyrate	Pineapple
Butyric acid	Ethyl	Ethyl butyrate	Strawberry

PROCEDURE

1. Prepare *n*-propyl acetate by putting 6 drops of *n*-propyl alcohol in a clean, dry 10-cm test tube.
2. Add 2 drops of glacial acetic acid.
3. Add 1 drop of concentrated sulfuric acid, agitate the tube to mix the contents, and place the test tube in a boiling-water bath for 3 minutes.

 **SAFETY ALERT**

> Both glacial acetic acid and concentrated sulfuric acid will vigorously attack tissue. Avoid contact and wash any contacted area immediately with large amounts of cool water.

4. When the heating is completed, remove the test tube from the bath, and add 20 drops of water to the test tube contents. Agitate to mix.
5. Cautiously note the odor of the product, using the method described in Step 6 of Part B. Record the odor in Table 22.6.

 **DISPOSAL**   Test tube contents in container labeled "Exp. 22, Used Chemicals"

6. Repeat Steps 1 to 5 three times, using the following pairs of alcohols and acids: isoamyl alcohol plus glacial acetic acid, ethyl alcohol plus butyric acid, and *n*-butyl alcohol plus butyric acid.
7. Note and record the odor of the produced esters. Don't worry about having the "correct" odor for a particular ester. Descriptions of odor often vary with the individual, and some odors are just simply hard to describe.

REPORT

A. Synthesis of Aspirin

The percentage yield of a synthesis is defined by the following equation:

$$\% \ \text{Yield} = \frac{\text{actual yield}}{\text{theoretical yield}} \times 100 \qquad \text{Eq. 22.5}$$

In Equation 22.5, the actual yield is the mass of product isolated, and the theoretical yield is the mass of product that was theoretically possible based on the amount of reactants used. In this synthesis, salicylic acid with a molecular weight of 138 was reacted with an excess of acetic anhydride.

Because the acetic anhydride was used in excess, the amount of aspirin product theoretically possible is determined by the amount of salicylic acid (the limiting reactant) used. Aspirin has a molecular weight of 180. According to Equation 22.2, 1 mol salicylic acid will react to produce 1 mol aspirin. Thus, if 138 g of salicylic acid were reacted, the amount of aspirin that would theoretically be produced (the theoretical yield) would be 180 g. The theoretical yield corresponding to any amount of reacted salicylic acid can be calculated by multiplying the number of grams reacted by the factor 180 g aspirin/138 g salicylic acid. Thus, the theoretical yield of aspirin is given by the following equation:

$$\text{Theor. yield} = \text{grams of salicylic acid reacted} \times \frac{180 \text{ g aspirin}}{138 \text{ g salicylic acid}} \qquad \text{Eq. 22.6}$$

1. Use the mass of salicylic acid used and recorded in Table 22.1 and Equation 22.6 to calculate the theoretical yield of aspirin. Record the value in Table 22.2.
2. Use the mass of the empty watch glass and the watch glass plus aspirin recorded in Table 22.1 to calculate the mass of aspirin you isolated. Record this mass in Table 22.2.
3. Use the results of Steps 1 and 2 and Equation 22.5 to calculate the percentage yield for your synthesis of aspirin. Record the calculated value in Table 22.2.

B. Synthesis of Oil of Wintergreen

1. List in Table 22.3 at least two commercial products in which you have detected the odor of oil of wintergreen.

C. Qualitative Test for Phenols

1. On the basis of the data in Table 22.4, decide which compounds contain a phenol group.
2. List your conclusions in Table 22.5.
3. Also indicate in Table 22.5 those compounds that, according to molecular formulas given in this experiment, should give a positive phenol test.

Experiment 22 ▪ Pre-Lab Review

The Synthesis of Aspirin and Other Esters

1. Are any specific safety alerts given in the experiment? List any that are given.

    _____

    _____

    _____

2. Are any specific disposal directions given in the experiment? List any that are given.

    _____

    _____

    _____

3. What substances or objects are to be weighed accurately in this experiment?

    _____

    _____

4. What test result in Part C indicates the presence of a phenol group? _____

    _____

    _____

5. Explain the proper technique to use to cautiously note the odor of a substance.

    _____

    _____

    _____

Experiment 22 ▪ Data & Report Sheet

The Synthesis of Aspirin and Other Esters

A. Synthesis of Aspirin

TABLE 22.1 (data)

Mass of salicylic
acid used _____

Mass of watch glass
plus aspirin _____

Mass of empty
watch glass _____

TABLE 22.2 (report)

Theoretical yield _____

Mass of aspirin
isolated _____

Percentage yield _____

B. Synthesis of Oil of Wintergreen

TABLE 22.3 (data and report)

Odor of prepared ester _____

Commercial products containing oil of wintergreen _____

_____

C. Qualitative Test for Phenols

TABLE 22.4 (data)

Sample Tested	Solution Color
Salicylic acid	_____
Commercial aspirin	_____
Prepared aspirin	_____
Commercial oil of wintergreen	_____

TABLE 22.5 (report)

Sample Tested	Contains a Phenol Group	Should Give a Phenol Test
Salicylic acid	_____	_____
Commercial aspirin	_____	_____
Prepared aspirin	_____	_____
Commercial oil of wintergreen	_____	_____

D. Synthesis of Some Fragrant Esters

TABLE 22.6 (data and report)

Alcohol and Acid Reacted	Product Odor
n-Propyl alcohol + acetic acid	_____
Isoamyl alcohol + acetic acid	_____
Ethyl alcohol + butyric acid	_____
n-Butyl alcohol + butyric acid	_____

QUESTIONS

1. Your prepared aspirin probably gave a positive test for phenol. How can this be explained?

   a. The aspirin structure contains a phenol group.
   b. It is contaminated with salicylic acid.
   c. It is contaminated with acetic anhydride.
   d. It is contaminated with sulfuric acid.

   Explain your answer: _____

   _____

   _____

2. Suppose in Step 10 of Part A, a student washed the solid aspirin with hot water instead of ice water. How would this influence the percentage yield of the isolated aspirin compared to that obtained if ice water had been used?

   a. Increase it.   b. Decrease it.   c. Would not influence it.

   Explain your answer: _____

   _____

   _____

3. Suppose in Step 12 of Part A, a student suspended the synthesized aspirin using hot water instead of ice water. How would this influence the percentage yield of the isolated aspirin compared to that obtained if ice water had been used?

   a. Increase it.   b. Decrease it.   c. Would not influence it.

   Explain your answer: _____

   _____

   _____

4. What is the correct formula for the ester produced when ethyl alcohol, $CH_3CH_2—OH$, is

    reacted with butyric acid ($CH_3CH_2CH_2—\overset{\overset{\displaystyle O}{\|}}{C}—OH$)?

    a. $CH_3—\overset{\overset{\displaystyle O}{\|}}{C}—CH_2CH_2CH_2CH_3$

    b. $CH_3—\overset{\overset{\displaystyle O}{\|}}{C}—O—CH_2CH_2CH_2CH_3$

    c. $CH_3CH_2CH_2—\overset{\overset{\displaystyle O}{\|}}{C}—O—CH_2CH_3$

    Explain your answer: _____

    _____

5. Which of the following is a use that could probably be made of all esters prepared in Part D?

    a. Coloring agents   b. Flavoring agents   c. Analgesics

    Explain your answer: _____

    _____

    _____

Identifying Functional Groups in Unknowns

In this experiment, you will

- Review the chemical properties of some common functional groups.

- Use tests based on chemical properties to identify the functional groups in two unknown compounds.

INTRODUCTION

When a new organic compound is synthesized in a laboratory or isolated from a natural source, one of the first things that is often done is to determine what functional groups are present in the compound.

In this experiment, the common functional groups shown in Table 23.1 will be studied. All chemical procedures and tests used in this experiment have been used in previous experiments, so only a brief review is included here.

TABLE 23.1 Functional Groups Studied in This Experiment

Class	Functional Group	Example
Alkane		$CH_3(CH_2)_4CH_3$
Alkene	$\diagup C = C \diagdown$	$CH_3CH = CHCH_3$
Alcohol	$-\overset{\displaystyle |}{\underset{\displaystyle |}{C}} - OH$	$CH_3CH_2CH_2 - OH$
Phenol	$\bigcirc\!\!\!\!\!-OH$	$\bigcirc\!\!\!\!\!-OH$
Aldehyde	$-\overset{\displaystyle O}{\overset{\|}{C}} - H$	$CH_3CH_2 - \overset{\displaystyle O}{\overset{\|}{C}} - H$
Ketone	$-\overset{\displaystyle |}{\underset{\displaystyle |}{C}} - \overset{\displaystyle O}{\overset{\|}{C}} - \overset{\displaystyle |}{\underset{\displaystyle |}{C}}-$	$CH_3 - \overset{\displaystyle O}{\overset{\|}{C}} - CH_3$
Carboxylic acid	$-\overset{\displaystyle O}{\overset{\|}{C}} - OH$	$CH_3CH_2 - \overset{\displaystyle O}{\overset{\|}{C}} - OH$
Amine	$- NH_2$	$CH_3CH_2CH_2 - NH_2$

This experiment is more complex than previous experiments because nearly all the functional groups studied before are included. Because of the number of functional groups involved, it is very useful to carry out the identifying chemical tests in a systematic way and in a specific order. A convenient way to do this is to follow flowcharts such as those given later on the Data and Report Sheet (Charts 23.1 and 23.2). The boxes of these charts have been provided with blanks in which you will record the results of the tests with known compounds and with your unknown compounds. The results of these tests will allow you to identify the functional groups in your unknown compounds.

EXPERIMENTAL PROCEDURE

It is useful to determine the solubility characteristics of an unknown before the functional group tests are performed because most tests are performed on solutions of the compounds, and suitable solvents must be selected. You will receive two unknowns; one will be identified as being water-soluble, and the other will be identified as being water-insoluble. The procedures followed for each type are outlined in Charts 23.1 (water-soluble) and 23.2 (water-insoluble). Be sure to follow the correct chart and record your observations as directed in the lettered boxes of the proper chart.

A. Water-Soluble Unknown

a. Carboxylic Acid, Phenol, and Amine pH Tests

Water-soluble carboxylic acids and phenols dissolve in water to form acidic solutions with pH values in the range of 2 to 4. Remember that distilled water is usually acidic with a pH of 5 to 6 because of dissolved carbon dioxide (CO_2). Thus, if a carboxylic acid or phenol is dissolved, the pH of the water solution will be lower than the pH of distilled water. Water-soluble amines dissolve in water to form basic solutions with pH values greater than that of distilled water, and usually greater than 7.

PROCEDURE

1. Obtain a water-soluble unknown from the stockroom and record its identification (ID) number in Table 23.2 of the Data and Report Sheet.
2. Add 10 drops of distilled water to 3 drops of glacial acetic acid in a clean 10-cm test tube. Agitate the tube to mix the contents.

SAFETY ALERT

Glacial acetic acid will vigorously attack tissue. If you contact it, wash the contacted area with cool water.

      Use wide-range pH paper to measure the pH of the test tube contents. Record the pH in Chart 23.1, Box A, in the blank labeled "Standard."
3. Repeat Step 2, using 3 drops of triethylamine in place of the acetic acid. Record the pH in Chart 23.1, Box C, in the blank labeled "Standard."

4. Repeat Step 2, using 3 drops (or 0.1 g if solid) of your water-soluble unknown in place of the acetic acid. Record the pH on the blank labeled "Unknown" in the box that matches the result. That is, if the pH is in the same range as a solution of acetic acid, record the pH of the unknown in Box A. If the pH is in the same range as triethylamine, record the pH of the unknown in Box C. If the pH is about the same as distilled water, record the value in Box B.

DISPOSAL Test tube contents in container labeled "Exp. 23, Used Chemicals Part A"

b. Ferric Chloride Test for Phenols

Carboxylic acids and phenols can be differentiated by using the ferric chloride test. Phenols form colored complexes (pink, green, or, often, violet) when mixed with ferric chloride. Carboxylic acids do not form colored complexes with ferric chloride.

PROCEDURE

1. Put 2 drops of liquified phenol into a clean 10-cm test tube.

Add 20 drops of distilled water and 2 drops of 1% ferric chloride solution to the test tube and agitate to mix the contents. Record the result (color) in Chart 23.1, Box D, in the blank labeled "Standard."

2. If your water-soluble unknown gave a water solution with a pH of 2 to 4, repeat Step 1 with 2 drops (or 0.1 g if solid) of your unknown in place of the liquified phenol. Record the results in the blank labeled "Unknown," in the box that matches the test result.

DISPOSAL  Test tube contents in container labeled "Exp. 23, Part A"

c. Ceric Nitrate Test for Alcohols

Alcohols react with yellow ceric nitrate solution to form a red-colored complex. Phenols also react with ceric nitrate but form a dark brown precipitate. Other functional groups do not change the color of a ceric nitrate solution.

PROCEDURE

1. Put 3 drops of ethanol into a clean 10-cm test tube. Add 8 drops of ceric nitrate solution to the test tube and agitate to mix. Note

the color and record it in Chart 23.1, Box F, in the blank labeled "Standard."

2. If your unknown has not yet been identified as a carboxylic acid, phenol, or amine, repeat Step 1 with 3 drops (or 0.1 g if solid) of your unknown in place of the ethanol. Note the resulting color and record it in the blank labeled "Unknown" in Box F or G, according to which one corresponds to your result.

DISPOSAL Test tube contents in container labeled "Exp. 23, Used Chemicals Part A"

d. Chromic Acid Test for Alcohols

Primary and secondary alcohols can be oxidized by chromic acid. Tertiary alcohols and most phenols are resistant to oxidation by this reagent. The disappearance of the red-orange chromic acid color and the appearance of the blue-to-green colored Cr^{3+} ions indicate that oxidation has taken place.

PROCEDURE

1. Put 1 drop of ethanol into a clean 10-cm test tube. Add 20 drops of acetone and 1 drop of chromic acid solution to the test tube and agitate to mix.

SAFETY ALERT

> Acetone is very flammable. Be sure no open flames are nearby before you do this test. Chromic acid is a carcinogen and will also attack tissue vigorously. If contact is made with it, immediately wash the contacted area with soap and water.

Note the color of the test tube contents and record it in Chart 23.1, Box H, in the blank labeled "Standard."

2. If your water-soluble unknown is an alcohol (if it gave a red color with ceric nitrate), repeat Step 1 with 1 drop of your unknown in place of the ethanol. Record the observed color in the blank labeled "Unknown," in Box H or I, depending on which one matches your result.

DISPOSAL Container labeled "Exp. 23, Used Chromic Acid Part A"

e. Tollens' Test for Aldehydes and Ketones

Aldehydes and ketones can be differentiated on the basis of the ease with which they are oxidized. Aldehydes are readily oxidized by Tollens' reagent to form a black precipitate of silver or a silver mirror on the walls of clean glass containers. Ketones do not react with Tollens' reagent.

PROCEDURE

1. Put 15 drops of dilute (6 M) sodium hydroxide solution (NaOH) into a clean 10-cm test tube and allow it to stand for at least 1 minute.

SAFETY ALERT

> Dilute sodium hydroxide solution will vigorously attack tissue and is especially damaging to the eyes. If contact is made with this material, immediately wash the contacted area with large amounts of cool water. If the eyes are contacted, inform your instructor.

2. After 1 minute, empty the NaOH solution from the test tube and, without rinsing the tube, add 15 drops of 0.1 M silver nitrate solution $Ag(NO_3)$.

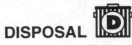 **DISPOSAL**   Discarded NaOH solution in sink

3. Add 6 M ammonia (NH_3 or NH_4OH) to the test tube dropwise, while agitating the tube. Continue the addition and agitation until the precipitate that forms initially is dissolved (about 15 drops will be required).
4. Add 5 drops of butyraldehyde to the test tube and agitate to mix the contents. Set the tube aside for 3 minutes, then observe it and record your observations in Chart 23.1, Box J, in the blank labeled "Standard."
5. If your water-soluble unknown is an aldehyde or a ketone (see Box G), repeat Steps 1 to 4 above, using 5 drops, or 0.1 g if solid, of your unknown in place of the butyraldehyde. Note and record your observations in the blank labeled "Unknown" in Box J or K, whichever matches your result.

 **DISPOSAL**   Test tube contents in sink, flush with cool water

B. Water-Insoluble Unknown

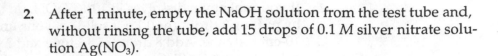

Several of the procedures used in Part A must be modified slightly when the unknown is not water-soluble. In addition, a water-insoluble unknown may also be an alkane or an alkene, compounds not included in Part A.

a. Solubility in Base

Higher-molecular-weight carboxylic acids and phenols that are water-insoluble in water will usually dissolve in a 10% solution of sodium hydroxide (NaOH).

PROCEDURE

1. Obtain a water-insoluble unknown from the stockroom and record its identification number in Table 23.2.
2. Put an amount of solid benzoic acid that would fill this circle (○) into a clean 10-cm test tube. Add 10 drops of 10% sodium hydroxide solution (NaOH), and agitate the tube to mix the contents. Note the solubility behavior of the solid and record it in Chart 23.2, Box L, in the blank labeled "Standard."

 **SAFETY ALERT**

> The 10% solution of sodium hydroxide will vigorously attack tissue and is very damaging to the eyes. If contact is made, immediately wash the contacted area with cool water and, if the eyes are involved, inform your instructor.

3. Repeat Step 2 with 3 drops (or 0.1 g if solid) of your water-insoluble unknown. Note and record your observations of solubility in the blank labeled "Unknown" of Box L or M, according to which box matches your result.

4. If your unknown dissolves in 10% NaOH, do the ferric chloride test by putting 2 drops (or 0.1 g if solid) of your unknown into a clean 10-cm test tube. Add 8 drops of acetone and 2 drops of 1% ferric chloride solution ($FeCl_3$). Agitate the tube and record the results (color) in the blank labeled "Unknown" in Box N or O of Chart 23.2, depending on which box matches your results.

b. Solubility in Acid

Water-insoluble amines usually dissolve in 6 M hydrochloric acid (HCl) because the amines are converted into salts that have a greater solubility than the amines themselves.

PROCEDURE

1. Put 3 drops of N,N-dimethylaniline into a clean 10-cm test tube. Add 10 drops of dilute (6 M) hydrochloric acid (HCl) to the tube and agitate the contents. Note the solubility behavior and record it in Chart 23.2, Box P, in the blank labeled "Standard."

SAFETY ALERT

> N,N-dimethylaniline is toxic and a skin irritant. Avoid contact, but if contact occurs, wash contacted area with soap and water. Dilute hydrochloric acid will attack tissue. If contact occurs, wash contacted area with cool water.

2. If your water-insoluble unknown is not a carboxylic acid or phenol, repeat Step 1 using 3 drops (or 0.1 g if solid) of your unknown in place of the N,N-dimethylaniline. Note the solubility characteristics and record your observation in the blank labeled "Unknown" in Box P or Q of Chart 23.2, depending on which box matches the characteristics of your unknown.

DISPOSAL  Test tube contents in container labeled "Exp. 23, Used Chemicals Part B"

c. Ceric Nitrate Test for Alcohols

This test was also done in Part A, but acetone must now be used as a solvent because of the water-insoluble nature of your unknown.

1. If your unknown has not yet been identified as a carboxylic acid, phenol, or amine, put 3 drops (or 0.1 g if solid) of unknown, 8 drops of acetone, and 8 drops of ceric nitrate solution into a clean 10-cm test tube.
2. Agitate the contents of the test tube and note the resulting color. Record your observation (color) in the blank labeled "Unknown" of Box R or S of Chart 23.2. Record in the box that matches your test results.
3. If your unknown gives a positive alcohol test (red color in Box R), perform the chromic acid test by putting 1 drop (or 0.1 g if solid) of your unknown into a clean 10-cm test tube and adding 20 drops of acetone and 1 drop of chromic acid. Agitate the tube contents and observe the color that results. Record your observations in the blank

labeled "Unknown" in Box T or U of Chart 23.2. Once again, record in the box that matches your test results.

 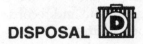

d. DNP Test for Aldehydes and Ketones

A characteristic property of the carbonyl group is the reaction with 2,4-dinitrophenylhydrazine (DNP) to form yellow-orange precipitates. This readily detectable reaction is characteristic of both aldehydes and ketones.

PROCEDURE

1. Put 10 drops of DNP solution into a clean 10-cm test tube.

   Add 2 drops of acetone to the test tube and agitate to mix. Note the changes that occur and record your observations in the blank labeled "Standard" in Box V of Chart 23.2.
2. If you have narrowed down the identity of your unknown to Box S (aldehyde, ketone, alkane, or alkene), repeat Step 1 using 2 drops (or a few crystals if solid) of your unknown in place of the acetone. Record your observations in the blank labeled "Unknown" of Box V or W in Chart 23.2. Choose the box that matches your results.
3. If your unknown gives a positive DNP test (Box V), perform the Tollens' test on a sample (see Part A-e). Record your observations in the blank labeled "Unknown" in Box X or Y of Chart 23.2. Match the results to the box chosen.

e. Bromine Test for Alkanes and Alkenes

Alkanes and alkenes are water-insoluble compounds and so were not included in Part A. Alkenes react rapidly with bromine solutions to cause the red-orange color of the bromine to disappear almost instantly when the two are mixed. Alkanes, however, react much more slowly, so bromine retains its red-orange color for a few minutes when mixed with alkanes.

PROCEDURE

1. Put 10 drops of hexene into a clean, dry 10-cm test tube and add 3 drops of 1% bromine solution.

   Note and record in the blank labeled "Standard" in Box Z of Chart 23.2 the color immediately after mixing.

2. If the identity of your unknown has been narrowed down to an alkane or an alkene (Box W), repeat Step 1 using 10 drops (or 0.1 g if solid) of your unknown in place of the hexene. Note and record your observations in the blank labeled "Unknown" in Box Z or Z' of Chart 23.2. Choose the box that matches your results.

DISPOSAL 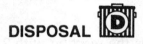 Test tube contents in container labeled "Exp. 23, Used Chemicals Part B"

REPORT

Characterization of Unknowns

1. Use the information recorded in Chart 23.1 to determine the functional group in your water-soluble unknown. Record the identity of the functional group in Table 23.2.
2. Use the information recorded in Chart 23.2 to determine the functional group in your water-insoluble unknown. Record the identity of the functional group in Table 23.2.

Experiment 23 ▪ Pre-Lab Review

Identifying Functional Groups in Unknowns

1. Are any specific safety alerts given in the experiment? List any that are given.

   _____

   _____

   _____

   _____

2. Are any specific disposal directions given in the experiment? List any that are given.

   _____

   _____

   _____

3. What result indicates that a water-soluble unknown is an amine? _____

   _____

   _____

4. What functional group in addition to alcohols will react with ceric nitrate?

   _____

5. Which type of alcohol fails to react with chromic acid? _____

   _____

6. What result of a Tollens' test indicates the presence of an aldehyde functional group in an unknown?

   _____

7. What test should be done to determine if a water-insoluble unknown contains a carbonyl (aldehyde or ketone) functional group?

   _____

8. Why is a test done for alkenes in Part B of the experiment but not in Part A?

   _____

Experiment 23 ▪ Data & Report Sheet

Identifying Functional Groups in Unknowns

A. Water-Soluble Unknown

CHART 23.1 (data)

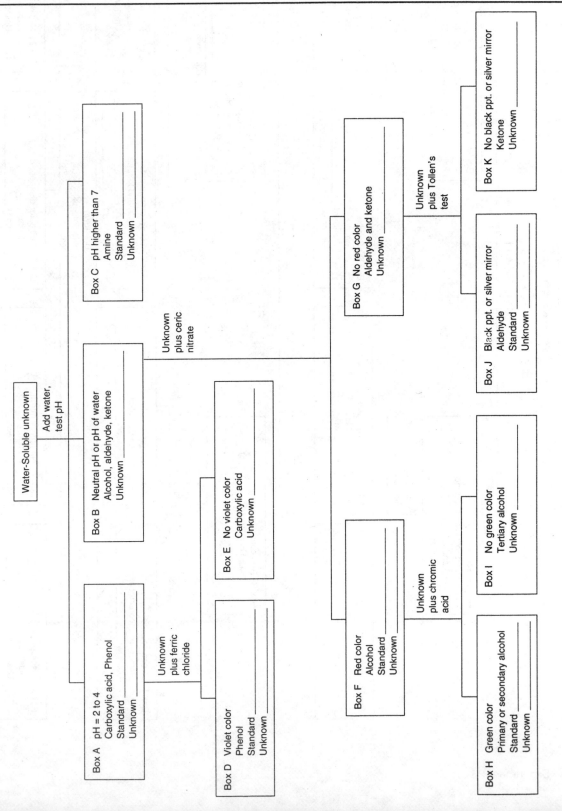

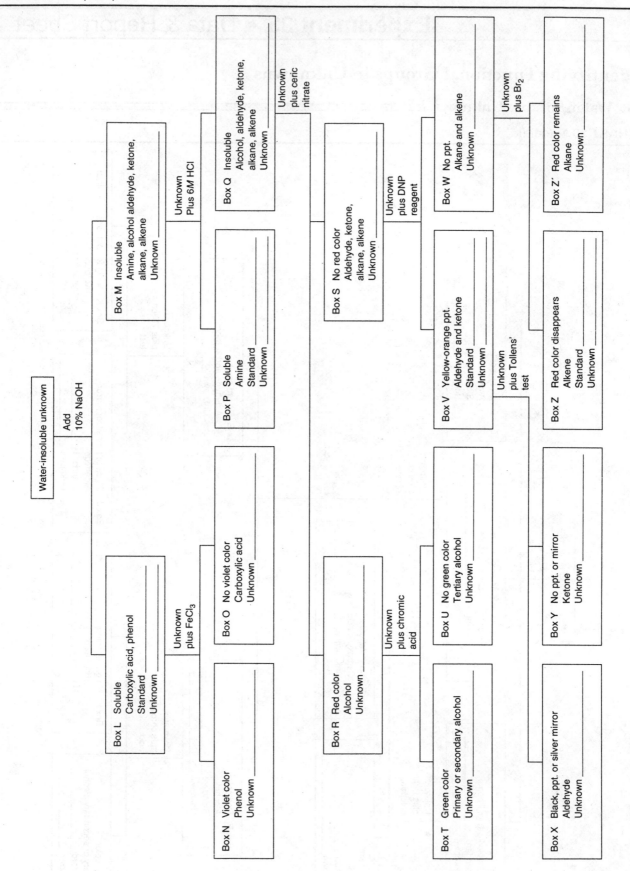

TABLE 23.2 (report)

Water-Soluble Unknown	Water-Insoluble Unknown
ID number _____	ID number _____
Functional group _____	Functional group _____

QUESTIONS

1. A water-soluble unknown is dissolved in water, and the pH of the resulting solution is 5.9. What would you do next?

    a. Classify the unknown as a carboxylic acid or phenol.
    b. Check the pH of the water used as a solvent.
    c. Ask for a new unknown.

    Explain your answer: _____

    _____

    _____

2. A water solution of a water-soluble unknown has a pH the same as water. The unknown gives no red color in a ceric nitrate test and no black precipitate or silver mirror in a Tollens' test. What functional group would you conclude was in the unknown?

    a. Alcohol   b. Aldehyde   c. Ketone   d. Not enough information is given to identify the functional group.

    Explain your answer: _____

    _____

    _____

    _____

3. A student concludes that a water-soluble unknown contains an alcohol functional group and does a chromic acid test on a sample. The color of the tested solution is green. How should the identity of the functional group be reported?

    a. A primary alcohol   b. A secondary alcohol   c. A tertiary alcohol
    d. A primary or secondary alcohol

    Explain your answer: _____

    _____

    _____

    _____

4. A water-insoluble unknown is soluble in 10% NaOH solution. What single additional test would enable a student to identify the functional group in the unknown?

    a. Determining the solubility in 6 M HCl
    b. Ferric chloride test
    c. Ceric nitrate test

Explain your answer: _____

_____

_____

5. A water-insoluble unknown is not soluble in 10% NaOH or 6 M HCl and gives no precipitate when tested with DNP reagent. Which of the following functional groups might be in the unknown?

    a. Alkane   **b.** Alkene   **c.** Aldehyde   **d.** Ketone
    e. More than one answer is correct.

Explain your answer: _____

_____

_____

_____

Synthetic Polymers

In this experiment, you will

- Prepare several synthetic polymers of commercial interest.

- Study some characteristics of polymers.

INTRODUCTION

Polymers are substances composed of huge molecules that contain repeating structural units. Reactions in which small molecular units called **monomers** are united to form polymers are called **polymerizations.**

Natural as well as synthetic polymers exist; both types exert a profound influence on our lives. Some important *natural polymers* are rubber (isoprene monomers); starch, glycogen, and cellulose (glucose monomers); proteins (amino acid monomers); and nucleic acids (nucleotide monomers).

On the basis of the polymerization reaction that takes place during their formation, *synthetic polymers* are classified into two categories: addition polymers and condensation polymers. *Addition polymers* are formed by a reaction in which identical monomers add to one another. The most common addition polymers are based on monomers that contain a vinyl grouping. The formation of these vinyl polymers is represented in Equation 24.1 where X is different in different polymers:

$$n \; CH_2{=}CH \quad \rightarrow \quad {+}CH_2{-}CH{+}_n \qquad \text{Eq. 24.1}$$
$$\underset{\displaystyle X}{|} \qquad\qquad\qquad \underset{\displaystyle X}{|}$$

a vinyl monomer   polymer

Some commercially important vinyl polymers of this type are polyethylene (where X is H), polypropylene (X is CH_3), polystyrene (X is ⬡), and poly (vinyl chloride), where X is Cl. In addition, Teflon and Lucite (Plexiglas) are important polymers with the following vinyl monomers: the Teflon monomer is $CF_2{=}CF_2$, and the monomer of Lucite is $CH_2{=}C{-}CH_3$.
$$\begin{array}{c} | \\ C{=}O \\ | \\ O{-}CH_3 \end{array}$$

Condensation polymers result when two or more di- or polyfunctional monomers react. In the process, a small molecule (often water) is eliminated. The polyfunctional monomers frequently used include glycols (di-alcohols), diamines, and dicarboxylic acids. Some well-known condensation polymers are nylon (a polyamide), Dacron (a polyester), Bakelite, and polyurethane. Most natural polymers may also be classified as condensation polymers. The widely used polymers rayon, cellophane, and cellulose acetate are modified forms of cellulose.

A. Preparation of Plexiglas

Plexiglas or *Lucite* is an addition polymer resulting from the polymerization of methyl methacrylate in the presence of benzoyl peroxide which initiates the reaction.

$$n \; CH_2{=}C \begin{matrix} CH_3 \\ | \\ | \\ C{=}O \\ | \\ O{-}CH_3 \end{matrix} \xrightarrow{\text{benzoyl peroxide}} \left[CH_2{-}\underset{\underset{O-CH_3}{\underset{|}{\overset{|}{C=O}}}}{\overset{CH_3}{\underset{|}{\overset{|}{C}}}} \right]_n \qquad \text{Eq. 24.2}$$

methyl methacrylate   polymer

PROCEDURE

1. Set up a boiling-water bath made from a 250-mL beaker.
2. Place 5 mL of methyl methacrylate in a clean, dry 15-cm test tube.
3. Your instructor will add 5-6 grains of benzoyl peroxide to the tube.

SAFETY ALERT

> Benzoyl peroxide is an explosive but poses little risk in the very small amounts used in this experiment.

4. Use a ring stand and clamp to suspend the test tube in a boiling-water bath without letting the test tube rest on the bottom of the bath. Stir the methyl methacrylate until the peroxide dissolves.
5. Heat the tube in the bath without stirring for 1 ½ hours. Go on to other parts of the experiment while the reaction proceeds.
6. After 1 ½ hours, remove the flame and allow the polymer to cool.
7. Examine the product in the tube or, if the instructor permits, break the tube by wrapping it in cloth and hitting it with a clamp. Carefully peel away any glass that sticks to the polymer.
8. Discard the test tube whether it is broken or not.
9. Record the appearance and general properties of the polymer in Table 24.1.

DISPOSAL

Broken test tube pieces in broken-glass receptacle
Polymer in wastebasket

B. Preparation of Glyoxal–Resorcinol Polymer

Bakelite, a condensation polymer widely used to manufacture molded plastic articles such as handles for skillets and kitchen utensils, is prepared from phenol and formaldehyde (Equation 24.3):

$$nH\!-\!\overset{\text{O}}{\overset{\|}{\text{C}}}\!-\!H \; + \; n\,C_6H_5\!-\!OH \;\longrightarrow\; \text{a portion of the polymer} \; + \; nH_2O \qquad \text{Eq. 24.3}$$

formaldehyde   phenol

a portion of the polymer

A similar polymeric material can be prepared from a less toxic aldehyde, called glyoxal, and resorcinol, a dialcohol.

$$H\!-\!\overset{\text{O}}{\overset{\|}{\text{C}}}\!-\!\overset{\text{O}}{\overset{\|}{\text{C}}}\!-\!H$$

glyoxal   resorcinol

PROCEDURE

1. Use a plastic weighing dish and a centigram or electronic balance to weigh out a sample of resorcinol with a mass of 1 g.
2. Put the resorcinol sample into a clean 15-cm test tube.
3. Use a 10-mL graduated cylinder to add 1.6 mL of 40% glyoxal solution to the test tube. Put the test tube into the boiling-water bath. Allow the tube to remain in the bath until all the solid has dissolved.

SAFETY ALERT

> Resorcinol and glyoxal are both skin and eye irritants. If contact is made with either compound, wash the contacted area thoroughly with soap and water.

4. After the solid has dissolved, remove the tube from the bath, agitate it to mix the contents well, and add 1 drop of 3 M hydrochloric acid (HCl) to the tube. Agitate immediately after adding the HCl to ensure that the contents are well mixed.
5. Return the test tube to the boiling-water bath and allow it to remain until the solid polymer forms (usually about 1 to 3 minutes).
6. After the solid polymer forms, remove the test tube from the bath and set it aside to cool for about 30 minutes. Go on to other parts of the experiment while it cools.
7. Examine the solid product in the test tube or, if your instructor permits, break the tube by wrapping it in a cloth and hitting it with a clamp or other heavy object. Carefully peel away any glass that sticks to the polymer.
8. Discard the test tube whether it is broken or not.
9. Record the appearance and general properties of the polymer in Table 24.1.

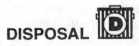

DISPOSAL Broken test tube pieces in broken glass container
Polymer in wastebasket

C. Preparation of Nylon

Nylon, a polyamide condensation polymer, can be formed in principle by reacting a diamine with a dicarboxylic acid. The two monomers form amide linkages to produce the polymer chain. We will use a carboxylic acid chloride rather than a carboxylic acid because the reaction between an acid chloride and amine is more rapid than that between an acid and amine. However, commercial nylon 6,6 (the 6,6 indicates that each monomer contains six carbon atoms) is produced from adipic acid and hexamethylenediamine. The polymer melts at 260 to 270°C and is drawn into fibers from the molten state. After cooling and further stretching, the fiber is spun into thread that dyes readily, resists wear, tolerates moisture, and has high strength. The reaction we will use is given in Equation 24.4:

$$n\, Cl-\overset{O}{\overset{\|}{C}}\!\!-\!\!(CH_2)_4\overset{O}{\overset{\|}{C}}\!\!-\!\!Cl + n\, H-\overset{H}{\overset{|}{N}}\!\!-\!\!(CH_2)_6\overset{H}{\overset{|}{N}}\!\!-\!\!H \rightarrow \left[\overset{O}{\overset{\|}{C}}\!\!-\!\!(CH_2)_4\overset{O}{\overset{\|}{C}}\!\!-\!\!\overset{H}{\overset{|}{N}}\!\!-\!\!(CH_2)_6\overset{H}{\overset{|}{N}}\right]_n + n\, HCl$$

adipoyl chloride   hexamethylenediamine   amide linkage   Eq. 24.4

PROCEDURE

1. Into a 50-mL beaker, pour 10 mL of a 5% solution of hexamethylene-diamine.
2. Carefully pour on top of this solution (down the side of the beaker) 10 mL of a 5% solution of adipoyl chloride (the solvent, hexane, will not mix with water).
3. Note the immediate formation of polymer at the interface of the two immiscible liquids.

SAFETY ALERT

Hexane is flammable. Be certain no open flames are nearby when you do this part of the experiment. Both hexamethylenediamine and adipoyl chloride are strong skin irritants. If contact is made with either one, wash the contacted area with soap and water.

FIGURE 24.1
The formation of nylon

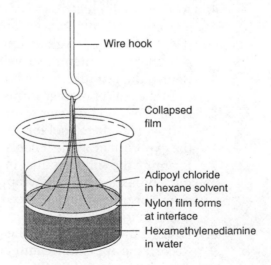

4. Use a wire hook made from a paper clip to grasp the polymer film in the center and gently lift it to form a rope about 12 inches long (Figure 24.1).

5. Cut or break the rope near the liquid surface, and rinse it several times with water.

6. Lay it on a piece of filter paper to dry.
7. Examine the nylon after it has dried and describe its appearance in Table 24.1.
8. Use a glass rod and vigorously stir the remaining portion of the carboxylic acid chloride–amine mixture.
9. Pour off the liquid and wash the polymer thoroughly in running tap water.

DISPOSAL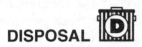

Liquid remaining after Step 8 in container labeled "Exp. 24, Used Chemicals Part C"
Washed nylon in wastebasket

D. Preparation of Glyptal Resin

Glyptal resins are polyesters formed by a condensation reaction between phthalic acid (the acid anhydride is generally used) and glycerol. The reaction is represented in Equation 24.5. These polymers, which find wide use as commercial surface coatings, are thermosetting and must be shaped by some sort of a mold. Before you begin the synthesis, make a mold capable of holding about 6 mL of liquid. Use aluminum foil formed around a small beaker, a cupcake mold, or some other suitable material. A coin may be placed on the bottom of the mold if you wish.

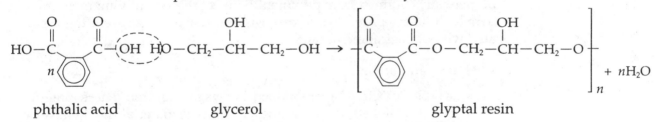

phthalic acid   glycerol   glyptal resin

Eq. 24.5

PROCEDURE

1. Place together in a 15-cm test tube 6 g of phthalic anhydride, 0.3 g of sodium acetate, and 2.4 mL of glycerol. This fills the test tube about ½ full, but it can be used safely.

2. Carefully heat the mixture by moving a *low* flame back and forth along the filled length of the test tube until all the solid melts. Avoid heating the bottom of the tube.
3. Continue heating until the melt appears to boil.

4. Maintain this gentle boiling for 5 minutes. Sufficient heating is necessary to produce a nonsticky product. Excessive heating will turn the material into a brittle, dark amber product.
5. Set the prepared mold in a hood, and pour the hot liquid into it. Allow it to cool until nearly the end of the laboratory period.
6. Peel the mold from the product.
7. Note and describe the appearance of the product in Table 24.1.
8. Discard the test tube used in the preparation.

DISPOSAL  Used test tube in broken-glass receptacle
Solid polymer in wastebasket

E. A Study of Some Polymer Characteristics

The large molecular size of polymers causes most of them to be insoluble in most solvents, especially highly polar water. However, interactions between water molecules and functional groups attached to large polymer chains can increase the solubility of the polymers in water. One interaction of this type that has importance in the behavior of biopolymers such as proteins is hydrogen bonding. Interactions between functional groups on different polymer chains can also result in cross-linking between the chains. Cross-linking can result in a change in characteristics for a polymer. For example, raw natural rubber consists of long polymer chains with only weak interactions between the chains. The resulting rubber is sticky and not very strong. However, when the rubber molecules are cross-linked in a process called **vulcanization,** sulfur atoms form cross-links between the polymer chains to give rubber the useful characteristics we all recognize.

Most naturally occurring membranes such as those that surround living cells are composed of polymeric materials. Some nonbiological membranes can be formed from polymers. Some of these nonliving membranes demonstrate the properties of osmosis and dialysis that are often associated with biological membranes.

a. Hydrogen Bonding

Various types of "slime" are marketed as toys for children. These materials consist of cross-linked natural gums. A synthetic material that behaves in a way similar to the natural gums is poly (vinyl alcohol). The monomers of poly (vinyl alcohol) are vinyl alcohol molecules. The structure of the monomer and polymer are shown below:

$$\begin{array}{cc} \text{OH} \quad \text{H} & \text{OH} \quad \text{H} \\ | \quad\quad | & | \quad\quad | \\ \text{CH}{=}\text{CH} & {+}\text{CH}{-}\text{CH}{+}_n \\ \text{vinyl alcohol} & \text{poly (vinyl alcohol)} \end{array}$$

The hydroxyl functional groups that protrude from the long chains of poly (vinyl alcohol) will form hydrogen bonds with water and cause poly (vinyl alcohol) to have a significant solubility in water. In addition, the hydroxyl groups hydrogen bond with borax to form weak cross-links between polymer chains that distinctly change the characteristics of the polymer. Both of these hydrogen-bonding interactions are represented below:

HC—O—H····O⟨H⟩ (left diagram)

HC—H

H⟨O⟩····H—O—CH

HC—H

hydrogen bonding between
water and poly (vinyl alcohol) chains

HC—O—H····O

HC—H

H—CH

HC—O····H—O

H

(with boron center B cross-links)

H—O—H····O—CH

H—CH

H—CH

O····H—O—CH

H

hydrogen bond cross-links between
poly (vinyl alcohol) chains

PROCEDURE

1. Use your 10-mL graduated cylinder to put 10.0 mL of 4% poly (vinyl alcohol) into a clean 50-mL beaker.

2. Rinse the 10-mL graduated cylinder several times with water to remove any residual poly (vinyl alcohol), then measure 2.0 mL of 4% borax solution. Leave the borax in the cylinder until it is needed.

3. Investigate the characteristics of the poly (vinyl alcohol) solution in the beaker and record them in Table 24.2.

4. Add the 2.0 mL of borax solution to the beaker and stir the mixture rapidly with a glass stirring rod. Continue the stirring until the liquid poly (vinyl alcohol) solution has all formed into a gel.

5. It is safe to remove the gel from the beaker with your hands. Remove the cross-linked polymer from the beaker and investigate its properties. Try molding it, stretching it slowly, stretching it rapidly, and dropping a ball of it a short distance onto the lab bench. Record the observed characteristics in Table 24.2.

Cross-linked poly (vinyl alcohol) can be taken with you and used to entertain yourself if you wish. It will stay in good condition for weeks if it is stored in a plastic food-storage bag. If left out, it will dry out. When it is time to dispose of it, it can be put into an ordinary waste receptacle. In the lab, put it into the wastebasket.

DISPOSAL

b. Osmosis

Poly (sodium acrylate) is sometimes called a super absorber because of its osmotic interaction with water. The monomer is a sodium salt of acrylic acid, as shown below. The structure of the polymer is also shown.

$$\underset{\text{sodium acrylate}}{\overset{\displaystyle \overset{O}{\underset{\displaystyle}{\|}}\;C—O^-Na^+}{CH_2{=}CH}} \qquad \underset{\text{poly (sodium acrylate)}}{\overset{\displaystyle \overset{O}{\|}\;C—O^-Na^+}{+CH_2—CH+_n}}$$

According to one explanation of the water-absorbing properties of this polymer, small crystals of the polymer can be thought of as being surrounded by a membrane of the polymer with the carboxylic acid salt portions enclosed inside the structure. Thus, the insides of the crystals have a high concentration of Na$^+$ and carboxylate ions. When the crystals are put into pure water, the water osmotically flows toward the high concentration of ions and causes the solid to swell.

PROCEDURE

1. Use a centigram or electronic balance to weigh out a 0.2-g sample of poly (sodium acrylate).
2. Put the sample into a clean, dry 100-mL beaker.
3. Use a 50- or 100-mL graduated cylinder to measure out 50 mL of distilled water. Rapidly pour all the water into the breaker that contains the sample of polymer, then invert the beaker.
4. Observe and record in Table 24.2 the characteristics of the material in the beaker.
5. Add 5 mL of saturated sodium chloride solution to the mixture in the beaker. Observe and record in Table 24.2 the results of adding the saltwater to the mixture.

DISPOSAL 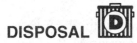 Beaker contents in sink, flushed down drain with running water

---

REPORT

The report for all parts of the experiment is completed when Tables 24.1 and 24.2 are completed.

Experiment 24 ▪ Pre-Lab Review

Synthetic Polymers

1. Are any specific safety alerts given in the experiment? List any that are given.

    _____

    _____

    _____

2. Are any specific disposal directions given in the experiment? List any that are given.

    _____

    _____

    _____

3. In Part A, what procedure is to be performed only after you get permission from your lab instructor?

    _____

    _____

4. In Part A, how long do you heat the reaction mixture?

    _____

    _____

5. Which polymer forms as a film at the interface between two immiscible solutions?

    _____

    _____

6. What do you do with the excess nylon formed in Part C?

    _____

    _____

7. For which polymer must you prepare a mold?

    _____

    _____

8. Which polymer can be given undesirable properties by either too little or too much heating?

    _____

Experiment 24 ▪ Data & Report Sheet

Synthetic Polymers

A–D Preparations

TABLE 24.1 (data and report)

Prepared Polymer	Appearance and General Properties
Plexiglas	_____
Glyoxal–resorcinol	_____
Nylon	_____
Glyptal resin	_____

E. A Study of Some Polymer Characteristics

TABLE 24.2 (data and report)

Characteristics of poly (vinyl alcohol) soln. before cross-linking	_____
Characteristics and behavior of poly (vinyl alcohol) soln. after cross-linking	_____
Characteristics of poly (sodium acrylate) plus water mixture	_____
Results of adding saltwater to the poly (sodium acrylate) plus water mixture	_____

QUESTIONS

1. Which of the following monomers would most likely be involved in the formation of an addition polymer?

   a. $HO-CH_2-CH_2-OH$ **b.** $H_2N-\overset{\overset{O}{\|}}{C}-NH_2$ **c.** $CH_2{=}CH-CH_3$

   Explain your answer: _____

   _____

   _____

2. Ethylene glycol, $HO-CH_2-CH_2-OH$, and terephthalic acid, $HO-\overset{\overset{O}{\|}}{C}-\langle \bigcirc \rangle-\overset{\overset{O}{\|}}{C}-OH$,

react to form a polyester that in fiber form is called Dacron or in sheet form called Mylar. Which polymer did you prepare that is closely related to this one?

   a. Glyptal resin   **b.** Bakelite   **c.** Nylon

Explain your answer: _____

_____

3. Which of the following formulas represents the polyester described in Question 2?

   a. $+O-CH_2-CH_2-O+_n$

   b. $\left[O-CH_2-CH_2-O-\overset{\overset{O}{\|}}{C}-\langle\bigcirc\rangle-\overset{\overset{O}{\|}}{C} \right]_n$

   c. $\left[O-\overset{\overset{O}{\|}}{C}-\langle\bigcirc\rangle-\overset{\overset{O}{\|}}{C}-O \right]_n$

Explain your answer: _____

_____

4. The solution of poly (vinyl alcohol) that was cross-linked in Part E was 96% water. What characteristics of the cross-linked product were indicative of this fact?

   a. The flow characteristics   **b.** The way the product felt to the touch

   c. The color of the product   **d.** More than one answer is correct.

Explain your answer: _____

_____

_____

5. How could the behavior of the poly (sodium acrylate)–water mixture when saltwater was added be explained?

   a. Water moved osmotically from an area of higher Na^+ concentration to an area of lower concentration.

   b. Water moved osmotically from an area of lower Na^+ concentration to an area of higher concentration.

   c. Na^+ ions moved from an area of lower concentration to an area of higher concentration.

Explain your answer: _____

_____

_____

Experiment 25

Dyes, Inks, and Food Colorings

In this experiment, you will

- Separate the component dyes of food colorings using paper chromatography.
- Determine the effect of solvent variation on the chromatographic separation of black ink components.

- Prepare three azo dyes.
- Compare several different methods of dyeing fabrics.

INTRODUCTION

A **dye** is a colored substance with the ability to stain such materials as paper, leather, plastics, and natural or synthetic fibers. **Food colorings** and **inks** are examples of dyes. For a colored substance to act as a dye, it must adhere to the substrate (the material to be colored).

Natural dyes obtained from flowers, berries, roots, and bark have been used through most of recorded history. The association of the color purple with royalty comes from the high cost and rarity of Tyrian purple, a dye derived from the sea snail *Murex brandairs.* However, today, most dyes are composed of synthetic organic chemicals, and almost all colors are available. More than 1000 different dyes are now produced commercially in the United States.

The colors of dyes result from a selective absorption of light in the visible region of the spectrum. The ability to absorb such light is usually imparted to a dye molecule by a group of atoms called a **chromophore.** The chromophore contains double bonds and is a part of a larger molecule containing at least five or six double bonds. The most important chromophores are the nitro ($-NO_2$), nitroso ($-NO$), vinyl ($-CH=CH_2$),

carbonyl ($-\overset{\overset{\displaystyle O}{\|}}{C}-$), and azo ($-N=N-$) groups. In addition to chromophores, dyes usually contain groups known as **auxochromes,** which increase the color intensity of a dye. Typical auxochromes contain such groups as $-NH_2$, $-NR_2$, $-OH$, $-OR$, $-SO_3H$, and $-COOH$. The chemical formulas for some of the dyes used in this experiment are given in the following sections. In each formula, note the presence of an extended chain of conjugated bonds (alternating single and double bonds) and auxochrome groups.

EXPERIMENTAL PROCEDURE

A. Separation of Food Colorings

Horizontal paper chromatography, a modification of the ascending technique used in Experiment 5, Part C, is useful for some separations. This technique will be demonstrated by separating the component dyes of commercial food coloring.

PROCEDURE

1. Cut a piece of 5.5-cm filter paper as shown in Figure 25.1.
2. Fold the cut portion down to form a wick.
3. At the three locations indicated by dots, use the capillary tubes that are provided to place small spots of three different food colors. Purple, orange, and chartreuse work well. The spots should be no larger than 3 mm in diameter.
4. While the spots are drying, put 3 mL of ethanol, 1 mL of distilled water, and 2 drops of glacial acetic acid into a clean 10-cm test tube.

FIGURE 25.1
Preparation of filter paper

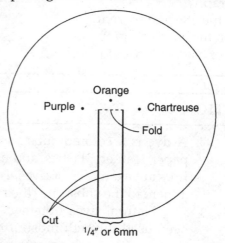

SAFETY
ALERT

Ethanol is flammable. Make certain there are no open flames nearby when you do this part of the experiment. Glacial acetic acid will vigorously attack tissue. If you contact the acid, wash the contacted area with cool water.

5. Clean a crucible, which has a height of about 3 cm, with soap and water. Dry the crucible with a towel and pour the mixture prepared in Step 4 into it.
6. After the spots have dried, place the filter paper (wick down) on the top of the crucible. The liquid in the crucible should reach about 0.5 cm up the wick.
7. Invert a large beaker over the crucible and paper to minimize solvent evaporation.
8. When the first of the colored components nears the outer edge of the paper, remove the paper and allow it to dry.
9. Record the color of each component of each of the food colorings in Table 25.1 and attach the chromatogram to your report.

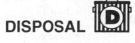

DISPOSAL

Contents of crucible in sink; flush with tap water

B. Separation of Black Ink Components

Most inks are complex mixtures containing several dye components. In this experiment, ascending paper chromatography is used to separate the dye components of Paper Mate's black ink no. 863-11 (not all black inks separate as well). Three different solvents will be used to determine the effect of solvent variation on the quality of separation. Some suggested solvents are pure water, 50% water/50% acetone, and pure acetone.

PROCEDURE

1. Carry out the following procedure with each of the following three solvents: 30 drops of distilled water, a mixture of 15 drops of distilled water plus 15 drops of acetone, and 30 drops of pure acetone.
2. Put the solvent in a 15-cm test tube.
3. Obtain a 1-cm wide by 15-cm long strip of chromatography paper.
4. Use a felt-tip pen to put a small spot of black ink on the paper 3 cm from the end.
5. Put the paper strip into the test tube and adjust the stopper so the end just dips into the solvent (Figure 25.2). The spot of ink should be between 1 and 2 cm above the solvent surface.
6. Allow the solvent to move up the paper until it reaches a height that is within 2 cm of the stopper.
7. Remove the filter paper and allow it to dry.
8. Report in Table 25.2 the different colors you can detect in the ink.
9. Record the solvent used on each chromatogram and attach the chromatograms to the report sheet.

FIGURE 25.2
Separation of ink components

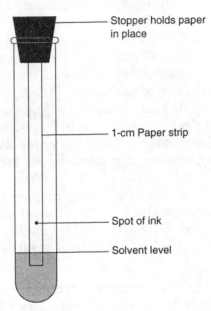

— Stopper holds paper in place

— 1-cm Paper strip

— Spot of ink

— Solvent level

DISPOSAL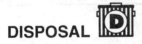

Chromatography solvent in container labeled "Exp. 25, Used Chromatography Solvent Part B"

C. Separation of Soft-Drink Food Colorings

Dyes are used in many processed foods to add visual appeal. The dyes used for such purposes are carefully monitored and must have FDA approval. In this part of the experiment, you will use ascending paper chromatography to separate the components of the dyes used to color powdered soft-drink mixes. The technique was used in Part C of Experiment 5 of this manual.

PROCEDURE

1. Make five spotting capillaries by heating the middle of glass melting-point capillaries in the flame of a burner until the glass softens and then stretching the softened glass to form a narrow neck. Bend the narrowed glass until it breaks to form the tip of the spotting capillary. Your instructor will demonstrate the technique.

2. Obtain a sheet of chromatography paper from the reagent area.
3. Obtain a ruler from the stockroom, measure the chromatography paper, then trim it with a pair of scissors so the dimensions match those given in Figure 25.3(a). Note that the paper is wider at the top than at the bottom.
4. Use a pencil to draw a line across the sheet of paper 1.5 cm from the bottom as shown in Figure 25.3(b).
5. Measure from the edge of the paper and use a pencil to mark an X on the line of the paper at the following distances from the edge: 1.0 cm, 2.0 cm, 3.0 cm, and 4.0 cm.
6. Take a clean, dry, plastic weighing dish to the reagent area and put a 1-drop sample of each of the four powdered drink solutions in separate corners of the weighing dish. Identify the samples with numbers, letters, or names as indicated on the sample containers.
7. Use one of the spotting capillaries prepared in Step 1 and practice spotting samples by dipping the narrow end of the capillary into a test tube of distilled water and then touching the narrow tip to a piece of filter paper. Practice until you can consistently form spots that are 2 to 4 mm in diameter. Be sure they are no more than 4 mm in diameter (the size of this circle ◯).
8. Use the remaining four capillaries to put spots of the samples of soft-drink mix on the paper prepared in Steps 2 to 5. Use a different capillary for each sample. Label under the X's of each spot with the number, letter, or name that identifies the sample. Allow the spots to dry.
9. While the sample spots are drying, prepare a chromatography chamber as follows: Use your 10-mL graduated cylinder to measure 3.0 mL of ethanol. Put the ethanol into a clean, dry 250-mL beaker and cover the top of the beaker with a watch glass.

SAFETY ALERT

Ethanol is flammable. Make certain no open flames are nearby when you use this material.

FIGURE 25.3
Preparation of
Chromatography Paper
(*not* actual size)

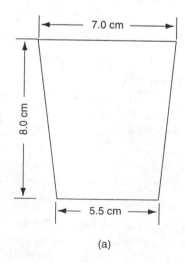

|← 7.0 cm →|

8.0 cm

|← 5.5 cm →|

(a)

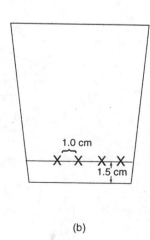

1.0 cm

X X X⌐X
1.5 cm

(b)

10. Add 2 drops of glacial acetic acid to the ethanol in the beaker and swirl the beaker to mix the contents.

11. Put the beaker into a location where it can remain undisturbed. Carefully put the spotted paper into the beaker with the spotted end down. The paper should rest evenly on the bottom of the beaker. It might be necessary to push the paper down into the beaker to position it properly. Do this carefully and quickly so the liquid is not splashed onto the paper but begins to move up the paper uniformly. The liquid level in the beaker should be below the line on the chromatography paper. Replace the watch glass cover.

12. Allow the chromatogram to develop undisturbed until the solvent front gets to within about 1 cm of the top of the paper. When the development is completed, remove the paper from the beaker and set it aside to dry. Dispose of the chromatography solvent.

DISPOSAL 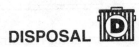 Chromatography solvent in sink; flush with running tap water

13. Record in Table 25.3 the color of each component of each sample of drink mix (use only the number of blanks you need). Attach the chromatogram to the Data and Report Sheet.

D. Preparation of Azo Dyes

The most common dyes are the azo dyes that contain the chromophore $-N=N-$. These dyes are formed by two separate reactions: (1) diazotization of a primary aromatic amine to give a diazonium salt (Equation 25.1) and (2) coupling the resulting diazonium salt with a phenol or aromatic amine in the presence of NaOH (Equation 25.2).

$$O_2N-\!\!\bigcirc\!\!-NH_3 \; + \; NaNO_2 \; + \; HCl \; \rightarrow \; O_2N-\!\!\bigcirc\!\!-N\equiv N^+Cl^- \qquad \text{Eq. 25.1}$$

p-nitroaniline
(a primary aromatic amine)

(a diazonium salt)

$$O_2N-\!\!\bigcirc\!\!-N\equiv N^+Cl^- \; + \; \text{(β-naphthol)} \; + \; NaOH \; \rightarrow \; O_2N-\!\!\bigcirc\!\!-N=N-\text{(para red)}$$

β-naphthol
(a phenol)

para red
(an azo dye)

Eq. 25.2

Azo dyes with intense yellow, orange, red, blue, or green colors can be prepared by attaching different substituent groups to the aromatic benzene rings.

The procedures of this part of the experiment and parts E and F will be done cooperatively by groups of 4 students. Each of the 4 members of a group will do one of the procedures listed below. All four members of each group will individually do the indicated steps that involve the soaking, dyeing, etc. of cloth samples, and will submit the dyed cloth samples with their reports. All group members will complete and submit the data and report tables for all three parts.

     Procedure D-a (steps 1-5)
     Procedure D-b (steps 1-5)
     Procedure E (steps 1-5), All group members do steps 3-5.
     Procdure F (steps 1-14), All group members do steps 2-4, 7-9, and 11-14.

PROCEDURE

a. Preparation of a Diazonium Salt

1. Place 0.1 g of *p*-nitroaniline into a 50-mL beaker.
2. Add 20 ml of water and 5 ml of concentrated hydrochloric acid (HCl).

SAFETY ALERT | Concentrated hydrochloric acid is very corrosive and will attack tissue vigorously. If you contact it, wash the contacted area with cool water.

3. Stir well for 2 to 3 minutes and filter the mixture if any solid remains undissolved. Collect the filtrate (the liquid) in a small flask and put the flask into an ice-and-water bath for 5 minutes.
4. Add 1.5 mL (30 drops) of 1.0 M sodium nitrite solution ($NaNO_2$).
5. Stir the resulting diazonium salt mixture well and leave it in the ice–water bath for use later in the preparation of azo dyes (see Part D-b below).

b. Preparation of Azo Dyes

1. Prepare separate solutions of β-naphthol, phenol, and resorcinol as follows: Mix together in a 15-cm test tube 0.1 g of the solute (β-naphthol, phenol, or resorcinol), 12 mL of water, and 1 mL (20 drops) of dilute (6 M) sodium hydroxide (NaOH).

SAFETY ALERT | Solid phenol is very corrosive and will attack tissue. All three solid solutes should be used with care. If contact occurs, wash the contacted area with soap and water.

     Warm the mixture containing β-naphthol in a hot-water bath and stir until the solid dissolves. Phenol and resorcinol will dissolve without heating.

2. After the solute has dissolved, stir each solution, place it in an ice-and-water bath, and cool it to 5°C.
3. Pour one-half of each solution into separate 15-cm test tubes and leave them in the cold bath for later use (Part F).
4. Add 9 mL of the cold diazonium salt mixture prepared earlier to the remaining half of each solution.
5. Stir each solution, and observe the formation of the azo dye in each case. Record the dye colors in Table 25.4. Note that in some cases the azo dye precipitates from the solution.

DISPOSAL Test tube azo dye contents in container labeled "Exp. 25, Used Chemicals Part D"

E. Direct Dyeing

A number of different methods must be used to apply dyes to fabrics because of the variety of substances now used as textile fibers. These substances range from nonpolar polyethylene to medium-polarity cotton to highly polar wool. For a dye to be effective, it must penetrate the fiber and remain fast (firmly fixed) during washing and cleaning. The methods used to accomplish this provide a means for classifying dyes.

Direct dyes adhere firmly to fibers without the aid of supplemental chemicals. Wool and silk are easily dyed by the direct method. These fibers have many polar sites and bind strongly to polar dyes such as malachite green.

$$(CH_3)_2-N-\bigcirc-C=\bigcirc=\overset{+}{N}(CH_3)_2Cl^-$$

malachite green

Martius yellow, picric acid, eosin, and methyl orange are also highly polar dyes and thus dye directly to wool and silk.

PROCEDURE

1. Place 10 mL of malachite green solution in a 100-mL beaker and add 40 mL of water and a boiling chip.
2. Heat the solution until it just begins to boil gently. Do not heat the solution strongly.
3. Remove the burner and put a piece of wool and a strip of cotton cheesecloth (or a single strip of special test fabric) into the hot dye solution.
4. After 2 minutes, use a glass stirring rod to remove the fabrics from the dye and rinse them in cool water.
5. Note whether or not the color is reasonably fast (does not wash away) and record your observations in Table 25.5. If fabric test strips are available, note also the results for other fabrics in the strip. Attach the dyed sample to the Data and Report Sheet.

DISPOSAL Used dye solution in container labeled "Exp. 25, Used Chemicals Part E"

F. Developed and Ingrain Dyeing

A superior method for applying azo dyes to cotton was patented in England in 1880 and is called *developed dyeing*. The cotton is soaked in an alkaline solution of a phenol and is then placed in a cold diazonium salt solution. The azo dye develops directly on the fiber. The reverse process, called *ingrain dyeing*, was introduced in 1887. In this process, cotton is impregnated with an amine, which is then diazotized and developed by immersion in a solution of the phenol. The first ingrain dye, primuline red, was obtained by coupling the diazonium salt of primuline with 2-naphthol.

primuline

PROCEDURE

1. Place 5 mL of primuline solution and 95 mL of water in a 250-mL beaker.
2. Soak three pieces of cotton cheesecloth (or three strips of test fabric) for 5 minutes in the resulting solution at a temperature just below the boiling point.
3. Rinse the cloth well in about 500 mL of water contained in a large beaker.

DISPOSAL   Rinse water in sink

4. After rinsing, squeeze the cloth between paper towels or filter paper to remove excess water.
5. Prepare a diazotizing bath by mixing together in a 100-mL beaker 3.0 mL (60 drops) of 1.0 M sodium nitrite solution ($NaNO_2$) and 50 mL of water.
6. Cool the solution to 5°C in an ice bath and add 2 mL of concentrated hydrochloric acid.

SAFETY ALERT

> Remember, concentrated hydrochloric acid is very corrosive.

7. Put the cloth samples impregnated with primuline into this diazotizing bath and leave them for about 5 minutes.
8. Remove the cloth samples from the diazotizing bath and rinse them in a beaker containing about 500 mL of water.

DISPOSAL   Rinse water in sink

9. Remove the samples from the beaker and press out excess water as before.
10. Use three small beakers and mix 25 mL of water with each of the β-naphthol, phenol, and resorcinol solutions saved from Part D-b.
11. Mix well and place one piece of cloth in each solution.
12. Let the color develop for about 5 minutes.
13. Remove the cloth pieces, rinse them in cool water, and remove the excess water.
14. Record the color and fastness of the dyes in Table 25.6. Attach the dyed samples to the Data and Report Sheet.

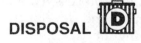

DISPOSAL   Used dye solutions in container labeled "Exp. 25, Used Chemicals Part F"

Experiment 25 ▪ Pre-Lab Review

Dyes, Inks, and Food Colorings

1. Are any specific safety alerts given in the experiment? List any that are given.

_____

_____

_____

2. Are any specific disposal directions given in the experiment? List any that are given.

_____

_____

_____

3. In Part A, what is the maximum spot size for the food colors put on the filter paper?

_____

4. What container is used for the solvent in Part A? _____

_____   _____

5. In which part of the experiment do you make spotting capillaries? How many do you make?

_____

6. What do you use to mark the chromatography paper in Part C? _____

_____

7. Describe how you get samples of powdered drink solutions in Part C.

_____

_____

8. What is the maximum spot size for samples put on the chromatography paper in Part C?

_____

9. Which parts of the experiment are done cooperatively by groups of students?

_____

_____

10. In Part D, what solutions must be kept cold until they are used? _____

_____

11. What solutions are used in Part F but prepared in Part D? _____

_____

Experiment 25 ▪ Data & Report Sheet

Dyes, Inks, and Food Colorings

A. Separation of Food Colorings

TABLE 25.1 (data and report)

Color of Sample	Colors of Separated Components
_____	_____
_____	_____
_____	_____

B. Separation of Black Ink Components

TABLE 25.2 (data and report)

Solvent Used	Colors of Separated Components
Water	_____
Acetone	_____
Acetone–water mix	_____

C. Separation of Soft-Drink Food Colorings

TABLE 25.3 (data and report)

Sample Color	Component Colors
_____	_____
	_____
_____	_____
	_____
_____	_____
	_____
_____	_____
	_____

ATTACH CHROMATOGRAMS BELOW

D. Preparation of Azo Dyes

TABLE 25.4 (data and report)

Phenol Compound Used	Color of Resulting Dye	Did Dye Precipitate?
β-Naphthol	_____	_____
Phenol	_____	_____
Resorcinol	_____	_____

E. Direct Dyeing

TABLE 25.5 (data and report)

Dye	Material	Color Fastness
Malachite green	wool	_____
Malachite green	cotton	_____

F. Developed and Ingrain Dyeing

TABLE 25.6 (data and report)

Dye Components	Color of Cotton	Color Fastness
Primuline + β-naphthol	_____	_____
Primuline + phenol	_____	_____
Primuline + resorcinol	_____	_____

ATTACH DYED SAMPLES BELOW

QUESTIONS

1. On the basis of the data recorded in Table 25.2, which solvent is best at separating black ink components?

   a. Water   b. Acetone   c. 50/50 acetone and water mix.

   Explain your answer: _____

   _____

   _____

2. Why would it be unwise to use an ink pen to mark the reference lines and points on the chromatography paper in Part C?

   a. The ink might dissolve in the solvent and disappear.
   b. The ink components might move with the solvent.
   c. The ink might spread as it was applied to the paper.

   Explain your answer: _____

   _____

   _____

3. What was the total number of different dyes used to make the four colors of soft-drink mixes studied in Part C?

   a. 1   b. 2   c. 3   d. 4

   Explain your answer: _____

   _____

   _____

A Study of Carbohydrates

In this experiment, you will

- Carry out several of the more common carbohydrate reactions.

- Identify an unknown sugar and the sugars present in some common foods.

INTRODUCTION

Carbohydrates are a class of naturally occurring organic compounds that undergo a variety of chemical reactions. Reactions typical of aldehydes, ketones, alcohols, hemiacetals, and acetals are characteristic of various members of this compound class.

Monosaccharides are the simplest carbohydrates and are the fundamental units that make up more complicated carbohydrates. Monosaccharides, sometimes called *simple sugars,* are classified as aldoses or ketoses on the basis of whether their characteristic carbonyl group is present in the form of an aldehyde or a ketone. The linking together of two monosaccharide units results in carbohydrates called **disaccharides. Polysaccharides** are carbohydrates composed of many monosaccharide units linked together. The characteristic linkage of most disaccharides and polysaccharides is an acetal. The formation of an acetal linkage produces water; thus, the addition of water will break acetal linkages (hydrolysis). The relationships between the various carbohydrates are illustrated by the following hydrolysis reactions. Note that the reactions are acid-catalyzed:

$$\text{polysaccharides} + H_2O \xrightarrow{H^+} \text{disaccharides} + H_2O \xrightarrow{H^+} \text{monosaccharides} \qquad \text{Eq. 26.1}$$

starch, cellulose   sucrose,   glucose, fructose,
lactose, maltose   galactose, xylose

EXPERIMENTAL PROCEDURE

A. Fermentation

During fermentation, certain carbohydrates are converted into CO_2 gas and ethanol. This transformation is brought about by several enzymes.

$$C_6H_{12}O_6 \xrightarrow{\text{enzymes}} 2CH_3CH_2-OH + 2CO_2 \qquad \text{Eq. 26.2}$$

glucose   ethanol

Yeast contains enzymes capable of fermenting glucose as well as many other sugars.

PROCEDURE

1. Obtain an unknown sugar solution and record the unknown identification (ID) number in Table 26.1 of the Data and Report Sheet.

2. Carry out the following procedure and those in Parts B to D with 4% solutions of the sugars fructose, glucose, lactose, sucrose, xylose, and your unknown and 10% solutions of gelatin dessert and honey.

3. Label eight 10-cm test tubes with the identity of the solutions and add 3 mL of the appropriate sugar solution to each tube.

4. Add 3 mL of yeast suspension to each tube. (Shake the yeast bottle before using.)

5. Invert once to mix and place the tubes in a 250-mL beaker that is half full of warm water (45 to 50°C). Take care not to overheat the bath; higher temperatures will deactivate the yeast. The bath does not have to be heated after the tubes are put in.

6. Allow the tubes to remain undisturbed for 1 hour. Gas (CO_2) will form, bubble off and create a foam on the top of some of the tubes within an hour.

7. While this test is proceeding, go on to other tests.

8. Record in Table 26.1 which samples produced CO_2 and which did not.

 **DISPOSAL** Test tube contents in sink

B. Benedict's Test

Benedict's reagent is a mild, alkaline oxidizing agent that will oxidize all monosaccharides, whether they are aldoses or ketoses, and some disaccharides. Sugars are classified as reducing or nonreducing, depending on whether or not they react with the Benedict's reagent. A color change from blue to orange, red, or dark brown indicates a reaction has taken place.

$$\underset{\substack{\text{blue Benedict's}\\\text{solution}}}{\text{reducing sugar}} + Cu^{2+} \xrightarrow{\text{heat}} \text{oxidized sugar} + \underset{\substack{\text{red-colored}\\\text{precipitate}}}{Cu_2O} \qquad \text{Eq. 26.3}$$

PROCEDURE

1. Label eight 10-cm test tubes and place 20 drops of the appropriate solution in each tube.

2. Add 20 drops of Benedict's solution to each tube and agitate to mix the contents.

3. Place all the tubes in a boiling-water bath at the same time.

4. Heat them for 2 minutes after the water starts boiling again.

5. Note and record in Table 26.2 the color of the agitated contents of each tube after the 2-minute boiling is complete.

 **DISPOSAL** Test tube contents in container labeled "Exp. 26, Used Chemicals Part B"

C. Barfoed's Test

Barfoed's reagent is a slightly acidic solution containing Cu^{2+} ions, but it is a weaker oxidizing agent than the alkaline Benedict's solution. Barfoed's reagent will oxidize monosaccharides (and produce the red-colored precipitate Cu_2O) but will not oxidize disaccharides. Thus, Barfoed's reagent serves to distinguish between monosaccharides and disaccharides.

PROCEDURE

1. Label eight 10-cm test tubes and place 20 drops of the appropriate solution in each tube.
2. Add 20 drops of Barfoed's reagent to each tube and agitate to mix the contents.
3. Place all the tubes in a boiling-water bath at the same time.
4. Heat them for 2 minutes after the water starts boiling again.
5. At the end of 2 minutes, note and record in Table 26.3 the appearance of any red precipitate.

DISPOSAL  Test tube contents in container labeled "Exp. 26, Used Chemicals Part C"

D. Seliwanoff's Test

Hexoses (sugars containing six carbon atoms) are dehydrated and form hydroxymethylfurfural when heated with HCl. Ketohexoses (such as fructose) and disaccharides (such as sucrose) that contain a ketohexose yield larger amounts of this product and react faster than aldohexoses. These differences are the basis of the Seliwanoff test for ketoses in which resorcinol is used to form a red product with the generated hydroxymethylfurfural. Other sugars produce gray, yellow, or faintly pink colors.

$$\text{Ketohexoses and disaccharides composed of ketohexoses} \xrightarrow[\text{heat}]{\text{HCl, resorcinol}} \text{red color} \qquad \text{Eq. 26.4}$$

PROCEDURE

1. Label eight 10-cm test tubes.
2. Place 2 mL of Seliwanoff's reagent in each tube.

SAFETY ALERT

> Seliwanoff's reagent is 6 *M* in hydrochloric acid, and will vigorously attack tissue. If contact occurs, wash the contacted area with cool water.

3. Add 1 drop of each solution to the appropriate tube and mix well.
4. Place all the samples in a boiling-water bath at the same time.
5. Heat them for 3 minutes after the boiling starts again.
6. Note and record in Table 26.4 the resulting color of each solution.

DISPOSAL  Test tube contents in container labeled "Exp. 26, Used Chemicals Part D"

E. Aniline Acetate Test

The furfural produced by the action of HCl on pentoses forms a bright pink color with aniline acetate. Hydroxymethylfurfural derived from hexoses gives only a slight pink color.

PROCEDURE

1. Perform the following test in a fume hood.
2. Place 2 mL of xylose solution, 10 mL of distilled water, and 10 mL of concentrated HCl into a 125-mL flask.

3. Boil the resulting solution for 1 minute over a burner located in a fume hood.
4. Discontinue heating and place a filter paper over the mouth of the flask.
5. Moisten the paper with 3 drops of aniline acetate solution.
6. Observe and record in Table 26.5 any color changes in the paper.
7. Repeat the test using your unknown in place of the xylose solution and record the results.

DISPOSAL

Flask contents in container labeled "Exp. 26, Used Chemicals Part E"
Used filter paper in wastebasket

REPORT

A to E. Positive Results

1. Look through the data of Tables 26.1 through 26.5 and decide what experimental result constitutes a positive test in each case.
2. Record a summary of these characteristic positive tests in Table 26.6.

A to E. Summary of Sample Behavior

1. Use the positive results summarized in Table 26.6 and the test results for each sample recorded in Tables 26.1 through 26.5 to decide which samples gave positive results for each test.
2. Record these results in Table 26.7 by placing an X in the blanks to indicate which samples gave positive results.

A to E. Identification of Unknowns

1. Your unknown contained a single sugar. Use the results in Table 26.7 to identify the sugar.
2. Record your conclusion in Table 26.8.
3. Also identify and record the name of the sugar found in the gelatin dessert.
4. Honey contains two sugars; one is glucose. Identify and record the name of the other one.

NAME _____ SECTION _____ DATE _____

Experiment 26 ▪ Pre-Lab Review

A Study of Carbohydrates

1. Are any specific safety alerts given in the experiment? List any that are given.

    _____

    _____

    _____

2. Are any specific disposal directions given in the experiment? List any that are given.

    _____

    _____

    _____

3. In Part A, what observation is used to detect fermentation in the samples?

    _____

4. In Parts B through E, what type of results will you be observing and recording?

    _____

    _____

5. Which carbohydrate tests require the use of a boiling-water bath? _____

    _____

    _____

6. Which carbohydrate test should be performed in a fume hood? _____

    _____

Experiment 26 ▪ Data & Report Sheet

A Study of Carbohydrates

A. Fermentation

TABLE 26.1 (data)

Unknown ID Number _____

Sample	Test Results (CO$_2$ Gas)
Fructose	_____
Glucose	_____
Lactose	_____
Sucrose	_____
Xylose	_____
Unknown	_____
Gelatin dessert	_____
Honey	_____

B. Benedict's Test

TABLE 26.2 (data)

Sample	Test Results (Color)
Fructose	_____
Glucose	_____
Lactose	_____
Sucrose	_____
Xylose	_____
Unknown	_____
Gelatin dessert	_____
Honey	_____

C. Barfoed's Test

TABLE 26.3 (data)

Sample	Test Results (Red Precipitate)
Fructose	_____
Glucose	_____
Lactose	_____
Sucrose	_____
Xylose	_____
Unknown	_____
Gelatin dessert	_____
Honey	_____

D. Seliwanoff's Test

TABLE 26.4 (data)

Sample	Test Results (Color)
Fructose	_____
Glucose	_____
Lactose	_____
Sucrose	_____
Xylose	_____
Unknown	_____
Gelatin dessert	_____
Honey	_____

E. Aniline Acetate Test

TABLE 26.5 (data)

Sample	Test Results (Color)
Xylose	_____
Unknown	_____

A to E. Positive Results

TABLE 26.6 (report)

Test	Positive Result
Fermentation	_____
Benedict's test	_____
Barfoed's test	_____
Seliwanoff's test	_____
Aniline acetate test	_____

A to E. Summary of Sample Behavior

TABLE 26.7 (report)

Sample	Fermentation	Benedict's	Barfoed's	Seliwanoff's	Aniline acetate
Fructose	_____	_____	_____	_____	_____
Glucose	_____	_____	_____	_____	_____
Lactose	_____	_____	_____	_____	_____
Sucrose	_____	_____	_____	_____	_____
Xylose	_____	_____	_____	_____	_____
Unknown	_____	_____	_____	_____	_____
Gelatin dessert	_____	_____	_____	_____	_____
Honey	_____	_____	_____	_____	_____

A to E. Identification of Unknown

TABLE 26.8 (report)

Unknown ID Number _____

Sample	Sugars Present
Unknown	_____
Gelatin dessert	_____
Honey	_____

QUESTIONS

1. An unknown consists of a single pure sugar. It gives a positive Benedict's test, but is negative for all others. What is the sugar?

   a. Sucrose   **b.** Lactose   **c.** Xylose

   Explain your answer: _____

   _____

   _____

2. An unknown mixture of two sugars is known to contain xylose. The mixture gives positive results for all five tests. What other sugar could be present?

   a. Glucose   **b.** Fructose   **c.** Lactose

   Explain your answer: _____

   _____

   _____

3. Which one of the following tests would allow you to differentiate between fructose and sucrose?

   a. Fermentation   **b.** Benedict's test   **c.** Seliwanoff's test

   Explain your answer: _____

   _____

   _____

4. An unknown consisting of a single sugar gives a negative fermentation test. Which one of the following tests would allow you to decide which of the five sugars used in this experiment you had?

   a. Seliwanoff's test   **b.** Benedict's test   **c.** Barfoed's test

   Explain your answer: _____

   _____

   _____

Preparation of Soap
by Lipid Saponification

In this experiment, you will

- Prepare a soap by saponifying a lipid.
- Prepare a detergent.
- Compare several properties of your prepared soap and detergent.

INTRODUCTION

Soaps are molecules containing a very long alkyl group, which is soluble in nonpolar substances (fats and oils), and an ionic end (the salt of a carboxylic acid), which is soluble in water.

$$\underbrace{CH_3CH_2CH_2CH_2CH_2CH_2CH_2CH_2CH_2CH_2CH_2CH_2CH_2CH_2CH_2}_{\substack{\text{long hydrocarbon chain} \\ \text{(soluble in nonpolar substances)}}}\underbrace{\overset{\overset{\textstyle O}{\|}}{C}-O^-Na^+}_{\substack{\text{ionic end} \\ \text{(soluble in water)}}}$$

Detergents are structurally similar to soaps but differ primarily in the water-soluble portion. Several examples are given below:

$$CH_3(CH_2)_nCH_2-O-\overset{\overset{\textstyle O}{\|}}{\underset{\underset{\textstyle O}{\|}}{S}}-O^-Na^+ \qquad \text{a sodium alkyl sulfate}$$

$$CH_3(CH_2)_nCH\overset{\overset{\textstyle CH_3}{|}}{}-\bigcirc\!\!\!\!\!-\overset{\overset{\textstyle O}{\|}}{\underset{\underset{\textstyle O}{\|}}{S}}-O^-Na^+ \qquad \text{a sodium alkylbenzene sulfonate}$$

$$CH_3(CH_2)_nCH_2-\overset{\overset{\textstyle O}{\|}}{C}-O-CH_2-\overset{\overset{\textstyle CH_2OH}{|}}{\underset{\underset{\textstyle CH_2OH}{|}}{C}}-CH_2OH \qquad \text{a non-ionic detergent}$$

The cleaning action of both soaps and detergents results from their ability to emulsify or disperse water-insoluble materials (dirt, oil, grease, and so on) and hold them in suspension in water. This ability comes from the molecular structure of soaps and detergents. When a soap or detergent is added to water that contains oil or other water-insoluble materials, the soap or detergent molecules surround the soil droplets. The alkyl group dissolves in the oil, while the ionic end dissolves in water. As a result, the oil droplets are dispersed throughout the water and may be rinsed away.

A. Preparation of a Soap

Soaps are prepared by the alkaline hydrolysis (saponification) of fats and oils:

$$
\begin{array}{l}
CH_2-O-\overset{\overset{\displaystyle O}{\|}}{C}-R \\[2mm]
CH-O-\overset{\overset{\displaystyle O}{\|}}{C}-R' \;+\; 3NaOH \;\rightarrow\; \\[2mm]
CH_2-O-\overset{\overset{\displaystyle O}{\|}}{C}-R''
\end{array}
\quad
\begin{array}{l}
CH_2OH \\[2mm]
CH-OH \;+\; \\[2mm]
CH_2OH
\end{array}
\quad
\begin{array}{l}
Na^{+}{}^{-}O-\overset{\overset{\displaystyle O}{\|}}{C}-R \\[2mm]
Na^{+}{}^{-}O-\overset{\overset{\displaystyle O}{\|}}{C}-R' \quad Eq.\ 27.1 \\[2mm]
Na^{+}{}^{-}O-\overset{\overset{\displaystyle O}{\|}}{C}-R''
\end{array}
$$

fat or oil   glycerol   soaps

This reaction, the basis of the soap industry, also provides a commercial source of glycerol that is useful as an antifreeze and a moisturizing agent, and is necessary in the manufacture of nitroglycerine.

PROCEDURE

1. Put 2.0 mL of vegetable oil and 3.0 mL of dilute (6 M) sodium hydroxide solution (NaOH) into a 100-mL beaker.

SAFETY ALERT

> The dilute sodium hydroxide solution is corrosive and will vigorously attack tissue. Wash any contacted area of skin with cool water. Inform your instructor if your eyes are involved.

2. Cover the beaker with a watch glass and carefully boil the contents over a low flame of a burner. There will be a significant amount of spattering inside the beaker, and steam will escape from the pouring spout of the beaker. Make certain the spout is directed away from you. The boiling rate is more easily controlled if you hold the burner in your hand and direct the flame against the bottom of the beaker as needed to maintain a steady but slow rate of boiling.
3. Heat the beaker until most of the water has evaporated and the contents are the consistency of a thick foam (approximately 3 to 5 minutes). Take care not to char the contents.
4. After the heating is completed, remove the heat source and allow the beaker to cool for 10 minutes before you remove the watch glass.
5. While the beaker is cooling, prepare a concentrated solution of sodium chloride by mixing together 15 g of solid sodium chloride (NaCl) and 55 mL of distilled water. Stir the mixture to dissolve the solid.
6. After the heated beaker is cool, add 20 mL of the sodium chloride solution prepared in Step 5 to the beaker. Loosen the solid in the beaker by stirring with a spatula. Stir the mixture well, but do not

beat it into a froth. Decant the liquid from the solid. Use a wire screen to keep the floating soap from leaving the beaker.

 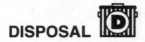

7. Repeat Step 6 again. Then, add the final 20 mL of sodium chloride solution and stir, but this time filter the mixture through a funnel lined with 2 or 3 layers of cheesecloth and allow the solid soap to collect in the cheesecloth.
8. Remove the cheesecloth from the funnel and twist it to remove excess liquid from the soap. Save the soap for use in Part C of the experiment.

B. Preparation of a Detergent

You will prepare a sodium alkyl sulfate called sodium dodecylsulfate by reacting dodecyl alcohol (dodecanol) with sulfuric acid. The resulting sulfate is converted to the sodium salt by a reaction with sodium hydroxide. Equations for the reactions are:

$$CH_3(CH_2)_{10}CH_2-OH + H_2SO_4 \rightarrow CH_3(CH_2)_{10}CH_2-O-\overset{\overset{\displaystyle O}{\|}}{\underset{\underset{\displaystyle O}{\|}}{S}}-OH + H_2O$$

dodecanol	sulfuric acid	dodecylsulfate	Eq. 27.2

$$CH_3(CH_2)_{10}CH_2-O-\overset{\overset{\displaystyle O}{\|}}{\underset{\underset{\displaystyle O}{\|}}{S}}-OH + NaOH \rightarrow CH_3(CH_2)_{10}CH_2-O-\overset{\overset{\displaystyle O}{\|}}{\underset{\underset{\displaystyle O}{\|}}{S}}-O^-Na^+ + H_2O$$

dodecylsulfate	sodium dodecylsulfate	Eq. 27.3

PROCEDURE

1. Place 2 mL of dodecanol into a 100-mL beaker.
2. With stirring, slowly add 1 mL (20 drops) of concentrated sulfuric acid (H_2SO_4).

> Concentrated sulfuric acid is very corrosive and will rapidly and vigorously attack tissue. If you contact it, immediately wash the contacted area with cool water.

3. Swirl to mix and let the mixture stand for an additional 10 minutes.
4. Fill a 100-mL beaker one-third full of ice, add about 5 g of sodium chloride (NaCl), and mix thoroughly. Add water to bring the total volume of the mixture to 40 mL.
5. Mix 2 mL of dilute (6 M) sodium hydroxide with 5 mL of water in a small beaker. Mix well, then add 4 drops of phenolphthalein indicator.

Dilute sodium hydroxide is corrosive and will vigorously attack tissue. It is especially damaging to eyes. Wash any contacted area with cool water. Inform your instructor if your eyes are involved.

6. Carefully pour the sodium hydroxide solution (from Step 5) into the dodecanol–sulfuric acid mixture (from Step 3). Stir until the pink color disappears or for 2 minutes, whichever comes first. A large amount of solid detergent will form.
7. Pour the ice–salt bath prepared in Step 4 into the detergent-containing mixture from Step 6. Stir vigorously for 1 minute.
8. Filter the precipitated detergent mixture through 2 or 3 layers of cheesecloth in a funnel mounted on a ringstand.
9. Wash the collected detergent twice with 10-mL portions of ice-cold water.

DISPOSAL  Filtrate of Step 8 and wash water of Step 9 in sink

10. Remove the cheesecloth from the funnel, squeeze excess water from the solid detergent, and save the detergent for use in Part C.

C. Properties of Soaps and Detergents

As we mentioned earlier, soaps and detergents are similar in some ways but different in others. This will become apparent to you as you study the following characteristics: emulsifying properties, behavior in hard water, alkalinity, and ability to react with mineral acids.

a. Emulsifying Properties

As a result of their molecular structures, soaps and detergents are both capable of emulsifying or dispersing oils and similar water-insoluble substances.

PROCEDURE

1. Prepare a soap solution and a detergent solution by dissolving about 1 g of your damp products from Parts A and B in separate 60-mL samples of boiling distilled water. These solutions will be used in all remaining parts of this experiment.
2. Place 4 drops of mineral oil into each of three separate 15-cm test tubes.
3. Add 5 mL of distilled water to one of the tubes, 5 mL of soap solution to another, and 5 mL of detergent solution to the third.
4. Close the mouth of each test tube with a stopper, parafilm or kitchen wrap, then shake each tube up and down vigorously for about 30 seconds.
5. Observe the degree of oil emulsification in each tube as indicated by the presence of suds and the absence of oil droplets in the liquid, or the absence of oil scum on the inside of the test tube.
6. Record your results in Table 27.1 of the Data and Report Sheet, using the following scale: ++ = good emulsifier; + = fair emulsifier; − = poor emulsifier.

DISPOSAL  Test tube contents in sink

b. Behavior in Hard Water

The sodium and potassium salts of most carboxylic acids are water-soluble. However, the calcium, magnesium, and iron salts are not. Thus, when soaps are placed in hard water that contains such ions, an insoluble, curdy solid forms. Most of us have seen these results in the form of a bathtub ring. This process removes soap ions from solution and decreases the cleaning effectiveness of soaps.

$$2CH_3(CH_2)_n\!-\!\overset{\displaystyle O}{\overset{\|}{C}}\!-\!O^-Na^+ + Ca^{2+} \rightarrow [CH_3(CH_2)_n\!-\!\overset{\displaystyle O}{\overset{\|}{C}}\!-\!O^-]_2Ca^{2+} + 2Na^+ \quad \text{Eq. 27.4}$$

a soluble soap   an insoluble soap

The calcium, magnesium, and iron forms of most detergents are more soluble in water than the corresponding soap compounds. Consequently, detergents function almost as well in hard water as they do in soft water.

PROCEDURE

1. Place 5 mL of soap solution in each of three 15-cm test tubes.
2. Place 5 mL of detergent solution in each of three 15-cm test tubes.
3. Add 2 mL of 1% calcium chloride ($CaCl_2$) to one soap-containing tube and one detergent-containing tube. Repeat this process, using 1% magnesium chloride ($MgCl_2$) and 1% iron (III) chloride ($FeCl_3$) solutions.
4. Mix the contents by inverting each tube and note whether or not a precipitate forms. Indicate the amount of precipitate by the following scale: ++ = very large; + = large; − = little or none.
5. Record your results in Table 27.2.
6. Add 4 drops of mineral oil to each tube, cover the mouth of the tube and shake the mixture vigorously.
7. Observe and record (Table 27.2) the emulsifying ability of the soap or detergent in each tube as indicated primarily by the amount of suds formed. Use the following scale: ++ = heavy suds; + = light suds; − = few or no suds.

DISPOSAL  Test tube contents in sink

c. Alkalinity

Soaps undergo a hydrolysis reaction in water. As a result, soap solutions tend to be alkaline. Detergent solutions, on the other hand, are more neutral.

PROCEDURE

1. Test small samples of your soap and detergent solutions with red litmus paper. Remember, the paper color will remain pink in acid, and will become blue in base.
2. Test small samples (20 drops) of your soap and detergent solutions by adding 2 or 3 drops of phenolphthalein indicator. Phenolphthalein is pink at pHs higher than about 8.
3. Record your results in Table 27.3.

d. Reaction with Mineral Acid

Soaps, the sodium salts of fatty acids, are water soluble, but the fatty acids themselves are not. A soap can be converted into the fatty acid by means of a reaction with a strong mineral (nonorganic) acid.

$$CH_3(CH_2)_n - \overset{\overset{\textstyle O}{\textstyle \|}}{C} - O^-Na^+ + H^+ \rightarrow CH_3(CH_2)_n - \overset{\overset{\textstyle O}{\textstyle \|}}{C} - OH + Na^+ \quad \text{Eq. 27.5}$$

a soluble soap　　　　　　　　an insoluble fatty acid

Thus, the acidification of a soap solution causes the fatty acid to precipitate. Acidification of detergents, on the other hand, produces acids that are often water soluble.

PROCEDURE

1. Place 25 mL of soap solution into a 100-mL beaker.
2. Add dilute (6 M) hydrochloric acid dropwise until the well-mixed beaker contents turn blue litmus paper pink.

SAFETY ALERT

> Dilute hydrochloric acid is corrosive and will attack tissue. Wash any contacted area with cool water.

3. With occasional stirring, cool the mixture in an ice bath for 3 or 4 minutes.
4. Note and record the formation of any precipitate in Table 27.4.
5. Repeat Steps 1 and 2, using your detergent solution. The solution may turn blue litmus to red without adding any HCl. Note and record whether a precipitate is present.

DISPOSAL　Beaker contents in sink

6. Your instructor might want you to submit your prepared soap and detergent with your report. If that is the case, let the remainder of your prepared soap and detergent dry in your locker until the next lab period. Then, put the samples into small, labeled test tubes and submit them to your instructor with your report. If the samples are not turned in, discard them.

DISPOSAL　Unused soap and detergent solutions in sink
Prepared soap and detergent in wastebasket

REPORT

C. Properties of Soaps and Detergents

1. Summarize the characteristics of soaps and detergents on the basis of the data recorded in Tables 27.1 through 27.4. Do this by writing in the blanks of Table 27.5 the appropriate term from the following: *soap*, *detergent*, or *about equal*.

Experiment 27 ▪ Pre-Lab Review

Preparation of Soap by Lipid Saponifiction

1. Are any specific safety alerts given in the experiment? List any that are given.

   _____

   _____

2. Are any specific disposal directions given in the experiment? List any that are given.

   _____

   _____

3. In Part A, what suggestion is given for controlling the rate of boiling of the reaction mixture?

   _____

4. In Part A, when do you stop heating the reaction mixture? _____

   _____

   _____

5. In Part B, how long should the dodecanol–sulfuric acid–sodium hydroxide mixture sit after the addition of sulfuric acid is completed?

   _____

6. In Part C, what procedure and observation are used to estimate the emulsifying properties of soaps and detergents?

   _____

   _____

7. What color is litmus paper in
   a. alkaline solutions? _____

   b. acidic solutions? _____

8. What color is phenolphthalein in
   a. alkaline solutions? _____

   b. acidic solutions? _____

Experiment 27 ▪ Data & Report Sheet

Preparation of Soap by Lipid Saponification

C. Properties of Soaps and Detergents

a. Emulsifying Properties

TABLE 27.1 (data)

Test Tube Contents	Emulsifying Ability
Oil + water	_____
Oil + soap solution	_____
Oil + detergent solution	_____

b. Behavior in Hard Water

TABLE 27.2 (data)

Test Tube Contents	Amount of Precipitate	Emulsifying Ability
Soap solution + $CaCl_2$	_____	
Soap solution + $MgCl_2$	_____	
Soap solution + $FeCl_3$	_____	
Detergent solution + $CaCl_2$	_____	
Detergent solution + $MgCl_2$	_____	
Detergent solution + $FeCl_3$	_____	
Soap solution + $CaCl_2$ + oil		_____
Soap solution + $MgCl_2$ + oil		_____
Soap solution + $FeCl_3$ + oil		_____
Detergent solution + $CaCl_2$ + oil		_____
Detergent solution + $MgCl_2$ + oil		_____
Detergent solution + $FeCl_3$ + oil		_____

c. Alkalinity

TABLE 27.3 (data)

Sample	Reaction to Red Litmus	Reaction to Phenolphthalein
Soap solution		
Detergent solution		

d. Reaction with Mineral Acid

TABLE 27.4 (data)

Sample	Result of Acidification
Soap solution	
Detergent solution	

TABLE 27.5 (report)

Best emulsifier	
Forms largest amount of precipitate with Ca^{2+}, Mg^{2+}, and Fe^{3+}	
Best emulsifier in presence of Ca^{2+}, Mg^{2+}, and Fe^{3+}	
Forms most basic solution	
Forms a precipitate when H^+ is added	

QUESTIONS

1. In an ion-exchange water softener, hard-water ions such as Ca^{2+}, Mg^{2+}, and Fe^{3+} are replaced by sodium ions, Na^+. How do the added Na^+ ions influence the behavior of soaps and detergents?

   a. Na^+ ions decrease their cleaning effectiveness.
   b. Na^+ ions increase their cleaning effectiveness.
   c. Na^+ ions have no effect on their cleaning effectiveness.

   Explain your answer: _____

   _____

2. Which would be the best emulsifier of soil if soft water (no Ca^{2+} ions, Mg^{2+} ions, and so on) were used?

   a. A soap   **b.** A detergent   **c.** Both would be about the same.

   Explain your answer: _____

   _____

3. The soil found in most dirty ovens consists primarily of fats or oils. Some popular oven cleaners are solutions of strong base. After the cleaner is applied, it is allowed to sit for some time. Much of the baked-on grease and oil then wipes away because it has been partially converted into which of the following?

   a. Soap   **b.** Detergent   **c.** Carbon   **d.** CO_2 gas

   Explain your answer: _____

   _____

   _____

Isolation of Natural Products: Trimyristin and Cholesterol

In this experiment, you will

- Gain experience in the use of several techniques used to isolate chemical substances from their natural sources.
- Isolate trimyristin, a fat, from nutmeg.
- Isolate the steroid cholesterol from egg yolk.

INTRODUCTION

Living organisms are amazing in their ability to specifically and efficiently synthesize complicated organic compounds. Many of the familiar organic materials we use daily are natural products, including substances used as (1) flavorings (peppermint, spearmint, cinnamon), (2) foods (carbohydrates, fats, proteins), (3) polymers (rubber, cotton, silk), (4) building materials (wood), (5) drugs (penicillin, streptomycin), and (6) vitamins.

EXPERIMENTAL PROCEDURE

A. Isolation of Trimyristin from Nutmeg

Naturally occurring animal or vegetable fats and oils are **glycerides** (triesters of glycerol and long-chain fatty acids). The fatty acids differ in chain length and degree of unsaturation, but they commonly contain 12 to 18 carbon atoms. Glycerides containing a high percentage of saturated acids are solids at room temperature and are called **fats.** Those composed of a higher percentage of unsaturated acids have lower melting points, are liquids at room temperature, and are called **oils.**

Nutmeg is somewhat unique because its total content of fats and oils is essentially composed of a single glyceride called trimyristin. The structure of this compound is given below, together with that of glycerol and a general glyceride:

$$
\begin{array}{ccc}
CH_2-OH & CH_2-O-\overset{\displaystyle O}{\overset{\displaystyle \|}{C}}-R & CH_2-O-\overset{\displaystyle O}{\overset{\displaystyle \|}{C}}-(CH_2)_{12}CH_3 \\[2mm]
CH-OH & CH-O-\overset{\displaystyle O}{\overset{\displaystyle \|}{C}}-R' & CH-O-\overset{\displaystyle O}{\overset{\displaystyle \|}{C}}-(CH_2)_{12}CH_3 \\[2mm]
CH_2-OH & CH_2-O-\overset{\displaystyle O}{\overset{\displaystyle \|}{C}}-R'' & CH_2-O-\overset{\displaystyle O}{\overset{\displaystyle \|}{C}}-(CH_2)_{12}CH_3 \\[2mm]
\text{glycerol} & \text{a glyceride} & \text{trimyristin}
\end{array}
$$

357

We note that the fatty acids contained in trimyristin are all the same—a condition that is not true for all glycerides, as indicated by the general formula. In addition, the fatty acids of trimyristin are saturated. As expected, trimyristin is a solid at room temperature.

The long carbon chains of the fatty acids of fats and oils impart hydrocarbonlike properties to the molecules. The preferred solubility of fats and oils in nonpolar solvents such as hexane is an example of these properties.

PROCEDURE

1. Use a plastic weighing dish and a centigram or electronic balance to accurately weigh out a sample of powdered nutmeg with a mass of about 1 g. Record the sample mass in Table 28.1 of the Data and Report Sheet.

2. Put the weighed sample into a 50-mL beaker and add 5 mL of hexane. Swirl the beaker to mix the contents and cover the mouth of the beaker with a watch glass to prevent evaporation.

 **SAFETY ALERT**

> Hexane is flammable. Make certain no open flames are nearby when you do this part of the experiment.

3. Allow the mixture to sit for 10 minutes with occasional swirling.

4. You are going to evaporate the hexane by one of two methods in step 6. Either a hot water bath or a fume hood with a good flow of air will be used. Check with your instructor about which method to use. If a hot water bath is required, set one up, using a 250-mL beaker (see Figure 9.1; disregard the test tubes) while your sample sits for the 10 minutes.

5. After the mixture has been allowed to sit for 10 minutes, filter the mixture through a dry filter paper and collect the filtrate (the liquid) in a dry, weighed 50-mL beaker. Record the empty beaker mass in Table 28.1.

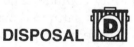 **DISPOSAL**

Filter paper containing nutmeg residue:
Allow to air dry, then put into a wastebasket

6. If you are going to use a hot-water bath to evaporate the hexane from the filtrate, leave the filtrate at your work area while you heat the water bath in the hood to boiling. Then, turn off the flame and use the hot water to evaporate the hexane from the filtrate in the weighed 50-mL beaker.

 **SAFETY ALERT**

> Under no circumstances should the beaker containing the hexane filtrate be brought near the hood while an open flame is present.

Use a buret clamp to hold the 50-mL beaker steady in the hot-water bath. If the fume hood is to be used, simply leave the uncovered beaker in the hood for 5 to 10 minutes.

7. When the hexane has all evaporated, remove the beaker from the hot-water bath or the hood and allow the contents to air dry and

cool (if it was heated) for another 3 minutes. Dry the outside of the beaker with a towel if it is wet.

8. Weigh the cool beaker and the contained crude trimyristin together on the same balance used in Step 1. Record the combined mass in Table 28.1.

9. Use a spatula to scrape the dried crude trimyristin from the bottom and sides of the beaker and put the product into a clean, dry 10-cm test tube.

10. Add 2.5 mL (50 drops) of acetone to the test tube to wash impurities from the crude trimyristin. Swirl the contents of the tube, then set it aside to allow the trimyristin to settle.

 **SAFETY ALERT**

> Acetone is flammable. Do not do this washing step if an open flame is nearby.

11. After the trimyristin has settled, use a plastic dropper to decant the liquid acetone from the test tube.

 **DISPOSAL**

Decanted acetone in sink, flushing it with running water

12. Add 20 drops of hexane to the trimyristin that remains in the test tube and agitate the tube to dissolve the solid. After the solid dissolves, pour the resulting solution onto a clean watch glass and let it evaporate.

 **SAFETY ALERT**

> Hexane is flammable. Make certain no open flames are nearby when you do this step.

13. Allow the trimyristin to air dry until it is powdery.

14. While the solid is drying, set up a 100-mL beaker water bath to use in the determination of the melting point of the product (see Figure 16.1). The melting point of pure trimyristin is 55 to 56°C.

15. After the solid is dry, put about 3 mm of it into a melting-point capillary. Measure and record the melting point in Table 28.1.

 **DISPOSAL**

Used melting-point capillary in broken-glass container
Surplus product in wastebasket

B. Isolation Of Cholesterol from Egg Yolk

Cholesterol is the best known and most abundant steroid in the body. It is a precursor for many of the important naturally occurring steroids, including male and female sex hormones, bile salts, and hormones of the adrenal glands.

cholesterol

The extreme insolubility of cholesterol in water sometimes causes it to come out of solution. In the bloodstream, this process contributes to the hardening of the arteries, and in the gallbladder it leads to the formation of gallstones. Cholesterol isolated from gallstones contains very small amounts of other steroids related to cholesterol.

PROCEDURE

1. Use a plastic weighing dish and a centigram or electronic balance to accurately weigh out a sample of cooked egg yolk with a mass of about 2 g. Record the mass in Table 28.3 of the Data and Report Sheet.
2. Put the weighed sample into a 100-mL beaker and add 5 mL of 2-propanol.

SAFETY ALERT

> 2-propanol is flammable. Make certain no open flames are nearby when you do this step.

Stir the mixture with a spatula to break up the egg yolk. Cover the mouth of the beaker with a watch glass to prevent evaporation.

3. Allow the mixture to sit for 10 minutes with occasional swirling while you set up a hot-water bath in a fume hood. Use a 250-mL beaker for the bath. If you previously set up a bath while doing Part A, the same bath can be used here.
4. After the mixture has been sitting for 10 minutes, filter it through three thicknesses of cheesecloth held in a funnel. Collect the filtrate in a weighed 50-mL beaker, the weight of which has been recorded in Table 28.3.

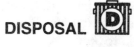

DISPOSAL   Solid residue in cheesecloth: Rinse with tap water, then put into a wastebasket.

5. Use a clamp to hold the 50-mL beaker in the water bath. Heat the bath to boiling and continue heating until all of the 2-propanol has been evaporated from the beaker.
6. When the evaporation is finished, remove the beaker from the water bath and allow it to cool for 5 minutes. Wipe the outside dry and weigh it with the contained cholesterol on the same balance used in Step 1. Record the mass in Table 28.3.
7. Add 20 drops of glacial acetic acid to the residue in the weighed beaker.

SAFETY ALERT

> Glacial acetic acid will vigorously attack tissue. Use it with care. If contact is made, wash the contacted area with cool water.

Swirl the beaker carefully to dissolve the solid residue from the sides and bottom of the beaker. Cover the mouth of the beaker with a watch glass and save the solution for use in Part C of the experiment.

C. The Lieberman–Burchard Test for Cholesterol

PROCEDURE

1. Put 4 drops of a 1% cholesterol solution into a dry 10-cm test tube. Put 4 drops of the cholesterol extract stored in the beaker in Step 7 of Part B into a second dry 10-cm test tube.
2. Add 4 drops of acetic anhydride and 1 drop of concentrated sulfuric acid to each of the test tubes.

3. If cholesterol is present in the test tubes, a characteristic color change occurs as a result of adding these reagents. An initial dark color changes to a blue-green color in 1 to 2 minutes. Note and record in Table 28.3 the color of the solutions 2 minutes after adding the acetic anhydride and sulfuric acid.

≡ CALCULATIONS AND REPORT

A. Isolation of Trimyristin from Nutmeg

1. Use the mass of the empty beaker and the mass of the beaker plus the isolated trimyristin recorded in Table 28.1 to calculate the mass of the isolated trimyristin. Record this mass in Table 28.2.
2. Use the mass of isolated trimyristin calculated in Step 1 and the mass of nutmeg recorded in Table 28.1 to calculate the percentage of trimyristin isolated. Record the percentage in Table 28.2.
3. Remember that the melting point of pure trimyristin is 55 to 56°C. Note the melting point of the isolated trimyristin recorded in Table 28.1 and classify the purity of the isolated product as *excellent*, *good*, or *poor*. Record your classification in Table 28.2 and explain the reasoning you used to arrive at the classification.

B. Isolation of Cholesterol from Egg Yolk

1. Use the mass data recorded in Table 28.3 to calculate the mass of cholesterol isolated. Record this mass in Table 28.4.
2. Use the mass calculated in Step 1 and the mass of egg yolk recorded in Table 28.3 to calculate the percentage of cholesterol isolated from the egg yolk. Record the calculated percentage in Table 28.4.

Experiment 28 ▪ Pre-Lab Review

Isolation of Natural Products:
Trimyristin and Cholesterol

1. Are any specific safety alerts given in the experiment? List any that are given.

   _____

   _____

   _____

2. Are any specific disposal directions given in the experiment? List any that are given.

   _____

   _____

   _____

3. What substances or objects are weighed accurately in Part A? _____

   _____

4. In Part A, what two alternative ways are suggested for evaporating hexane?

   _____

5. In Part A, what special condition is specified for the filter paper used to filter the hexane–nutmeg mixture?

   _____

6. How is the crude trimyristin washed in Part A? _____

   _____

   _____

7. In Part B, what substances or objects are weighed accurately? _____

   _____

8. How is solid egg yolk separated from the 2-propanol used to extract cholesterol in Part B?

   _____

   _____

   _____

9. What is done with the cholesterol after it is isolated in Part B? _____

_____

_____

10. In Part C, what result of the Lieberman–Burchard test indicates the presence of cholesterol?

_____

_____

Experiment 28 ▪ Data & Report Sheet

Isolation of Natural Products:
Trimyristin and Cholesterol

A. Isolation of Trimyristin from Nutmeg

TABLE 28.1 (data)

Mass of nutmeg sample _____

Mass of 50-mL beaker _____

Mass of crude, dry tri-
    myristin plus beaker _____

Melting point of isolated
    trimyristin _____

TABLE 28.2 (report)

Mass of isolated trimyristin _____

Percentage of trimyristin
    isolated _____

Purity of isolated trimyristin _____

Reason for purity
    classification _____

B. Isolation of Cholesterol from Egg Yolk

TABLE 28.3 (data)

Mass of egg yolk sample _____

Mass of 50-mL beaker _____

Mass of isolated, dry
    cholesterol plus beaker _____

Color of 1% cholesterol
    solution _____

Color of cholesterol
    extract solution _____

TABLE 28.4 (report)

Mass of isolated
    cholesterol _____

Percentage of cholesterol
    isolated _____

QUESTIONS

1. Suppose you make an error in Part A and record a 50 mL–beaker mass that is higher than correct. How will this error influence the final calculated percentage of trimyristin isolated?

   a. Error will increase it.   **b.** Error will decrease it.   **c.** Error will have no effect on it.

   Explain your answer: _____

   _____

   _____

2. Which of the following is a likely result of failing to wash your trimyristin properly as described in Part A, Steps 10 and 11?

   a. The percentage isolated will be reduced.
   b. The melting point of the isolated product will be lowered.
   c. The isolated product will be lighter in color.

   Explain your answer: _____

   _____

3. Suppose in Part A, Step 7, you fail to dry the beaker, and it has some water on the outside when it is weighed in Step 8. How will this influence the calculated percentage of isolated trimyristin?

   a. Increase it.   b.   Decrease it.   c.   Have no effect on it.

   Explain your answer: _____

   _____

   _____

4. Suppose in Part B, a student fails to weigh the empty 50-mL beaker before using it to collect the cholesterol. The mistake is not recognized until the student is cleaning up after the experiment has been completed. The student discards the cholesterol solution used in Part C, dries the beaker carefully, then weighs it and records the mass as the mass of the 50-mL beaker in Table 28.3. How is this likely to influence the calculated percentage of isolated cholesterol?

   a. Will probably make it too high.
   b. Will probably make it too low.
   c. Will probably not influence it.

   Explain your answer: _____

   _____

   _____

5. Suppose a Lieberman–Burchard test was done on a sample of the filtrate collected in Step 4 of Part B. Would you expect a positive result for the test?

   a. Yes   b. No

   Explain your answer: _____

   _____

   _____

Amino Acids and Proteins

In this experiment, you will

- Separate mixtures of amino acids and identify the amino acids in an unknown, using paper chromatography.

- Isolate the protein casein.
- Observe several chemical properties of amino acids and proteins.

INTRODUCTION

Proteins are essential constituents of all living cells, and are vital for proper cellular structure and function. Proteins are polymers composed of many amino acid units joined by amide linkages.

$$H_2N-\underset{\underset{R}{|}}{CH}-\overset{\overset{O}{||}}{C}-OH \qquad -HN-\underset{\underset{R}{|}}{CH}-\overset{\overset{O}{||}}{C}-NH-\underset{\underset{R'}{|}}{CH}-\overset{\overset{O}{||}}{C}-NH-\underset{\underset{R''}{|}}{CH}-\overset{\overset{O}{||}}{C}-$$

an amino acid   a protein segment

EXPERIMENTAL PROCEDURE

A. Separation of Amino Acids Using Paper Chromatography

The technique of paper chromatography was described in detail in Part C of Experiment 5. Only a brief review is given here. If you need more detail, refer to Experiment 5.

In paper chromatography, a solvent is allowed to move up a strip of paper by capillary action. Solutions that contain the components to be separated have been placed on the paper in the form of spots. As the solvent moves through the sample spots, components that are more strongly attracted to the solvent than to the paper tend to dissolve in the solvent and move with it. Components that are more strongly attracted to the paper also move up the paper but more slowly than those components that are attracted to the solvent. As a result of the different rates of moving up the paper, the components are separated.

The identification of components can often be accomplished by using R_f values. This quantity is the ratio of the distance the component moved, divided by the distance the solvent moved during the same time:

$$R_f = \frac{\text{distance traveled by component}}{\text{distance traveled by solvent front}} \qquad \text{Eq. 29.1}$$

Amino acids, the components separated in this part of the experiment, are not only the building blocks of proteins but also occur in an uncombined form in numerous foods. Molecules of amino acids contain a carboxyl group and an amino group. Most amino acids of interest are α-amino acids, in which the amino group is attached to the α carbon and occupies the α position:

$$H_2N - \underset{\underset{R}{|}}{CH} - \overset{\overset{O}{||}}{C} - OH$$

amino group ↗   ↖ α carbon

an α-amino acid

Amino acids are colorless, and their migration on paper cannot be followed visually. However, upon treatment with ninhydrin, most of them form an intensely blue-colored product. This reaction, given below, is used to detect the amino acids after they have migrated and after the paper has dried:

ninhydrin   amino acid   blue product

$$ + \quad H_2N - \underset{\underset{R}{|}}{CH} - \overset{\overset{O}{||}}{C} - OH \quad \rightarrow \quad \text{(blue product)} \quad + \quad R - CH \quad + \quad CO_2 \quad + \quad H_2O $$

Eq. 29.2

Since a nitrogen atom is the only part of the blue product obtained from the amino acid, all amino acids containing a primary amino group react to give the same product. Proline, an amino acid containing a secondary amino group, reacts somewhat differently and yields a yellow-colored product.

PROCEDURE

1. Obtain an unknown amino acid mixture and record its identification (ID) number in Table 29.1.
2. Make four spotting capillaries by heating the middle of glass melting-point capillaries in the flame of a burner until the glass softens and then stretching the softened glass to form a narrow neck. Bend the narrowed glass until it breaks to form the tip of the spotting capillary. Your instructor will demonstrate the technique.

3. Obtain a sheet of chromatography paper from the reagent area.
4. Obtain a ruler from the stockroom, measure the chromatography paper, then trim it with a pair of scissors so the dimensions match those given in Figure 29.1(a). Note that the paper is widest at the top.
5. Put the trimmed chromatography paper on a piece of clean scratch paper to protect it from contamination by chemicals on the bench top.

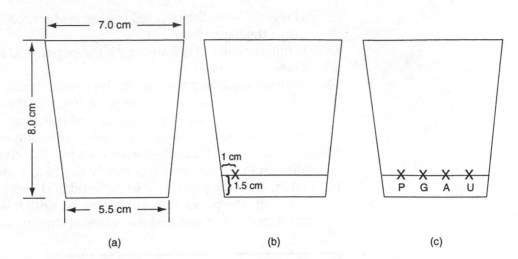

FIGURE 29.1
Preparation of
chromatography paper
(*not* actual size)

7.0 cm

8.0 cm

1 cm

X

1.5 cm

5.5 cm

X X X X
P G A U

(a) (b) (c)

6. Use a pencil (not a pen) to draw a line across the trimmed sheet of paper that is 1.5 cm from the bottom, as shown in Figure 29.1(b).

7. Measure from the edge of the paper and use a pencil to mark an X on the line of the paper that is 1 cm from the left edge. Mark three more Xs at 1-cm intervals, as shown in Figure 29.1(c). Use a pencil to label the paper with an identifying letter below each X as given below:

Sample	Identifying Letter
Proline	P
Glycine	G
Asparagine	A
Unknown	U

8. Use a pencil to mark one corner of a plastic weighing dish with a letter *P*. Mark a second corner with a *G*, and a third corner with an *A*. Take the marked weighing dish to the reagent area and put a 1-drop sample of 4% proline solution near the *P*, a 1-drop sample of 2% glycine solution near the *G*, and a 1-drop sample of 2% asparagine solution near the *A*.

9. Return to your work area and use a separate spotting capillary to put a spot of each amino acid solution obtained in Step 8 on the appropriate X of your chromatography paper. This is done by touching the spotting capillary to the drop of appropriate sample on the plastic weighing dish. The sample will be drawn into the capillary. Then, the capillary is touched to the chromatography paper, and liquid will leave the capillary and go onto the paper. Be sure to use a different capillary for each amino acid solution. Use a separate capillary to put a spot of your unknown solution on the X labeled with a *U*. Each spot should be no larger than 3 mm in diameter (about the size of this circle ◯). You might want to practice your spotting technique with your unknown and a piece of filter paper before you apply any spots to the X marks on your chromatography paper.

10. After spotting, allow the paper to air dry for 5 minutes.

11. While the spots are drying, put 3 mL of an 80% methanol solution in a 250-mL beaker and cover the mouth of the beaker with a watch glass.

12. After the spots are dry, put the spotted paper into the beaker, with the marked edge down. Cover the mouth of the beaker with a watch

glass and leave the covered beaker undisturbed until it is time to remove the paper (Step 14).

13. While the solvent is moving up the paper, you should do other parts of the experiment.

14. Without removing the watch glass, periodically check the level to which the solvent has risen up the paper in the beaker. When the solvent gets to within ½ cm of the top of the paper, remove the paper and use a pencil to mark the height of the solvent on the paper.

15. Put the chromatography paper on a piece of clean scratch paper and allow it to remain until it is nearly dry (4 to 8 minutes).

16. Then, hang the paper in a fume hood and spray it with ninhydrin solution. The paper should be sprayed until it is thoroughly moistened, but not so wet that the solution streams down the paper.

SAFETY ALERT

Ninhydrin will irritate skin and eyes. It will also stain tissue. Gloves should be worn when the spraying is being done.

17. After spraying, put the paper into a 100°C oven for 5 minutes. Be sure your paper is marked in some way so you can identify it.

18. Examine the dried paper for spots produced by the separated components. Measure and record in Table 29.1 the distance from the original sample position (the X) to the center of each spot. These distances are the component distances used in Equation 29.1. For the unknown, fill in the number of blanks corresponding to the number of spots on the developed chromatography paper.

19. Also measure and record the distance from the original sample position to the line indicating the height of the solvent on the paper. This value is the solvent distance of Equation 29.1.

DISPOSAL

Used solvent in beaker in container labeled "Exp. 29, Used Chromatography Solvent Part A"

B. Isolation of Casein

Casein, the principal protein of milk, can be precipitated by acidification of the milk. Some fat also precipitates but can be removed by washing the precipitate with alcohol.

PROCEDURE

1. Place 20 mL (20 g) of milk into a 125-mL flask.
2. Heat the milk to 40°C in a water bath heated with hot tap water.
3. Add 5 drops of glacial acetic acid and stir for about 1 minute.

SAFETY ALERT

Glacial acetic acid will vigorously attack tissue. If contact occurs, wash the contacted area with cool water.

4. Filter the resulting mixture through 4 layers of cheesecloth held in a funnel, and gently squeeze out most of the liquid.

5. Remove the solid (casein and fat) from the cheesecloth, place it into a 100-mL beaker, and add 10 mL of 95% ethanol.
6. Stir well to break up the product. Pour off the liquid (sink) and add 10 mL of a 1:1 ether–ethanol mixture to the solid.

SAFETY ALERT

> Ether is extremely flammable. No open flames should be used anywhere in the laboratory while this part of the experiment is being performed. Make certain no one is using ether before you begin using a flame for other parts of the experiment.

7. Stir well and filter through 4 layers of cheesecloth.

DISPOSAL Used 1:1 ether–ethanol solution in container labeled "Exp. 29, Used Ether–Ethanol Solution Part B"

8. Let the solid drain well, then scrape it onto a weighed filter paper and let it dry in the air.
9. Weigh the combined casein and filter paper and record the mass in Table 29.3 of the Data and Report Sheet. Some of the casein will be used in other parts of this experiment.

C. Amino Acid Structure and pH

Many amino acids are neutral in solution because they contain an equal number of acidic and basic functional groups. However, some amino acids have a side chain (R group) that contains an additional carboxylic acid group. A solution of such an acid would be acidic, and the term **acidic amino acid** is used to describe amino acids of this type. Glutamic acid is an example. **Basic amino acids** have an additional amino group in the side chain that causes solutions of the amino acid to be basic (alkaline). Lysine is an example.

In this part of the experiment, you will determine the pH of solutions of alanine, arginine, and aspartic acid.

PROCEDURE

1. Label four clean 10-cm test tubes a, b, c, and d.
2. Put 10 drops of distilled water into tube a. This tube will serve as a standard, or control. Put 10 drops of 4% alanine solution into tube b, 10 drops of 4% arginine into tube c, and 10 drops of 4% aspartic acid into tube d.
3. Determine the pH of the contents of each tube. Initially use wide-range pH paper to estimate the approximate pH of each sample. Remember to use only small pieces of pH paper (1.5 cm) for each measurement. Then, use the appropriate narrow-range pH paper to accurately determine the pH value. Record the accurately measured pH values in Table 29.5 of the Data and Report Sheet.

DISPOSAL 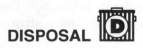 Tested amino acid solutions in sink

D. Amino Acids as Buffers

In solution amino acids exist in a dipolar form called the **zwitterion**. This zwitterion can react with both acids and bases:

$$H_3\overset{+}{N}-CH-\overset{O}{\overset{\|}{C}}-OH \quad \text{Eq. 29.3}$$
$$\overset{|}{R}$$

$$H_3\overset{+}{N}-CH-\overset{O}{\overset{\|}{C}}-O^- \xrightarrow{H^+}$$
$$\overset{|}{R}$$

dipolar ion
(zwitterion)

$$\xrightarrow{OH^-} H_2N-CH-\overset{O}{\overset{\|}{C}}-O^- + H_2O \quad \text{Eq. 29.4}$$
$$\overset{|}{R}$$

For this reason, solutions of amino acids make good buffers. In this experiment the buffering ability of alanine is demonstrated.

PROCEDURE

1. Label four 10-cm test tubes a, b, c, and d.
2. Place 2 mL of distilled water in tubes a and b and 2 mL of 4% alanine in tubes c and d.
3. Determine the pH of tubes a and c, using wide-range paper first and then narrow-range paper as you did in Step 3 of Part C. Record the wide-range values to the nearest whole pH unit and the narrow-range values to the tenth of a pH unit in Table 29.7.
4. Place the following reagents into the appropriate tubes. Be certain to use only NaOH and HCl solutions that come from containers labeled as 0.1%. These are specially-prepared very dilute solutions.

Tube	Reagent
a	10 drops of 0.1% NaOH
b	10 drops of 0.1% HCl
c	10 drops of 0.1% NaOH
d	10 drops of 0.1% HCl

5. Stir each mixture thoroughly with a clean, dry stirring rod.
6. Determine the pH of the resulting solutions by first using wide-range and then the appropriate narrow-range paper. Record the pH values as you did in Step 3.

DISPOSAL 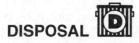 Tested amino acid solutions in sink

E. Color Tests of Proteins and Amino Acids

The presence of proteins in a solution is often detected by a general protein test or by specific tests that depend upon the presence of a specific amino acid. In this exercise, you will use a general color test (biuret test) and a specific test (Millon test) for proteins. Record the results in Table 29.8. Use (−) for no reaction; (+) for a weak, positive reaction; and (++) for a strong, positive reaction.

a. Biuret Test

The biuret test detects all proteins and requires the presence of two peptide linkages. When a protein in an alkaline solution is treated with copper sulfate, a pink-blue to violet color is formed.

PROCEDURE

1. Mix together in a 10-cm test tube 20 drops of 2% gelatin solution and 10 drops of dilute (6 M) sodium hydroxide (NaOH).

SAFETY ALERT

> The 6 M sodium hydroxide solution is corrosive and will attack tissue vigorously. If contact occurs, wash the contacted area with cool water.

2. Add 1 drop of 0.5% copper sulfate solution ($CuSO_4$) mix well, and record your observations in Table 29.8.
3. Mix together in a second test tube approximately 0.1 g of your isolated casein, 20 drops of water, and 10 drops of dilute (6 M) NaOH.
4. Mix well, then add 1 drop of 0.5% $CuSO_4$.
5. Record your observations.

DISPOSAL Test tube contents in container labeled "Exp. 29, Used Biuret Test Materials Part E"

b. Millon Test

Millon's reagent (a 2.2% solution of mercury nitrate, $Hg(NO_3)_2$, in 5.3 M nitric acid) is specific for the hydroxy–phenyl group. The positive test, indicated by a red solution or a red precipitate, is given only by proteins that contain tyrosine.

hydroxy–phenyl group   tyrosine

PROCEDURE

1. Add 5 drops of Millon's reagent to 2 mL of 2% gelatin solution in a 10-cm test tube, mix it, and heat it in a boiling-water bath for 1 to 2 minutes. Record your observations in Table 29.8.

SAFETY ALERT

> The Millon's reagent is toxic and corrosive. The reagent must be used with care. If contact occurs, wash the contacted area with soap and water.

2. Repeat Step 1 using about 0.1 g of your isolated casein suspended in 2 mL of water in place of the gelatin solution. Record the results.

 DISPOSAL Test tube contents in container labeled "Exp. 29, Used Millon Test Materials Part E"

F. Protein Denaturation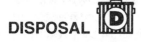

Proteins have rather complex structures in which the long chains of amino acids assume characteristic three-dimensional shapes. When a substance or condition disrupts the three-dimensional features of a protein's structure, the protein is said to be *denatured*. Denaturation usually causes a protein to lose its biological activity. In addition, it often makes the protein less soluble, leading to coagulation or precipitation. In this exercise, you will compare several common denaturing agents and processes.

PROCEDURE

1. Place 2 mL (40 drops) of 2% egg albumin in a 10-cm test tube to serve as a standard. The egg albumin is a protein.
2. Perform each of the following tests on an individual 2-mL sample of 2% egg albumin solution and record your observations in Table 29.10. Use (−) for no precipitation; (+) for slight precipitation (clouding of the solution); and (++) for heavy precipitation.

   a. *The effect of heat* Place a test solution in a boiling-water bath for 5 minutes. Compare the appearance of the tube with the unheated standard.

 DISPOSAL Test tube contents in sink

   b. *The effect of alcohol* Add 30 drops of isopropyl alcohol to a test sample, mix well, and compare to the standard.

 SAFETY ALERT Isopropyl alcohol is flammable. Make certain there are no open flames nearby when you do this test.

 DISPOSAL Test tube contents in sink

   c. *The effect of heavy metals* To separate test samples, add 2 drops of 5% HgCl$_2$ (mercury (III) chloride), 5% AgNO$_3$ (silver nitrate), and 5% Pb(C$_2$H$_3$O$_2$)$_2$ (lead acetate). Mix well and record your observations.

SAFETY ALERT The heavy metals are toxic and should be used with care. If contact occurs, wash the contacted area with soap and water.

 DISPOSAL Test tube contents in container labeled "Exp. 29, Used Heavy Metal Test Materials Part F"

A. Separation of Amino Acids Using Chromatography

1. Use the solvent distance and the component distances recorded in Table 29.1 and Equation 29.1 to calculate R_f values for the three amino acids and each component of the unknown. Record the R_f values in Table 29.2.
2. Use the R_f values of Table 29.2 to identify the component amino acids in your unknown. Record these results in Table 29.2.
3. Attach the chromatogram to the Data and Report Sheet in the indicated space below Table 29.1.

B. Isolation of Casein

1. Use the data in Table 29.3 to determine the mass of casein isolated. Record this result in Table 29.4.
2. Assume the 20 mL of milk weighed 20 g and calculate the weight–percentage of casein isolated from the milk:

$$\% \text{ Casein} = \frac{\text{grams of casein}}{\text{grams of milk}} \times 100$$

3. Record the calculated percentage in Table 29.4.

C. Amino Acid Structure and pH

1. Classify each amino acid tested as neutral, acidic, or basic on the basis of the data recorded in Table 29.5.
2. Record the classification in Table 29.6.

E. Color Tests of Proteins and Amino Acids

1. Use the test results recorded in Table 29.8 to decide whether or not gelatin and casein contain tyrosine.
2. Indicate your conclusions by writing either *yes* or *no* in each blank of Table 29.9.

Experiment 29 ▪ Pre-Lab Review

Amino Acids and Proteins

1. Are any specific safety alerts given in the experiment? List any that are given.

    _____

    _____

    _____

2. Are any specific disposal directions given in the experiment? List any that are given.

    _____

    _____

    _____

3. What should be used to mark the chromatography paper in Part A? _____

4. What technique is it suggested you might want to practice before you attempt to use it in Part A?

    _____

5. In Part A, what process is carried out in a fume hood? _____

    _____

6. How far should you allow the solvent to move up the paper strip in Part A?

    _____

7. How is fat removed from the precipitated casein in Part B? _____

    _____

    _____

8. Why do you need to weigh a piece of filter paper in Part B? _____

    _____

    _____

9. Describe the steps used to obtain an accurately measured pH value in Parts C and D.

    _____

    _____

10. What special precaution is given about the NaOH and HCl solutions used in Part D?

_____

_____

11. What change in a sample constitutes a positive result in each of the following tests?

Biuret test _____

Millon test _____

12. In Part F, what observation indicates that protein denaturation has taken place?

_____

_____

Experiment 29 ▪ Data & Report Sheet

Amino Acids and Proteins

A. Separation of Amino Acids Using Paper Chromatography

TABLE 29.1 (data)

Unknown ID number _____

Solvent distance _____

Sample	Component Distance
Proline	_____
Glycine	_____
Asparagine	_____
Unknown	_____
	_____
	_____

TABLE 29.2 (report)

Sample	R_f Value
Proline	_____
Glycine	_____
Asparagine	_____
Unknown	_____
	_____
	_____
Amino acids in unknown	_____

ATTACH CHROMATOGRAM BELOW

B. Isolation of Casein

TABLE 29.3 (data)

Mass of dry casein plus
filter paper _____

Mass of filter paper _____

TABLE 29.4 (report)

Mass of casein isolated _____

% casein isolated _____

C. Amino Acid Structure and pH

TABLE 29.5 (data)

Test Tube	Sample	Sample pH
a	Water	_____
b	Alanine solution	_____
c	Arginine solution	_____
d	Aspartic acid solution	_____

TABLE 29.6 (report)

Amino Acid	Classification
Alanine	_____
Arginine	_____
Aspartic acid	_____

D. Amino Acids as Buffers

TABLE 29.7 (data and report)

Test Tube	Test Tube Contents	Measured pH, Wide Range	Measured pH, Narrow Range
a	Water	_____	_____
a	Water + NaOH	_____	_____
b	Water + HCl	_____	_____
c	Alanine solution	_____	_____
c	Alanine solution + NaOH	_____	_____
d	Alanine solution + HCl	_____	_____

E. Color Tests of Proteins and Amino Acids

TABLE 29.8 (data)

Test Results

Test	Gelatin	Casein
Biuret	_____	_____
Millon	_____	_____

TABLE 29.9 (report)

	Gelatin	Casein
Contains tyrosine	_____	_____

F. Protein Denaturation

TABLE 29.10 (data and report)

Reagent or Condition	Results
Heat	_____
Isopropyl alcohol	_____
Heavy metals	
$HgCl_2$	_____
$AgNO_3$	_____
$Pb(C_2H_3O_2)_2$	_____

QUESTIONS

1. Why does ninhydrin stain the skin blue?

   a. Skin contains amino acids.
   b. Ninhydrin is blue-colored.
   c. Ninhydrin turns blue when warmed.

   Explain your answer: _____

   _____

2. Why would it be unwise to use an ink pen to mark the reference lines and points on the chromatography paper in Part A?

    a. The ink contains amino acids.
    b. The ink would move with the solvent.
    c. The ink would spread as it was applied onto the paper.

    Explain your answer: _____

    _____

3. Suppose you want to separate two amino acids by the technique described in Part A, but you find that they have identical R_f values. Which of the following would probably help solve the problem?

    a. Change solvents.
    b. Use longer pieces of paper and increase the spot migration distance.
    c. Use larger samples of amino acids.

    Explain your answer: _____

    _____

4. Suppose that while performing Part B, you weigh the casein and filter paper before they are completely dry. How will this error influence the reported percentage of casein?

    a. Increase it.   b. Decrease it.   c. Have no effect on it.

    Explain your answer: _____

    _____

5. Which of the following accounts for the smaller pH change that resulted when HCl was added to the alanine solution compared to the pH change when HCl was added to water? (See Table 29.7)

    a. Some of the added H^+ was removed from solution by the alanine.
    b. The added H^+ remained free in water and increased the pH.
    c. The added H^+ remained free in the alanine solution and decreased the pH.

    Explain your answer: _____

    _____

6. According to advertisements, a commercial hair dressing contains "protein." Which of the following would be the best test to perform on the product to check this claim?

    a. The biuret test   b. The Millon test

    Explain your answer: _____

    _____

7. In which of the following processes is protein denaturation likely to play an important role?

    a. In sterilizing instruments in boiling water   c. In disinfecting the skin with alcohol
    b. In preserving meat by freezing   d. More than one answer is correct.

    Explain your answer: _____

    _____

Experiment 30

Enzymes: Nature's Catalysts

In this experiment, you will

- Use enzymes to carry out a reaction of commercial importance.
- Observe some of the diverse types of reactions catalyzed by enzymes.

- Gain experience in the use of several different techniques used to detect or observe enzyme activity.

INTRODUCTION

Enzymes, the catalytic proteins found in all plants and animals, account for the remarkable efficiency that is characteristic of chemical reactions in living organisms. A given enzyme ordinarily catalyzes reactions of a very specific nature. However, many different enzymes are available, and thus many different reactions are catalyzed. Typical enzyme-catalyzed reactions of living systems are hydrolysis, polymerization, oxidation-reduction, dehydration, and transfer reactions.

EXPERIMENTAL PROCEDURE

A. Preparation of Cheese

Cheese is the solid portion (curd) of milk which has been separated from the liquid portion (whey). Cheese is prepared by adding an enzyme called *rennin* to milk. Rennin, isolated from calf stomachs or microbial sources, catalyzes the coagulation of milk and, hence, the formation of curd.

During cheesemaking, the curd is cut into small pieces, filtered to remove whey, and salted. It is then pressed in a cheese press to remove more whey and harden the cheese. The resulting block of cheese is wrapped to exclude air and allowed to age in order to develop its flavor.

Rennin is most effective in acidic solutions. Accordingly, the regular pasteurized milk used in this experiment will be treated with a starting culture that contains lactic acid-forming bacteria. Part of the lactose in the treated milk is converted to lactic acid, and the pH is lowered. The milk you will use has been treated by adding 20 mL of buttermilk to each quart of milk and allowing the mixture to stand for at least 10 minutes at 25°C.

PROCEDURE

1. Put 20 mL of treated milk into a 100-mL beaker and heat it with constant, slow stirring to 37°C. It is important to not overheat the milk, so carry this step out carefully.
2. When the temperature of the milk reaches 37°C, discontinue the heating and add 8 drops of 5% rennin solution to the warm milk. Stir the resulting mixture well.

3. Allow the mixture to stand undisturbed for 15 minutes or until the milk assumes a gel-like consistency.
4. Break the resulting curd into pieces, but don't stir it vigorously. Filter the mixture between two layers of cheesecloth (about 12 cm square) to remove as much of the liquid whey as you can.
5. Put the isolated curd into a previously weighed 50-mL beaker. Weigh the curd-containing beaker on the same centigram or electronic balance used to weigh the beaker. Record the masses in Table 30.1 of the Data and Report Sheet.

DISPOSAL  Liquid filtrate in sink
Isolated curd in wastebasket

B. Lipid Digestion

Lipases catalyze the hydrolysis of the ester linkages in glycerides and produce fatty acids and glycerol. These enzymes, which are widely distributed in nature, are especially important in the digestion of fats and oils. Since fatty substances are not water soluble, the enzymes can act only on the lipid surface that is available at the water–lipid interface. Consequently, **emulsifying agents,** such as the bile salts, often enhance the enzymatic action of lipases by dispersing the lipids and providing an increased surface area at the water–lipid interface.

Homogenized milk contains fat globules in a finely divided state (large total surface area) and thus serves as an excellent substrate for lipases. In this exercise, the action of pancreatin on milk fat (butterfat) will be investigated. *Pancreatin* is a very potent enzyme mixture that contains lipase. The milk sample to be used contains litmus, an acid–base indicator that shows a characteristic red color in acid and blue color in base. The accumulation of free fatty acids gradually reduces the pH of the milk. This process is indicated by a color change of the indicator.

PROCEDURE

1. Put 3 mL of litmus milk into each of two 10-cm test tubes.
2. Add 20 drops of 5% pancreatin suspension to one tube and 20 drops of distilled water to the other tube. The milk in both tubes should still be blue-colored.
3. Put both tubes into a warm-water bath maintained at 30 to 40°C. Maintain the temperature by adding hot tap water to the bath as needed. If necessary, use a plastic dropper to remove water from the bath before adding hot water.
4. Allow the tubes to remain in the bath for 30 minutes, then observe their color.
5. Record the observed color of each tube in Table 30.3 of the Data and Report Sheet.

DISPOSAL  Test tube contents in sink

C. Protein Digestion

Proteases are enzymes capable of cleaving the peptide linkages of proteins. Gelatin, a protein prepared from collagen, is readily hydrolyzed by pepsin and trypsin, two proteases of the digestive system. The ability of certain plant enzymes to hydrolyze gelatin is the basis for such warnings as "Do not use fresh or frozen pineapple" often found on the back of gelatin dessert packages. Hydrolysis of gelatin to smaller peptides and amino acids would destroy the colloidal properties of the protein and prevent the desired gel from forming. Canned pineapple is suitable for use in gelatin desserts because the heat treatment used in the canning process deactivates the proteolytic enzymes. Bromelain, the protease found in pineapple, is used commercially in some meat tenderizers.

The process of gel formation in gelatin provides us with a visual method for demonstrating the proteolytic properties of pepsin, trypsin, and bromelain. If gelation occurs in the presence of a proteolytic enzyme, it means that the enzyme is not active (that is, it may not be at the optimum pH). If gelation does not occur, it means the enzyme is active and is hydrolyzing the gelatin under those conditions. The heat stability and pH dependence of these digestive enzymes will also be shown.

PROCEDURE

1. Label four 15-cm test tubes and prepare their contents according to the following:

Concept	Tube	Tube Contents
Control	1	1 mL H_2O
Effect on gelatin	2	1 mL of 1% buffered pepsin solution
	3	1 mL of 1% buffered trypsin solution
	4	1 mL of 4% meat tenderizer

2. Add to each tube 1 mL (20 drops) of double-strength gelatin dessert solution (made according to package directions, but with one-half the normal amount of water and allowed to cool to about 50°C). The gelatin solution should be warm and in a liquid state. If it is not, check with your lab instructor.
3. Mix the contents well by agitation and allow the tubes to stand for 10 minutes.
4. After 10 minutes, put the tubes in a ice-and-water bath. Leave them in the ice–bath until complete gelation occurs in tube 1 (indicated by failure of the mixture to flow when the test tube is tilted almost to horizontal).
5. When tube 1 has gelled, examine tubes 2, 3, and 4, and record in Table 30.4 whether or not complete gelation has occurred. If complete gelation has not occurred, estimate the extent of gelation as slight, significant, and so on.

DISPOSAL  Test tube contents in sink

D. Sucrose Digestion

Upon hydrolysis, the disaccharide sucrose is converted into equal amounts of glucose and fructose. During digestion this process is catalyzed by *sucrase*, a hydrolase contained in the natural secretions of the small intestine. Yeast cells will serve as the enzyme source in this exercise. The reaction can be followed by using a positive Benedict's test that is characteristic of the products (glucose and fructose) but not of the reactant (sucrose).

PROCEDURE

1. Put 2 mL (40 drops) of 5% sucrose solution into a 10-cm test tube and 2 mL of distilled water (as a control) in another 10-cm test tube.
2. To each tube, add 2 mL (40 drops) of yeast suspension (swirl the bottle to mix the contents before you get your samples of yeast suspension).
3. Heat both tubes for 15 minutes in a water bath maintained at 38 to 40°C. Maintain the bath temperature by adding hot tap water.
4. After 15 minutes, remove the tubes from the bath and allow them to cool to room temperature. Then, test the contents of each tube by adding 8 drops of solution from the tube to a 10-cm test tube that contains 2 mL (40 drops) of Benedict's solution.
5. Heat the test tubes containing the Benedict's solution and samples in a boiling-water bath for 5 minutes.
6. Note and record in Table 30.5 the formation of any yellow-to-red precipitate, which indicates a positive Benedict's test.

DISPOSAL

Test tubes that contain Benedict's solution in container labeled "Exp. 30, Used Benedict's Test Materials Part D"
Other test tubes in sink

E. Hydrolysis of Urea

Urease is a plant and microbial enzyme that exhibits absolute specificity. It catalyzes the hydrolysis of urea and will not act on any other substrate.

$$H_2N - \overset{\overset{\displaystyle O}{\|}}{C} - NH_2 + H_2O \xrightarrow{\text{urease}} 2NH_3 + CO_2 \qquad \text{Eq. 30.1}$$
$$\text{urea}$$

The action of urease may be demonstrated by the detection of ammonia (NH_3) released during the hydrolysis reaction. Ammonia forms basic solutions in water and can be detected by the pH indicator phenol red. This indicator changes color from yellow-orange to reddish pink as the alkalinity of a solution increases. In this part of the experiment, you will observe the action of urease on urea.

PROCEDURE

1. Put 3 drops of 1% urease solution and 7 drops of distilled water into a clean 10-cm test tube.
2. Add 1 drop of phenol red indicator to the tube and mix well by agitation.
3. Add 4 drops of 5% urea solution to the test tube and mix.
4. Note and record in Table 30.6 the rapid color change that occurs.

DISPOSAL   Test tube contents in sink

F. Catalase from Potatoes

Catalase is an enzyme found in most living cells. It is highly specific and acts on no natural substrate other than hydrogen peroxide, which it causes to decompose into water and oxygen gas:

$$2H_2O_2 \xrightarrow{\text{catalase}} O_2 + 2H_2O \qquad \text{Eq. 30.2}$$

Catalase is one of the most active enzymes known; one molecule catalyzes the decomposition of 5 million H_2O_2 molecules per second.

Hydrogen peroxide is produced by a number of cellular oxidation reactions, and catalase is thought to be present in cells to prevent damage from an accumulation of H_2O_2.

PROCEDURE

1. Obtain a piece of freshly cut potato with approximate dimensions of 1 cm x 0.3 cm x 0.3 cm. You might have to cut the piece from slices of freshly peeled potato provided in the lab.

SAFETY ALERT

> Use caution to avoid cutting yourself if you cut your own piece of potato.

2. Put the piece of potato into a 10-cm test tube, and add 2 mL (40 drops) of 3% hydrogen peroxide (H_2O_2). Note and record in Table 30.7 the evolution of any gas.

DISPOSAL

Liquid test tube contents in sink
Rinse the piece of potato with water, then put in wastebasket

CALCULATIONS AND REPORT

A. Preparation of Cheese

1. Use the data in Table 30.1 and determine the mass of curd isolated. Record this result in Table 30.2.
2. Assume that milk has a density of 1.0 g/mL. Use this value, the volume of milk, and the mass of curd determined in Step 1 to calculate the percentage yield of curd:

$$\% \text{ Yield} = \frac{\text{mass of curd}}{\text{mass of milk}} \times 100$$

3. Record the calculated percent yield in Table 30.2.

Experiment 30 ▪ Pre-Lab Review

Enzymes: Nature's Catalysts

1. Are any specific safety alerts given in the experiment? List any that are given.

   _____

2. Are any specific disposal directions given in the experiment? List any that are given.

   _____

   _____

   _____

3. Why do you weigh an empty 50-mL beaker in Part A? _____

   _____

4. How do you maintain the temperature of the water bath used in Parts B and D?

   _____

5. In Part B, what observation indicates that fat has been hydrolyzed? _____

   _____

6. In Part C, what observation is used to indicate the activity of a protease enzyme?

   _____

7. In Part D, what observation indicates a positive result for a Benedict's test?

   _____

8. In Part E, what observation indicates that urea has been hydrolyzed? _____

   _____

9. What substance produced during urea hydrolysis is responsible for the observation you described in Question 8?

   _____

10. In Part F, what observation is used to detect the activity of catalase? _____

    _____

    _____

Experiment 30 ▪ Data & Report Sheet

Enzymes: Nature's Catalysts

A. Preparation of Cheese

TABLE 30.1 (data)

Mass of curd + beaker _____

Mass of empty beaker _____

TABLE 30.2 (report)

Mass of curd _____

% yield of curd _____

B. Lipid Digestion

TABLE 30.3 (data and report)

Tube Contents	Observations
Milk + water	_____
Milk + pancreatin	_____

C. Protein Digestion

TABLE 30.4 (data and report)

Tube Number	Tube Contents	Extent of Gelation
1	Gelatin + water	_____
2	Gelatin + buffered pepsin	_____
3	Gelatin + buffered trypsin	_____
4	Gelatin + meat tenderizer	_____

D. Sucrose Digestion

TABLE 30.5 (data and report)

Tube Contents	Result of Benedict's Test
Sucrose + yeast	_____
Water + yeast	_____

E. Hydrolysis of Urea

TABLE 30.6 (data and report)

Reactants	Observations
Urease + phenol red + urea	

F. Catalase from Potatoes

TABLE 30.7 (data and report)

Reactants	Observations
Potato + H_2O_2	

QUESTIONS

1. Suppose in Part A, Step 4, one student squeezes the curd more thoroughly than another student. How would this influence the calculated percentage yield of the two students?

   a. The thorough squeezer's percentage would be higher.
   b. The thorough squeezer's percentage would be lower.
   c. The thoroughness of squeezing should not influence the percentage.

   Explain your answer: _____

   _____

   _____

2. What observation recorded in Table 30.3 indicated that butterfat in the milk had been hydrolyzed?

   a. Milk color changed from pink to blue.
   b. Milk color changed from blue to pink.
   c. Milk color remained unchanged.

   Explain your answer: _____

   _____

   _____

3. Which of the following observations from Part C of the experiment indicates the greatest activity for the proteolytic enzyme such as pepsin or trypsin?

   a. Gelation occurs rapidly.
   b. Only partial gelation occurs during an hour.
   c. No gelation occurs during an hour.

   Explain your answer: _____

   _____

   _____

4. Suppose in Part E, Step 2, a student gets a red solution when 1 drop of phenol red indicator is added to the tube. Which of the following would explain this result.

    a. The test tube was dirty and contained a little hydrochloric acid.
    b. The test tube was dirty and contained a little sodium hydroxide.
    c. The test tube was dirty and contained a little urease.

    Explain your answer: _____

    _____

    _____

5. A student does Part F of the experiment and notices the evolution of a gas when the hydrogen peroxide was added. What is the gas?

    a. Water vapor given off when the reaction mixture got hot
    b. Oxygen produced by the reaction
    c. Hydrogen produced by the reaction
    d. More than one answer is correct.

    Explain your answer: _____

    _____

    _____

6. Which of the following would you expect to cause the evolution of a gas when added to hydrogen peroxide?

    a. A piece of raw carrot
    b. A piece of raw cabbage
    c. A piece of raw radish
    d. More than one answer is correct.

    Explain your answer: _____

    _____

    _____

Factors That Influence Enzyme Activity

In this experiment, you will

- Prepare a potato extract that contains the enzyme polyphenoloxidase.
- Study the specificity of polyphenoloxidase by attempting to react it with structurally similar compounds.

- Determine the effect of several variables on the activity of polyphenoloxidase.

INTRODUCTION

Enzymes serve one of the most important functions of all proteins. They catalyze the vital chemical reactions that occur in biological organisms. Enzymes speed up both simple and complex reactions that would otherwise take place much too slowly for life to continue.

Various factors of the cellular environment, such as temperature and pH, directly influence the rates of enzymatic reactions. The activity of an enzyme may be measured by monitoring the catalyzed reaction at fixed time intervals.

An **oxidase** is an enzyme that catalyzes the transfer of hydrogen from some compound to molecular oxygen. Polyphenoloxidase is a copper-containing enzyme that catalyzes the oxidation (hydrogen removal) of dihydroxy phenols to the corresponding quinones:

$$2 \underset{\text{catechol}}{\text{[OH, OH benzene ring]}} + O_2 \xrightarrow{\text{polyphenoloxidase}} 2 \underset{\text{benzoquinone}}{\text{[O, O quinone ring]}} + 2H_2O \qquad \text{Eq. 31.1}$$

This type of oxidation is accompanied by a color change since the product quinones are colored. The reaction commonly occurs in nature and is responsible for the browning of peeled potatoes and bruised fruit. The color change will be used in this experiment as a way to measure the extent of the reaction.

EXPERIMENTAL PROCEDURE

A. Preparation of an Enzyme Extract

The source of polyphenoloxidase for this experiment is a potato. Retain your potato extract until you have completed all parts of the experiment.

PROCEDURE

1. Obtain about 10 g of peeled potato.
2. Cut the potato into small pieces and place the pieces into a mortar.
3. Add 5 mL of distilled water and about 5 g of clean sand to the mortar.
4. Grind the mixture thoroughly and rinse it into a 100-mL beaker using 30 mL of 2% sodium fluoride (NaF) solution.

SAFETY ALERT

Sodium fluoride (NaF) is toxic and should be used with care. Be sure to wash any contacted area with cool water.

5. Let the mixture stand for 2 minutes and then pour it into a 100-mL beaker through 4 layers of cheesecloth held in a funnel. The liquid potato extract contains polyphenoloxidase.

DISPOSAL

Used cheesecloth and potato residue in funnel: rinse with tap water and put in wastebasket

B. Enzyme Specificity

Enzymes are often quite specific in their activity, catalyzing a reaction only if the proper substrate is present. In this exercise, you will study the specificity of polyphenoloxidase by providing various substrates that have structural similarities.

PROCEDURE

1. Prepare four labeled 10-cm test tubes with contents as follows:

Tube	Contents
a	20 drops of distilled water
b	20 drops of 0.01 M catechol solution
c	20 drops of 0.01 M phenol solution
d	20 drops of 0.01 M 1,4-cyclohexanediol solution

SAFETY ALERT

The three substrates—catechol, phenol, and 1,4-cyclohexanediol—are used in dilute solutions and so represent a minimal safety problem. But, all should be used with care because they are toxic.

The tube containing water will serve as a control in which no reaction takes place. The molecular structures of the experimental substrates are

catechol   phenol   1,4-cyclohexanediol

2. Place all the tubes into a water bath maintained at 37°C. Use a 250-mL beaker half filled with water for the bath.

3. Obtain four 10-cm test tubes and put 30 drops of potato extract into each of the four tubes. Place these tubes into the 37°C water bath.

4. After the tubes with extract have been in the water bath for 5 minutes, quickly empty the contents of one extract-containing tube into each labeled tube. Mix, and leave the labeled tubes in the bath for another 5 minutes.

5. After 5 minutes, remove the labeled tubes from the water bath. Compare the color of tubes b, c, and d to the color of tube a. The best way to detect and compare colors is to hold the tubes vertically over a piece of white paper, and look down into the tubes. Record in Table 31.1 the intensity of brown color that developed in each tube, using the following scale and symbols:

Most intense color (darkest):	Next most intense:	Least intense:	No color:
+++	++	+	−

If more than one tube shows the same intensity, assign them the same symbol. The development of a brown color is visual evidence of the oxidation process that was catalyzed by polyphenoloxidase.

DISABLE  **DISPOSAL**

Test tube contents in container labeled "Exp. 31, Used Chemicals Part B"

C. Substrate Concentration

The rate of an enzyme-catalyzed reaction increases as the substrate concentration increases until a limiting rate is reached. Beyond this point, the rate is independent of increases in the substrate concentration.

PROCEDURE

1. Obtain four 10-cm labeled test tubes, and check them to make sure they are identical in terms of their diameter and brand name.

2. Label the four test tubes to indicate the number of drops of catechol solution they will contain from the following list, then put the appropriate amount of catechol solution in each tube.

Tube	Contents
a	40 drops of 0.01 M catechol solution
b	20 drops of 0.01 M catechol solution + water
c	10 drops of 0.01 M catechol solution + water
d	5 drops of 0.01 M catechol solution + water

3. Use a dropper to add distilled water carefully to tubes b, c, and d. Add enough distilled water to raise the liquid level in each tube until it exactly matches the liquid level in tube a.

4. Place these four labeled tubes into a 37°C bath.

5. Obtain four more 10-cm test tubes and put 20 drops of potato extract into each one. Place these tubes into the 37°C bath for 5 minutes.

6. After the extract-containing tubes have been in the bath for 5 minutes, quickly empty the contents of one tube into each labeled tube. Mix and leave the labeled tubes in the water bath for another 2 minutes.

7. After 2 minutes, remove the labeled tubes from the water bath. Note and record in Table 31.2 the intensity of color that has developed in each tube. Use the scale and symbols given in Part B.

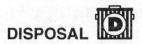

DISPOSAL Test tube contents in container labeled "Exp. 31, Used Chemicals Part C"

D. Enzyme Concentration

In all practical cases, the enzyme concentration is much lower than the concentration of the substrate. As a result, the rate of an enzymatic reaction is always directly dependent on the enzyme concentration.

PROCEDURE

1. Obtain three 10-cm test tubes, and check them to make certain they are identical in terms of their diameter and brand name.
2. Label the three test tubes to indicate the number of drops of potato extract they will contain from the following list, then put the appropriate amount of potato extract in each tube.

Tube	Contents
a	15 drops of potato extract
b	5 drops of potato extract + water
c	1 drop of potato extract + water

3. Use a dropper to add distilled water very carefully to tubes b and c. Add enough distilled water to raise the liquid level in each tube until it exactly matches the liquid level in tube a.
4. Place these three tubes into a 37°C water bath. If necessary check the bath temperature every few minutes to maintain the 37°C temperature.
5. Put 10 drops of 0.01 M catechol solution into each of three other 10-cm test tubes. Place these tubes into the water bath for 5 minutes.
6. After 5 minutes, quickly pour the catechol solution from one 10-cm tube into each of the labeled tubes. Agitate to mix and leave the labeled tubes in the water bath for another 5 minutes.
7. After 5 minutes, remove the labeled tubes from the water bath. Note and record in Table 31.3 the intensity of any color that develops. Use the scale and symbols given in Part B.

DISPOSAL Test tube contents in container labeled "Exp. 31, Used Chemicals Part D"

E. Effect of pH

The pH exerts a significant influence on enzyme activity. Most enzymes are active over a relatively narrow pH range. The pH at which the enzyme activity is at its maximum is called the **optimum pH** for the enzyme. Activity falls off sharply when the pH changes to a value on either side of the optimum value.

PROCEDURE

1. Prepare four labeled 10-cm test tubes with contents as follows:

Tube	Contents
a	20 drops of 0.1 M hydrochloric acid (HCl)
b	20 drops of pH 4 buffer
c	20 drops of pH 7 buffer
d	20 drops of 0.1 M sodium carbonate solution (Na_2CO_3)

These give pH values of approximately 1, 4, 7, and 10, respectively.

2. Add 10 drops of 0.01 *M* catechol solution to each tube.
3. Add 10 drops of potato extract to each tube.
4. Agitate to mix well, then place the tubes in a 37°C water bath.
5. Wait 10 minutes, then examine each tube for color changes in the solution. Record the results in Table 31.4, using the scale and symbols of Part B.

DISPOSAL Test tube contents in container labeled "Exp. 31, Used Chemicals Part E"

F. Effect of Temperature

A rough rule of thumb that holds for many chemical reactions is that a 10°C rise in temperature approximately doubles or triples the reaction rate. To a certain extent, this is also true for enzymatic reactions. After a certain point, however, an increase in temperature causes a decrease in the reaction rate because the enzyme begins to denature. The temperature of maximum enzyme activity is called the **optimum temperature.** Most enzymes of warm-blooded animals have optimum temperatures near the normal body temperature of about 37°C. In the following procedure, pairs of students should collaborate to maintain the 37°C and 70°C baths. One member of each pair should maintain one of the baths. Both members of each pair should make their own ice bath, and each member should perfom all parts of the experiment.

PROCEDURE

1. Label three 10-cm test tubes a, b, and c.
2. Add 10 drops of potato extract to each tube.
3. Place the tubes in the following temperature baths for 10 minutes:

Tube	Temperature
a	0°C (an ice bath)
b	37°C
c	70°C

Each bath should be maintained at those temperatures. Tube c will cool off especially fast if not maintained.
4. Put 10 drops of 0.01 *M* catechol solution into each of three other 10-cm test tubes. Place one of these tubes into each of the water baths. Leave them there until the labeled tubes have been in for the proper amount of time.
5. As quickly as possible, mix and agitate the contents of the two tubes in each bath. Leave the tube containing the final mixture in each temperature bath.
6. Wait 5 minutes, then examine each tube for the development of a color. Record the results in Table 31.5, using the scale and symbols of Part B.

DISPOSAL Test tube contents in container labeled "Exp. 31, Used Chemicals Part F"

G. Inhibitors

Inhibitors decrease or destroy the ability of an enzyme to catalyze reactions. Because enzymes are proteins, they are inhibited by substances that denature proteins. In addition, any agent that combines with a necessary cofactor can function as an inhibitor. In this exercise, the effect of the following inhibitors will be demonstrated:

Trypsin	An enzyme that digests other proteins (including other enzymes)
Phenylthiourea	Combines with the copper ion cofactor
Lead nitrate	A heavy metal denaturant

PROCEDURE

1. Label four 10-cm test tubes a, b, c, and d.
2. Add 10 drops of potato extract to each tube.
3. Add the following to the tubes:

Tube	Contents
a	10 drops of distilled water
b	10 drops of 5% trypsin suspension
c	10 drops of phenylthiourea (saturated solution)
d	10 drops of 5% lead nitrate ($Pb(NO_3)_2$) solution

SAFETY ALERT

> Phenylthiourea and lead nitrate are toxic and must be used with care. Wash any contacted skin with soap and water. Wash hands after lab before eating and so on.

4. Place the four tubes into a 37°C water bath and leave them there for a minimum of 10 minutes.
5. After 10 minutes, add 10 drops of 0.01 M catechol solution to each tube and mix well. Leave the labeled tubes in the water bath for another 5 minutes.
6. Examine the tubes for the development of a color in the solution. If a precipitate forms, allow it to settle before observing the color. Record the results in Table 31.6, using the scale and symbols of Part B.

DISPOSAL Test tube contents in container labeled "Exp. 31, Used Chemicals Part G"
Unused potato extract in sink

REPORT

Parts B to E

1. Plot the data recorded in Tables 31.2, 31.3, 31.4, and 31.5 on the appropriate graphs of the Data and Report Sheet.
2. Connect the resulting points in each graph with a smooth (but not necessarily straight) line to form a graphical representation of enzyme behavior.

Experiment 31 ▪ Pre-Lab Review

Factors That Influence Enzyme Activity

1. Are any specific safety alerts given in the experiment? List any that are given.

   _____

   _____

2. Are any specific disposal directions given in the experiment? List any that are given.

   _____

   _____

3. What observation is used in all parts of the experiment to indicate that polyphenoloxidase has acted on a substrate?

   _____

   _____

4. What serves as the source of polyphenoloxidase in all parts of the experiment?

   _____

5. In Parts C and D, what special directions are given about selecting test tubes to use?

   _____

   _____

6. In Parts C and D, how much distilled water is used to dilute the samples in some of the tubes?

   _____

   _____

7. In which part of the experiment is more than one water bath required? Why is more than one needed? What special directions are given?

_____

_____

_____

8. In Part G, what observation would indicate that an inhibitor had prevented polyphenol-oxidase from functioning as an enzyme?

_____

_____

Experiment 31 ▪ Data & Report Sheet

Factors That Influence Enzyme Activity

B. Enzyme Specificity

TABLE 31.1 (data and report)

Tube	Contents	Color Intensity
a	Water	_____
b	Catechol	_____
c	Phenol	_____
d	1,4-Cyclohexanediol	_____

C. Substrate Concentration

TABLE 31.2 (data)

Tube	Relative Substrate Concentration	Color Intensity
a	8	_____
b	4	_____
c	2	_____
d	1	_____

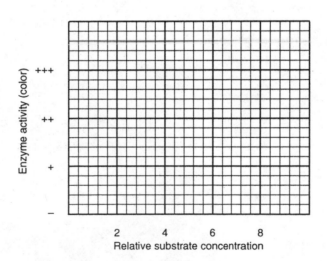

D. Enzyme Concentration

TABLE 31.3 (data)

Tube	Relative Enzyme Concentration	Color Intensity
a	15	_____
b	5	_____
c	1	_____

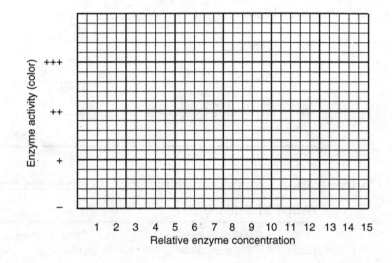

E. Effect of pH

TABLE 31.4 (data)

Tube	Contents	pH	Color Intensity
a	HCl	1	_____
b	Buffer	4	_____
c	Buffer	7	_____
d	Sodium carbonate	10	_____

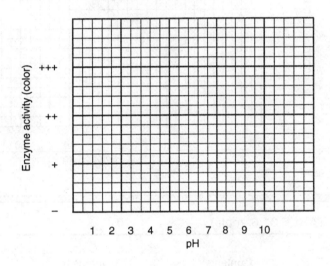

F. Effect of Temperature

TABLE 31.5 (data)

Tube	Approximate Temperature	Color Intensity
a	0°	_____
b	37°	_____
c	70°	_____

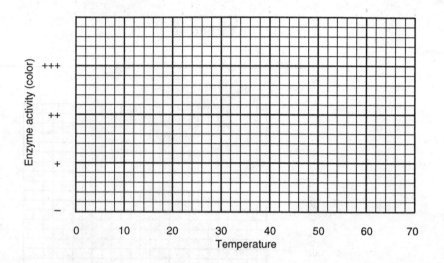

G. Inhibitors

TABLE 31.6 (data)

Tube	Contents	Color Intensity
a	Water	_____
b	Trypsin	_____
c	Phenylthiourea	_____
d	Lead nitrate	_____

QUESTIONS

1. According to the data recorded in Table 31.1, which of the following structural features of molecules appears to be necessary in a substrate for polyphenoloxidase?

   a. An OH group attached to a six-membered ring
   b. An OH group attached to an aromatic ring
   c. Two OH groups attached to an aromatic ring

   Explain your answer: _____

   _____

2. The results represented by the graph of enzyme activity versus substrate concentration indicate that

   a. The enzyme activity would probably increase if the relative substrate concentration were increased beyond 8.
   b. The enzyme activity would probably decrease if the relative substrate concentration were increased beyond 8.
   c. The enzyme activity would probably change very little if the relative substrate concentration were increased beyond 8.

   Explain your answer: _____

   _____

3. On the basis of the results represented by the graph of enzyme activity versus enzyme concentration, how would the enzyme activity change if the relative enzyme concentration were increased to a value of 20?

   a. Increase   b. Decrease   c. Remain essentially unchanged

   Explain your answer: _____

   _____

4. According to the graph of enzyme activity versus pH, which of the following pH values is closet to the optimum pH for polyphenoloxidase?

   a. 2   b. 3   c. 7   d. 10

   Explain your answer: _____

   _____

5. Which of the following processes is most likely to take place as a result of the influence of temperature on enzyme activity, and *not* because of enzyme denaturation?

   a. Preserving food by freezing
   b. Increasing the cooking rate of food by using a pressure cooker
   c. Preventing peeled fruit from turning brown by adding ascorbic acid (an easily oxidized substance)

   Explain your answer: _____

   _____

6. On the basis of the data in Table 31.6, which of the following functioned best as an enzyme inhibitor?

a. Trypsin   c. Lead nitrate
b. Phenylthiourea   d. More than one response is correct.

Explain your answer: _____

_____

7. In Part G, it is likely that a precipitate formed in some test tubes. It is likely that the precipitate was

a. denatured enzyme.
b. digested enzyme.
c. inhibitor combined with cofactor.

Explain your answer: _____

_____

_____

Experiment 32

Vitamin C Content of Foods, Part I

In this experiment, you will

- Determine the vitamin C content of an unknown.
- Compare the vitamin C content of related food samples.

INTRODUCTION

Vitamin C (ascorbic acid) is one of the more abundant vitamins; it is found in significant amounts in a wide variety of fresh fruits and vegetables including citrus fruits, cabbage, tomatoes, lettuce, strawberries, and broccoli. The currently accepted reference daily intake (RDI) of vitamin C is 60 mg/day.

Pure vitamin C is a colorless, water-soluble solid. It is somewhat stable in acidic solutions; but in neutral or alkaline solutions, it is rapidly destroyed by air oxidation. This tendency toward oxidation makes it difficult to retain the vitamin in preserved food. The vitamin is best retained in fresh foods by protecting the foods as much as possible from exposure to air. Therefore, the natural protective coverings of fresh fruits and vegetables should be left in place until just before the foods are eaten.

The ease with which vitamin C is oxidized forms the basis for the analysis procedure. The vitamin is titrated with a solution of *N*-bromosuccinimide (NBS), a mild oxidizing agent.

$$\text{vitamin C} + \text{N-bromosuccinimide} \rightarrow \text{dehydroascorbic acid} + \text{succinimide} + HBr$$

Eq. 32.1

The titration end point is detected by adding potassium iodide (KI) and starch to the mixture being titrated. When all the vitamin C has been oxidized, the addition of a slight excess of NBS causes the KI to be oxidized to free iodine (I_2), which forms a blue complex with starch:

$$2KI \xrightarrow{\text{NBS}} I_2$$

Eq. 32.2

$$I_2 + \text{starch} \rightarrow \text{blue complex}$$

Eq. 32.3

A. Standardization of Oxidizing Agent

The stock solution of *N*-bromosuccinimide is somewhat unstable, and it is therefore necessary to standardize it before use. This is accomplished by titrating a known amount of vitamin C.

PROCEDURE

1. Into a clean, dry 15-cm test tube, pour about 15 mL of standard vitamin C solution containing 0.50 mg/mL. Record this concentration in Table 32.1.
2. Pour about 150 mL of NBS solution into a clean, dry 250-mL beaker. This solution will be used for all parts of the experiment.
3. Use a 10-mL pipet and place 10.00 mL of vitamin C solution into a 125-mL flask. Record this volume in Table 32.1.

SAFETY ALERT

> The vitamin C solution contains 1.3% oxalic acid as a preservative. Oxalic acid is toxic. Be sure to use a pipet bulb. Do *not* pipet by mouth. Wash your hands after lab before eating and so on.

4. Add 5 mL of 4% potassium iodide solution (KI), 2 mL of 10% acetic acid solution, 5 drops of 4% starch indicator, and about 30 mL of distilled water to the flask.
5. Rinse a 25- or 50-mL buret with about 5 mL of the NBS solution. Then, fill the buret with NBS solution, adjust the liquid level to 0.00 or lower and record the initial buret reading in Table 32.1.
6. Titrate the vitamin C sample contained in the flask. The end point is reached when 1 drop of NBS solution causes the mixed contents of the flask to have a permanent light-blue color. Some titrated samples might require more than 25 mL of NBS to reach the end point. If you are using a 25-mL buret and you think your sample might require more than 25 mL, stop the titration before the liquid level in your buret reaches the bottom mark on the buret, and check with your lab instructor.
7. Record the final buret reading.

DISPOSAL Titrated samples in sink

B. Analysis of an Unknown

PROCEDURE

1. Obtain an unknown vitamin C solution from the stockroom and record its identification (ID) number in Table 32.4.
2. Rinse a 10-mL pipet with about 5 mL of your unknown solution. Then, use the pipet to transfer a 10.00-mL sample of your unknown vitamin C solution to a 125-mL flask. Record the sample volume.

3. Add 5 mL of 4% potassium iodide solution, 2 mL of 10% acetic acid solution, 5 drops of starch indicator, and about 30 mL of water to the flask.
4. Titrate the sample with NBS solution as described in Part A.
5. Record the initial and final buret readings in Table 32.3.

DISPOSAL Titrated samples in sink

C. Vitamin C Content of Orange Juice

In this part of the experiment, you will determine and compare the vitamin C content of fresh orange juice and reconstituted frozen orange juice. The vitamin C content of frozen orange juice will depend upon the brand selected. Some brands contain added vitamin C in addition to the natural amount.

PROCEDURE

1. Obtain 15 mL of fresh orange juice (you might have to squeeze it yourself) and filter it through 2 or 3 layers of cheesecloth to remove pulp and fibers.
2. Use a 10-mL pipet and transfer 10.00 mL of the juice to a 125-mL flask. Record the sample volume in Table 32.5.
3. Add about 30 mL of 1% oxalic acid solution, 5 mL of 4% potassium iodide solution, 2 mL of 10% acetic acid solution, and 5 drops of starch indicator to the flask.
4. Titrate the sample with NBS solution as described in Part A. The light-blue end point will be somewhat masked by the color of the juice and will appear as a dark coloration.
5. Record the initial and final buret readings.
6. Repeat Steps 1 to 5 with reconstituted frozen orange juice.
7. Rinse your 10-mL pipet thoroughly with tap water and then distilled water before you store it or return it to the stockroom.

DISPOSAL Titrated samples in sink

D. Vitamin C Content of Cabbage

In this part of the experiment, you will determine and compare the vitamin C content of raw cabbage and cooked cabbage.

PROCEDURE

a. Analysis of Raw Cabbage

1. Accurately weigh about 5 g of raw cabbage on a centigram or electronic balance and record the mass in Table 32.7.
2. Place the cabbage and 15 mL of 1% oxalic acid solution in a mortar and grind thoroughly.

SAFETY ALERT   Remember, the oxalic acid solution is toxic. Use it with care.

3. Filter the mixture through 2 or 3 layers of cheesecloth and collect the filtrate.
4. Rinse the mortar with an additional 15 mL of 1% oxalic acid, pour the solution through the same filter, and allow the filtrate to mix with that from Step 3.
5. Transfer the nearly 30 mL of filtrate to a 125-mL flask.
6. Add 5 mL of 4% potassium iodide solution, 2 mL of 10% acetic acid solution, and 5 drops of starch indicator to the flask.
7. Titrate the sample with NBS solution as described in Part A.
8. Record the initial and final buret readings.

DISPOSAL  Titrated samples in sink
Cabbage residues in wastebasket

b. Analysis of Cooked Cabbage

1. Accurately weigh about 5 g of raw cabbage on a centigram or electronic balance and record the mass in Table 32.7.
2. Place the cabbage in a 100-mL beaker, add 30 mL of distilled water, cover the beaker with a watch glass, and *gently* boil the mixture for 15 minutes. If necessary, add small amounts of distilled water so the beaker does not boil to dryness.
3. Pour off the cooking water and wash the cooked cabbage with 25 mL of distilled water. Pour off the wash water.
4. Transfer the boiled cabbage (solid only) to a mortar and repeat Steps 2 to 8 as you did for the raw cabbage. Record the data in Table 32.7.

DISPOSAL 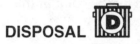 Titrated samples in sink
Cabbage residue in wastebasket

CALCULATIONS AND REPORT

A. Standardization of Oxidizing Agent

1. Use the volume and concentration of vitamin C solution in Table 32.1 to calculate the number of milligrams (mg) of vitamin C in the titrated sample. Record this result in Table 32.2.
2. Use the buret readings of Table 32.1 to determine the volume of NBS required to titrate the vitamin C sample. Record this result in Table 32.2.
3. Divide the volume of NBS used in the titration by the number of milligrams of vitamin C in the titrated sample. Record this number, which is the milliliters (mL) of NBS required to titrate 1 mg of vitamin C.

B. Analysis of an Unknown

1. Record the volume of unknown titrated in Table 32.4.
2. Use data from Table 32.3 to calculate and record the volume of NBS required in the titration.

3. Calculate and record the number of milligrams of vitamin C in the sample by dividing the volume of NBS used by the factor calculated in Step 3 of Part A of the Calculations and Report.
4. Divide the result of Step 3 by the sample volume to get the number of milligrams of vitamin C per milliliter of unknown. Record this result in Table 32.4.

C. Vitamin C Content of Orange Juice

1. Use data from Table 32.5 to calculate and record in Table 32.6 the volume of NBS required to titrate each sample.
2. Calculate the number of milligrams of vitamin C in each sample and the milligrams of vitamin C per milliliter of sample as you did in Steps 3 and 4 of Part B.
3. Record these results in Table 32.6.

D. Vitamin C Content of Cabbage

1. Use data from Table 32.7 to calculate and record in Table 32.8 the volume of NBS required to titrate each sample.
2. Calculate the number of milligrams of vitamin C in each sample as you did in Step 3 of Part B.
3. Divide the result of Step 2 by the sample mass to obtain the milligrams of vitamin C per gram of sample. Record this result in Table 32.8.

Experiment 32 ▪ Pre-Lab Review

Vitamin C Content of Foods, Part I

1. Are any specific safety alerts given in the experiment? List any that are given.

   _____

   _____

2. Are any specific disposal directions given in the experiment? List any that are given.

   _____

   _____

3. How many places after the decimal should be used to record

   a. a buret reading? _____   b. a pipet volume? _____

4. What observation indicates that the end point of a vitamin C titration has been reached?

   _____

   _____

5. What should you do if you are using a 25-mL buret, and it appears that a titration will require more than 25 mL of NBS to reach the end point?

   _____

   _____

6. Why is the fresh orange juice used in Part C filtered? _____

   _____

   _____

7. What problem in detecting the titration end point might arise in Part C?

   _____

   _____

   _____

8. What should you do to your pipet before you store it or return it to the stockroom?

_____

_____

9. What substances are weighed accurately in this experiment? _____

_____

10. What precautions are mentioned in Part D concerning the boiling of your cabbage sample?

_____

_____

Experiment 32 ▪ Data & Report Sheet

Vitamin C Content of Foods, Part I

A. Standardization of Oxidizing Agent

TABLE 32.1 (data)

Concentration of standard
  vitamin C solution (mg/mL) _____

Volume of standard vitamin C
  solution titrated _____

Initial buret reading _____

Final buret reading _____

TABLE 32.2 (report)

Number of mg vitamin C
  in titrated sample _____

Volume of NBS used in
  titration _____

Volume of NBS required to
  titrate 1 mg of vitamin C _____

B. Analysis of an Unknown

TABLE 32.3 (data)

Volume of unknown titrated _____

Initial buret reading _____

Final buret reading _____

TABLE 32.4 (report)

Unknown ID number _____

Volume of unknown titrated _____

Volume of NBS required in
  titration _____

Number of mg vitamin C in
  titrated sample _____

Number of mg vitamin C
  per mL of unknown _____

C. Vitamin C Content of Orange Juice

TABLE 32.5 (data)

Sample	Volume of Sample Titrated	Initial Buret Reading	Final Buret Reading
Fresh	_____	_____	_____
Reconstituted	_____	_____	_____

TABLE 32.6 (report)

Sample	Volume of NBS Required to Titrate Sample	Number of mg Vitamin C in Sample	Number of mg Vitamin C per mL of Sample
Fresh			
Reconstituted			

D. Vitamin C Content of Cabbage

TABLE 32.7 (data)

Sample	Sample Mass	Initial Buret Reading	Final Buret Reading
Fresh			
Cooked			

TABLE 32.8 (report)

Sample	Volume of NBS Required to Titrate Sample	Number of mg Vitamin C in Sample	Number of mg Vitamin C per Gram of Sample
Fresh			
Cooked			

QUESTIONS

1. A student makes a mistake while preparing a vitamin C sample for titration and adds the potassium iodide solution twice. How will the larger quantity of KI influence the amount of NBS needed to titrate the sample?

   a. Increase it.   **b.** Decrease it.   **c.** Will have no effect on it.

   Explain your answer: _____

   _____

   _____

2. A student finds that cooked cabbage contains less vitamin C per gram than raw cabbage. How might this result be explained?

    a. Vitamin C was destroyed by heat.
    b. Vitamin C dissolved into the cooking water.
    c. Vitamin C was destroyed by grinding the sample.
    d. More than one response is correct.

    Explain your answer: _____

    _____

    _____

3. Two students carry out a vitamin C analysis on raw and boiled cabbage (Part D). One follows the directions carefully, but the other fails to wash the boiled cabbage with water as directed in Step 3. How will the determined amounts of vitamin C per gram of boiled cabbage compare?

    a. The results will be the same.
    b. The washed sample will have a higher value.
    c. The washed sample will have a lower value.

    Explain your answer: _____

    _____

    _____

4. While doing a vitamin C titration, a student suddenly remembers that no starch indicator has been added. Upon the addition of indicator, the solution turns dark blue. What may the student conclude?

    a. The correct end point has not yet been reached.
    b. The correct end point has been reached.
    c. The correct end point has been passed.

    Explain your answer: _____

    _____

5. According to your experimental results, how much vitamin C would you get from a 4 oz. (120 mL) serving of fresh orange juice?

    a. Less than the RDI   b. More than the RDI   c. Almost exactly the RDI

    Explain your answer: _____

    _____

Vitamin C Content of Foods, Part II: Samples from Home

In this experiment, you will

■ Bring several related food samples from home.

■ Determine and compare the vitamin C content of the samples.

INTRODUCTION

This is a companion experiment to Experiment 32 where you first analyzed samples for their vitamin C content. Several factors make vitamin C an excellent subject for a second experiment: The analysis for vitamin C is straightforward and fairly accurate; articles about the health effects of vitamin C occur frequently in the media and create interest in the material; vitamin C is widely distributed in foods; and vitamin C is easily lost from foods because of its solubility in water and the ease with which it is oxidized.

There are numerous interesting comparisons that can be done on samples of your own choosing. Your instructor will give you some directions concerning the number of samples you will be required to analyze in this experiment. The number of analyses will, of course, influence the kind of comparison you choose to make. Some possible comparisons are the following:

- Compare several fruits and/or vegetables for their vitamin C content. Some good candidates are apples, bananas, peppers (both green and red), peaches, grapes, potatoes, grapefruit, lemons, and limes.
- Compare commercially available juices and drinks such as Kool-Aid and Tang.
- Compare different parts or regions of fruits and vegetables. The peelings versus the interior of potatoes or outer leaves versus inner leaves of cabbage are good examples.
- Compare preparation conditions. Fresh versus frozen, fresh versus canned, frozen versus canned, lemon juice concentrate from a freshly opened container versus that from an older container, or a whole radish versus a radish that was sliced the night before are good examples.

EXPERIMENTAL PROCEDURE

A. Standardization of Oxidizing Agent

The stock solution of *N*-bromosuccinimide is somewhat unstable, and it is therefore necessary to standardize it before use. This is accomplished by titrating a known amount of vitamin C.

PROCEDURE

1. Into a clean, dry 15-cm test tube, pour about 15 mL of standard vitamin C solution containing 0.50 mg/mL. Record this concentration in Table 33.1.

2. Pour about 150 mL of NBS solution into a clean, dry 250-mL beaker. This solution will be used for all parts of the experiment.

3. Use a 10-mL pipet and place 10.00 mL of vitamin C solution into a 125-mL flask. Record this volume in Table 33.1.

4. Add 5 mL of 4% potassium iodide solution (KI), 2 mL of 10% acetic acid solution, 5 drops of 4% starch indicator, and about 30 mL of distilled water to the flask.

5. Rinse a 25- or 50-mL buret with about 5 mL of the NBS solution. Then, fill the buret with NBS solution, adjust the liquid level to 0.00 or lower, and record the initial buret reading in Table 33.1.

6. Titrate the vitamin C sample contained in the flask. The end point is reached when one drop of NBS solution causes the mixed contents of the flask to have a permanent light-blue color.

7. Record the final buret reading.

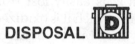
B. Vitamin C Content of Juices or Other Liquids

If your samples are liquids (juices, reconstituted drinks, cooking water, and so on), use this procedure. The vitamin C content can vary significantly between foods, so you might find it necessary to adjust the volume of your sample so that at least 10 mL of NBS is needed for the titration but that the total volume of your buret is not exceeded. Check with your instructor for further directions.

PROCEDURE

1. Obtain 15 mL of liquid and filter it through three layers of cheesecloth if it contains solids or pulp that would clog a pipet.

2. Use a 10-mL pipet and transfer 10.00 mL of the liquid to a 125-mL flask. Record the sample volume in Table 33.3, along with the identity of the sample.

3. Add about 30 mL of 1% oxalic acid solution, 5 mL of 4% potassium iodide solution, 2 mL of 10% acetic acid solution, and 5 drops of starch indicator to the flask.

4. Titrate the sample with NBS solution as described in Part A. The light-blue end point will be masked somewhat if your sample is colored. If this is the case, look for a distinct change in color to signal the end point.

5. Record the initial and final buret readings in Table 33.3 for each sample titrated.

C. Vitamin C Content of Solids

If your samples are solids, use the following procedure. Once again, it might be necessary to make adjustments in the amount of solid used or in the amount of liquid extract titrated in order to use at least 10 mL of NBS from the buret without exceeding the buret capacity.

PROCEDURE

1. Use a centigram or electronic balance to accurately weigh a sample of your solid with a mass of about 5 g. Record the mass in Table 33.5 of the Data and Report Sheet. Also record the identity of your sample.
2. Put the weighed sample and 15 mL of 1% oxalic acid solution into a mortar and grind them together thoroughly.

 **SAFETY ALERT**

Remember, the oxalic acid solution is toxic. Use it with care.

3. Filter the ground mixture through 3 layers of cheesecloth and collect the liquid filtrate in a clean 125-mL flask.
4. Rinse the mortar with an additional 15 mL of 1% oxalic acid, pour the rinse mixture through the same filter used in Step 3, and allow the filtrate to mix with that from Step 3 in the 125-mL flask.

5. Add 5 mL of 4% potassium iodide solution, 2 mL of 10% acetic acid, and 5 drops of starch indicator to the flask.
6. Titrate the sample with NBS solution as described in Part A.
7. Record the initial and final buret readings in Table 33.5.

CALCULATIONS AND REPORT

A. Standardization of Oxidizing Agent

1. Use the volume and concentration of vitamin C solution in Table 33.1 to calculate the number of milligrams (mg) of vitamin C in the titrated sample. Record this result in Table 33.2.
2. Use the buret reading of Table 33.1 to determine the volume of NBS required to titrate the vitamin C sample. Record this result in Table 33.2.
3. Divide the volume of NBS used in the titration by the number of milligrams of vitamin C in the titrated sample. Record this number, which is the milliliters (mL) of NBS required to titrate 1 mg of vitamin C.

B. Vitamin C Content of Juices or Other Liquids

1. Record the volume of sample titrated in Table 33.4.
2. Use data from Table 33.3 to calculate the volume of NBS solution required in the titration. Record the volume in Table 33.4.
3. Calculate and record the number of milligrams of vitamin C in the titrated sample by dividing the volume of NBS solution used by the factor calculated in Step 3 of Part A of Calculations and Report.
4. Divide the result of Step 3 by the sample volume to get the number of milligrams of vitamin C per milliliter of liquid. Record this result in Table 33.4.

C. Vitamin C Content of Solids

1. Use data from Table 33.5 to calculate the volume of NBS solution required to titrate each sample. Record the volume in Table 33.6 of the Data and Report Sheet.
2. Calculate the number of milligrams of vitamin C in each titrated sample as you did in Step 3 of Part B.
3. Divide the result of Step 2 by the sample mass recorded in Table 33.5 to obtain the milligrams of vitamin C per gram of sample. Record this result in Table 33.6.

Summary of Results

1. Write a brief summary of the results of your comparison in Table 33.7 of the Data and Report Sheet.

Experiment 33 ▪ Pre-Lab Review

Vitamin C Content of Foods,
Part II: Samples from Home

1. Are any specific safety alerts given in the experiment? List any that are given.

    _____

    _____

2. Are any specific disposal directions given in the experiment? List any that are given.

    _____

    _____

3. How many places after the decimal should be used to record

    a. a buret reading? _____ b. a pipet volume? _____

4. What observation indicates that the end point of a vitamin C titration has been reached?

    _____

    _____

5. Why are the liquid samples used in Part B filtered? _____

    _____

    _____

6. What problem in detecting the titration end point might arise in Part B?

    _____

    _____

    _____

7. What substances are weighed accurately in this experiment? _____

    _____

Experiment 33 ▪ Data & Report Sheet

Vitamin C Content of Foods,
Part II: Samples from Home

A. Standardization of Oxidizing Agent

TABLE 33.1 (data)

Concentration of standard
   vitamin C solution (mg/mL) _____

Volume of standard
   vitamin C solution titrated _____

Initial buret reading _____

Final buret reading _____

TABLE 33.2 (report)

Number of mg vitamin C
   in titrated sample _____

Volume of NBS used in
   titration _____

Volume of NBS required to
   titrate 1 mg of vitamin C _____

B. Vitamin C Content of Juices or Other Liquids

TABLE 33.3 (data)

Sample Identity	Sample Volume	Initial Buret Reading	Final Buret Reading
_____	_____	_____	_____
_____	_____	_____	_____
_____	_____	_____	_____
_____	_____	_____	_____
_____	_____	_____	_____

TABLE 33.4 (report)

Sample Identity	Sample Volume	Volume of NBS Required	mg Vit. C in Sample	mg Vit. C per mL of Sample

C. Vitamin C Content of Solids

TABLE 33.5 (data)

Sample Identity	Sample Mass	Initial Buret Reading	Final Buret Reading

TABLE 33.6 (report)

Sample Identity	Volume of NBS Required	mg Vit. C in Sample	mg Vit. C per Gram of Sample

Summary

TABLE 33.7

Appendix A Graphs and Graphing

Graphs provide a convenient way to represent the simultaneous behavior of the values of related quantities. In most graphs, this is done by using a grid containing perpendicular x and y axes. By convention, the x axis is horizontal, and the y axis is vertical. A single point on such a grid represents a specific value for each of two related quantities. The graph is formed by connecting the points with a line that can be either curved or straight. These characteristics are shown in Figure A.1.

FIGURE A.1
Graphical representation of values of related quantities

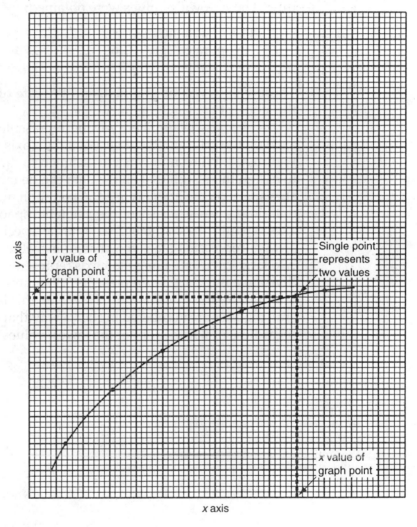

y axis

y value of graph point

Single point represents two values

x value of graph point

x axis

Some graphs represent experimentally measured relationships between quantities, while others represent the behavior of quantities related by a specific mathematical formula. The abilities to construct graphs and obtain information from them are important in the study of science. The techniques used are illustrated by the following example.

EXAMPLE

Some chemistry students were given the assignment of experimentally determining the rate at which acetone, a volatile liquid, evaporates at room temperature. To accomplish the task, they put a dish that contained some acetone on an electronic balance and determined the mass of the dish plus acetone at one minute intervals for a total of 6 minutes. The following data were collected:

Mass of dish + acetone (g):	51.33	51.07	50.83	50.60	50.37	50.15	49.93
Time (minutes):	0.0	1.0	2.0	3.0	4.0	5.0	6.0

The total mass of acetone evaporated at each time was obtained by subtracting the mass of the dish plus acetone at that time from the mass at zero time. The following values were obtained:

Mass of acetone evaporated (g):	0.26	0.50	0.73	0.96	1.18	1.40
Time (minutes):	1.0	2.0	3.0	4.0	5.0	6.0

These values will be graphed, and the rate of evaporation determined as the slope of the resulting line.

The mass evaporated will be represented along the y axis (vertical axis) on the grid, and the time along the x axis (horizontal axis) as shown in Figure A.2. When numbers are assigned to the x and y axes, it is important to assign them such that all the data can be included and most of the available space on the graph is used. A common mistake is to crowd the plotted data into a small corner of the available space. Also, the numbers on the axes should be assigned so that a specific distance along an axis always represents the same change in value. In Figure A.2, for example, two large divisions along the x axis represents 1.0 minutes, while each large division along the y axis represents 0.10 grams. The lowest values shown on both axes of this graph is zero. However, this is not always the case. In general, the axes are numbered in such a way that the data will fit on the graph regardless of whether or not the lowest values shown are zero.

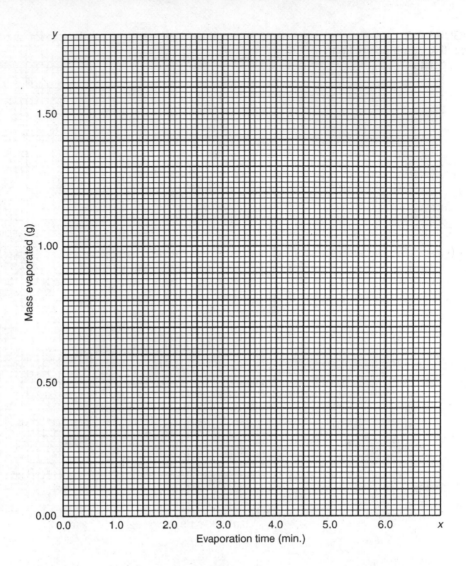

After numbers and labels are assigned to the axes, the positions of the points are determined. This is done by taking a time such as 1.0 minute and a corresponding mass that evaporated (0.26 g corresponds to the time of 1.0 min.) and locating their proper positions on the x and y axes. A line is then drawn vertically from the 1.0-min. point on the x axis and horizontally from the 0.26-g point on the y axis. The single graph point that represents both values is located at the intersection of these two lines, as shown in Figure A.3.

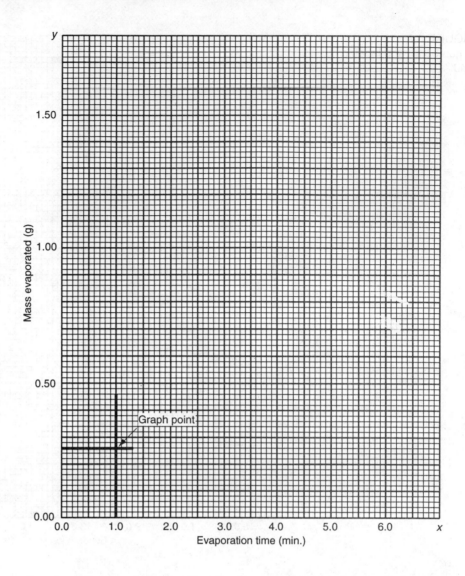

Graph points corresponding to the remaining pairs of data points (2.0 min. and 0.50 g, and so on) are obtained in the same way. The resulting graph points are connected by a smooth line, and the graph is completed as shown in Figure A.4. Once the graph is drawn, the value of one of the graphed quantities that corresponds to a value of the other quantity can be easily obtained. For example, let's determine the mass of acetone that would have evaporated after a time of 3.5 minutes had passed in the experiment. This is done by drawing a vertical line up from the 3.5-min. point on the x axis until the line intersects the line that connects the graph points. A horizontal line is then drawn to the left from this point of intersection until it meets the y axis. The value at the point of intersection with the y axis gives the mass of acetone that would have been evaporated after 3.5 minutes in the experiment. This process, represented by the dotted lines in Figure A.4, gives a mass of acetone evaporated value of 0.84 g.

FIGURE A.4
Determination of mass
evaporated from a
completed graph

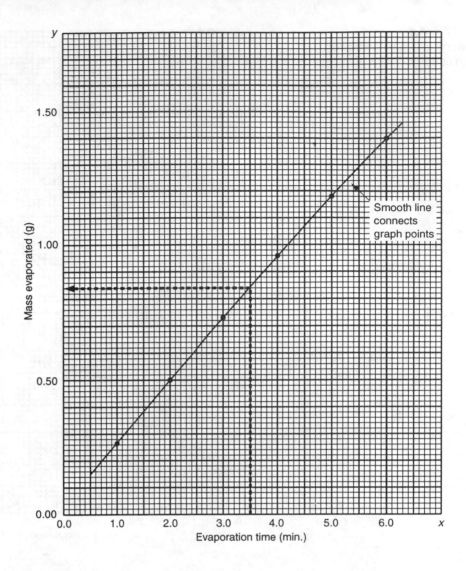

Useful information can often be obtained from the slope of a graph. In this example, the slope will give the rate of evaporation of acetone in units of grams per minute. The slope of a graph is the change in the quantity plotted on the y axis that occurs when the quantity plotted on the x axis changes. The slope is calculated by determining the value of the rise of the graph divided by the run of the graph. The rise of a graph is simply the change in the value of the quantity plotted on the y axis, and the run is the corresponding change in the value of the quantity plotted on the x axis.

The method for determining the slope is illustrated in Figure A.5. A horizontal line and a vertical line are drawn that both intersect the graph and also intersect each other (the dotted lines in Figure A.5). Both the points where the drawn lines intersect the graph correspond to a value of the quantity plotted on the x axis and the quantity plotted on the y axis. In Figure A.5, the values of the x and y quantities at the intersection of the horizontal line with the graph are denoted x_1 and y_1. This point is enclosed in a small square in Figure A.5, and corresponds to $x_1 = 1.6$ min., and $y_1 = 0.40$ g. The values of the x and y quantities at the intersection of the vertical line with the graph (enclosed in a small circle) are denoted x_2 and y_2. Their values are $x_2 = 5.0$ min., and $y_2 = 1.18$ g.

FIGURE A.5
Determination of the slope
of a graph

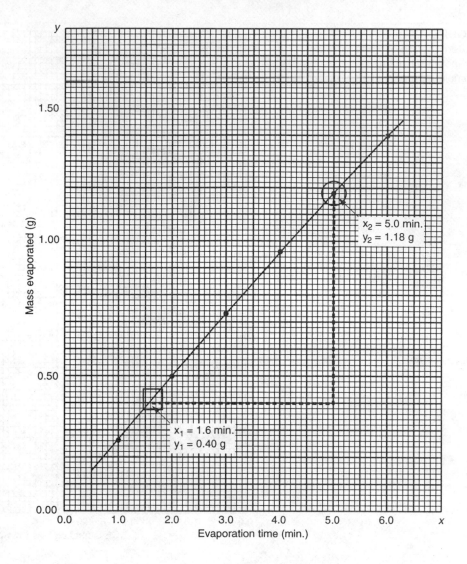

The rise of the graph is the change in the value of the y quantity, and will always be the value of y corresponding to the vertical dotted line (y_2) minus the value of y corresponding to the horizontal dotted line (y_1). The run of the graph is the change in the value of the x quantity, and will always be the value of x corresponding to the vertical dotted line (x_2) minus the value of x corresponding to the horizontal dotted line (x_1). Thus, the slope is calculated as follows:

$$\text{slope} = \frac{\text{rise}}{\text{run}} = \frac{y_2-y_1}{x_2-x_1} = \frac{1.18\,\text{g}-0.40\,\text{g}}{5.0\,\text{min}-1.6\,\text{min}} = \frac{0.78\,\text{g}}{3.4\,\text{min}} = \frac{0.23\,\text{g}}{\text{min}}$$

The slope gives the rate of evaporation of acetone as 0.23 g per minute. This value has a positive sign and is called a positive slope; y increases when x increases. In some cases, y decreases when x increases. In such cases, y_2 will have a smaller value than y_1, and the quantity $y_2 - y_1$ will have a negative value. This will cause the calculated slope to have a negative value, and the graph is said to have a negative slope. If the data consisting of the mass of dish + acetone had been plotted on the y axis, and the time on the x axis, a graph with a negative slope would have resulted. The slope of this graph would give the rate of change of the mass of the dish + acetone (actually the mass of acetone) with time in units of grams per minute. However, the negative slope indicates that the mass is decreasing with time.

Appendix B Equipment, Chemicals, Reagents, and Supplies

This appendix consists of three parts. Part I lists the locker equipment to which each student should have access in order to do the experiments. Part II contains preparation directions for the common acid and base solutions, such as 6 M NaOH, that are used in many of the experiments. Part III contains a listing of the chemicals, reagents (solutions), and other supplies needed for each experiment.

I. LOCKER EQUIPMENT

The listing below includes the locker equipment that we assume students will have available to do the experiments. Some of the more expensive items such as pipets and burets are listed as locker/stockroom items, indicating that students might have to check them out of the stockroom. These items are also listed later in Part III as equipment from the stockroom. The number in parentheses is the number of items we recommend be in the locker.

Glassware and Plasticware

Beakers, 50 mL (2)
Beakers, 100 mL (2)
Beakers, 250 mL (2)
Beakers, 400 mL (1)
Flasks, Erlenmeyer, 125 mL (3)
Flasks, Erlenmeyer, 250 mL (2)
Funnels, long stem (2)
Graduated cylinder, 10 mL (1)

Graduated cylinder, 50 mL (1)
Plastic droppers (6)
Polyethylene wash bottle, 250 mL (1)
Polyethylene storage bottle, 500 mL (1)
Test tubes, 15 cm (4)
Test tubes, 10 cm (12)
Test tubes, 7.5 cm (4)
Weighing dishes, plastic (4)

Ironware and Porcelain

Clamp, buret (1)
Clamp, pinch (1)
Clamp, test tube (1)
Crucibles with covers, 15 mL (2)

Crucible tongs (1)
Evaporating dishes, 75 mm (2)
Mortar and pestle, 50 mL (1)
Spatula, stainless (1)

Miscellaneous Items

Clay triangle (1)
Filter paper, 11 cm (1 pkg)
Litmus paper, blue (1 vial)
Litmus paper, red (1 vial)

Stirring rods, glass (2)
Test tube brush (1)
Test tube rack (1)
Wire gauze (1)

Locker/Stockroom Equipment

Buret, 25 or 50 mL (1)
Pipet, 10 mL (1)
Ringstand and ring (1)

Suction bulb, for pipet (1)
Thermometer, 150°C (1)

II. COMMON LABORATORY ACIDS AND BASES

The following acids and bases are commonly used in academic labs, and the experiments in this manual are no exception. When reference is made to concentrated solutions in the preparation directions, we are referring to the solutions normally provided by chemical supply houses. The directions are for the preparation of 1 L of solution. All preparations should be done in a fume hood because some of the concentrated solutions give off irritating and dangerous fumes.

$6\ M$ (dilute) acetic acid: Pour 350 mL of glacial acetic acid into 650 mL of distilled water.

$6\ M$ (dilute) ammonia (or ammonium hydroxide): Pour 400 mL of concentrated ammonia into 600 mL of distilled water.

$6\ M$ (dilute) hydrochloric acid: Pour 500 mL of concentrated hydrochloric acid into 500 mL of distilled water.

$6\ M$ (dilute) nitric acid: Pour 375 mL of concentrated nitric acid into 625 mL of distilled water.

$6\ M$ sodium hydroxide: Dissolve 240 g of solid sodium hydroxide in enough distilled water to give 1 L of solution. Use caution; the solution gets very hot.

$3\ M$ (dilute) sulfuric acid: Slowly pour (with stirring) 167 mL of concentrated sulfuric acid into 833 mL of distilled water. Use caution; the solution gets very hot and might spatter.

III. CHEMICALS, REAGENTS, AND SUPPLIES

Listed below are the chemicals (pure solids or liquids), reagents (solutions), and other supplies needed for each experiment. Also listed are any pieces of equipment that might not be included in student lockers and would be checked out of the stockroom. The amounts of chemicals, reagents, and miscellaneous supplies (including items from the stockroom) sufficient for ten students are given. You might want to add 10%–30% to these amounts to allow for student waste.

Directions are given for preparing all reagent solutions. In some cases, students will use only very small amounts of liquids. In these instances, the use of a dropper bottle is indicated. Directions for preparing solutions for dropper bottles assume a bottle capacity of 200 mL. It is also assumed that the common acids and bases will be available in the lab in the usual dilute (see Part II) and concentrated forms. When these substances are used, we recommend that a 250-mL or 500-mL reagent bottle be put in a hood together with a 50-mL beaker and three or four plastic droppers. The students should then be encouraged to pour small amounts into the beaker and use the droppers that are provided to get the amount they need. Solid chemicals should be put into jars or bottles with openings that are wide enough to allow students to get the amounts they need by dipping into the container with a spatula.

All chemicals and reagents should be clearly labeled, and the labels should be protected with transparent tape to prevent them from becoming unreadable. We use the following label format, where Chem. xxx is the university course number, and Exp. yy is the number of the experiment being done. A specific example from our program is given along with the general form.

Chem. xxx	Exp. yy	Chem. 111	Exp. 4
Chemical or reagent name		Ammonium thiocyanate	
Chemical or reagent formula*		NH_4SCN	
Concentration of solutions		$0.1\ M$	

*The formula is not always included. It is often omitted for organic compounds such as glycine, sucrose, or poly (vinyl alcohol).

When special disposal containers are needed in the lab, we have indicated the label that should be attached to each container. These labels should also be protected with transparent tape so they remain readable. The labels should be written as given because they are referred to specifically in the experiments and students will be looking for them in the form given in the experiments. The amount of material generated by ten students is indicated for each container. For containers, we clean up and use empty 5-pint or 1-gallon screw-cap bottles that are used by our chemical suppliers to ship acids, bases, and solvents to us. Directions and suggestions for disposing of the collected materials are included in the Instructor's Manual that is available to adopters of this lab text.

EXPERIMENT 1: MEASUREMENTS AND SIGNIFICANT FIGURES

Miscellaneous Supplies for Ten Students
Pennies: 100

Equipment from Stockroom for Ten Students
Several pair of scissors should be available.

EXPERIMENT 2: THE USE OF CHEMICAL BALANCES

Chemicals for Ten Students
Table salt (sodium chloride, NaCl): 25 g

Miscellaneous Supplies for Ten Students
Centigram balances: 3. More balances are useful if available.
Electronic balances: 2. More balances are useful if available.
Pennies: 100
Unknown masses: One for each student. We use ½-inch diameter metal rod sawed into ¾-inch to 1-inch lengths. Copper, aluminum, or zinc work well. We stamp each mass with an identifying number, then weigh the mass accurately by difference.

EXPERIMENT 3 THE USE OF VOLUMETRIC WARE AND THE DETERMINATION OF DENSITY

Miscellaneous Supplies for Ten Students
Rubber stoppers, solid no. 1: 10
Unknown liquids: 30-mL sample (one per student). Place the samples in small flasks or large test tubes. See the Instructor's Manual for preparation directions.

Equipment from Stockroom for Ten Students
Burets, 25 mL or 50 mL: 10
Pipets, 10 mL: 10
Suction bulbs: 10

EXPERIMENT 4: PHYSICAL AND CHEMICAL CHANGES

Chemicals for Ten Students
Copper carbonate ($CuCO_3$): 2 g
Iron chloride ($FeCl_3 \cdot 6H_2O$): 2 g
Magnesium (Mg), 5-cm strips of ribbon: 20
Sodium bicarbonate ($NaHCO_3$): 2 g
Sodium chloride (NaCl): 4 g

Reagents for Ten Students
Ammonium thiocyanate (NH_4SCN), 0.1 M: 2 mL (dropper bottle). Dissolve 1.6 g of NH_4SCN in 200 mL of distilled water.
Calcium nitrate ($Ca(NO_3)_2$), 0.1 M: 5 mL (dropper bottle). Dissolve 4.8 g of $Ca(NO_3)_2 \cdot 4H_2O$ in 200 mL of distilled water.
Hydrochloric acid (HC1), 6 M (dilute): 4 mL (reagent bottle, beaker, and droppers in hood)
Potassium ferrocyanide ($K_4Fe(CN)_6$), 0.1 M: 2 mL (dropper bottle). Dissolve 8.5 g of $K_4Fe(CN)_6 \cdot 3H_2O$ in 200 mL of distilled water.
Silver nitrate ($AgNO_3$), 0.1 M: 4 mL (dropper bottle). Dissolve 3.4 g of $AgNO_3$ in 200 mL of distilled water.

Disposal Containers
1. 120 mL/ten students: Chem. xxx, Exp. 4, Used Chemicals

EXPERIMENT 5: SEPARATIONS AND ANALYSIS

Chemicals for Ten Students
Acetone: 30 mL (dropper bottle in hood)
Ammonium chloride (NH_4Cl): 1 g
Calcium carbonate ($CaCO_3$): 2 g
Copper sulfate ($CuSO_4 \cdot 5H_2O$): 1 g (fine crystals)
Calcium carbonate/copper sulfate mixture: 11 g. Mix together equal masses of solid $CaCO_3$ and solid $CuSO_4 \cdot 5H_2O$ (fine crystals). Shake container vigorously to mix contents well.
Calcium carbonate/ammonium chloride mixture: 11 g. Mix together equal masses of solid $CaCO_3$ and solid NH_4Cl. Shake the container vigorously to mix the contents well.

Reagents for Ten Students
Ammonia (NH_3 or NH_4OH), 15 M (conc): 110 mL (reagent bottle, beaker, and droppers in hood)
Cobalt chloride ($CoCl_2$), 0.2 M: 2 mL (dropper bottle). Dissolve 9.5 g of solid $CoCl_2 \cdot 6H_2O$ in 200 mL of distilled water.
Copper chloride ($CuCl_2$), 0.2 M: 2 mL (dropper bottle). Dissolve 6.8 g of solid $CuCl_2 \cdot 2H_2O$ in 200 mL of distilled water.
Hydrochloric acid (HCl), 6 M (dilute): 30 mL (reagent bottle, beaker, and droppers in hood)
Iron (III) chloride ($FeCl_3$), 0.2 M: 2 mL (dropper bottle). Dissolve 10.8 g of solid $FeCl_3 \cdot 6H_2O$ in 200 mL of distilled water.
Mixed ions: 2 mL (dropper bottle). Dissolve 9.5 g of solid $CoCl_2 \cdot 6H_2O$, 6.8 g of solid $CuCl_2 \cdot 2H_2O$, and 10.8 g of solid $FeCl_3 \cdot 6H_2O$ in 200 mL of distilled water.
Nitric acid (HNO_3), 6M (dilute): 5 mL (reagent bottle, beaker, and droppers in hood).

Miscellaneous Supplies for Ten Students
Chromatography paper, 9 cm × 7 cm: 10. Cut from Whatman #1 chromatography paper, cat. no. 3001917.

Melting-point capillaries, both ends open: 60. Put several vials in the lab.
Unknowns: 4-drop sample (one per student). See instructor's manual for preparation instructions.

Equipment from Stockroom for Ten Students
Pencils: A few should be available for students who need one.
Rulers, plastic, 15 cm: 10
Scissors: A few pair should be available.

Disposal Containers
1. 120 mL/ten students: Chem. xxx, Exp. 5, Used Chemicals Part B
2. 40 mL/ten students: Chem. xxx, Exp. 5, Used Chromatography Solvent
3. 100 mL/ten students: Chem. xxx, Exp. 5, Used Ammonia
4. 4 mL/ten students: Chem. xxx, Exp. 5, Used Chromatography unknowns

EXPERIMENT 6:   CLASSIFICATION OF CHEMICAL REACTIONS

Chemicals for Ten Students
Copper (Cu) solid wire (16 gauge), 2-cm pieces: 10
Magnesium (Mg) ribbon, 3-cm pieces: 10
Potassium chlorate $KClO_3$: 5 g. Mix together 100 g of solid $KClO_3$ and 1 g of solid Fe_3O_4 (black) or Fe_2O_3 (red). Shake the container to make sure the contents are well mixed.
Zinc (Zn), small pieces (no larger than 4 mm): 10

Reagents for Ten Students
Ammonia (NH_3 or NH_4OH), 15 M (conc.): 3 mL (reagent bottle, beaker, and droppers in hood)
Copper nitrate ($Cu(NO_3)_2$), 0.1 M: 10 mL (dropper bottle). Dissolve 4.6 g of solid $Cu(NO_3)_2 \cdot 2\,\tfrac{1}{2}H_2O$ in 200 mL of distilled water.
Hydrochloric acid (HCl), 12 M (conc.): 3 mL (reagent bottle, beaker, and droppers in hood)
Hydrogen peroxide (H_2O_2), 3%: 5 mL (dropper bottle). Mix together 20 mL of 30% H_2O_2 and 180 mL of distilled water.
Iron (III) nitrate ($Fe(NO_3)_3$), 0.1 M: 6 mL (dropper bottle). Dissolve 8.1 g of solid $Fe(NO_3)_3 \cdot 9H_2O$ in 200 mL of distilled water.
Silver nitrate ($AgNO_3$), 0.1 M: 11 mL (dropper bottle). Dissolve 3.4 g of solid $AgNO_3$ in 200 mL of distilled water.
Sodium chloride (NaCl), 1 M: 6 mL (dropper bottle). Dissolve 11.7 g of solid NaCl in 200 mL of distilled water.
Sodium hydroxide (NaOH), 1 M: 6 mL (dropper bottle). Dissolve 8.0 g of solid NaOH in 200 mL of distilled water.
Sodium nitrate ($NaNO_3$), 1 M: 6 mL (dropper bottle). Dissolve 17.0 g of solid $NaNO_3$ in 200 mL of distilled water.
Yeast suspension: 4 mL (dropper bottle). Mix together ½ packet of dry yeast and 200 mL of distilled water. It is adviseable to make this fresh each lab day.

Miscellaneous Supplies for Ten Students
Wooden splints: 12

Disposal Containers
1. 10 mL/ten students: Chem. xxx, Exp. 6, Used Chemicals Part C
2. 35 mL/ten students: Chem. xxx, Exp. 6, Used Chemicals Part D

EXPERIMENT 7: ANALYSIS USING DECOMPOSITION REACTIONS

Miscellaneous Supplies for Ten Students

Barium chloride hydrate ($BaCl_2 \cdot nH_2O$): 5 g. Put a small container of solid $BaCl_2 \cdot 2H_2O$ in the lab. Be sure the label reads "Barium chloride hydrate ($BaCl_2 \cdot nH_2O$)."

Unknown solid $KClO_3$: 0.5-g sample (one per student). Each sample is made up of solid $KClO_3$ and solid Fe_2O_3 (red) or Fe_3O_4 (black). A variety of different mass percentages of $KClO_3$ can be used, but the $KClO_3$ should not be less than about 70% in any sample. When weighed amounts of the solids are mixed, be sure to shake the containers vigorously to make certain the solids are uniformly mixed.

Disposal Containers

1. 5 g/ten students: Chem. xxx, Exp. 7, Used Barium Chloride
2. 3 g/ten students: Chem. xxx, Exp. 7, Used Potassium Chlorate

EXPERIMENT 8: GAS LAWS

Reagents for Ten Students

Ammonia (NH_3 or NH_4OH), 15 M (conc.): 10 mL (reagent bottle, beaker, and droppers in hood)

Hydrochloric acid (HCl), 12 M (conc.): 10 mL (reagent bottle, beaker, and droppers in hood)

Miscellaneous Supplies for Ten Students

Cotton: 10 small balls. Put a large-mouthed jar full in the lab.

Food coloring: 1 mL (dropper bottle). A small dropper bottle containing 10 to 20 mL of red, blue, or green food coloring should be in the lab.

Ice, crushed: One 2- or 3-gallon bucket for the lab.

Equipment from Stockroom for Ten Students

Capillary tubing, glass, 5- or 6-mm o.d., 0.5-mm bore, 50-cm lengths: 10

Flexible tubing (to fit side arm of test tube), 4-cm pieces: 10

Glass tubing, 4-5 mm i.d., 6-7 mm o.d., 50 cm lengths: 5

One-hole rubber stoppers no. 4: 10

Side-arm test tubes, 25 × 200 mm: 10

We recommend that the above four items (excluding the glass tubing) be assembled into the Boyle's law apparatus shown in Figure 8.2 and issued to students as one item from the stockroom. This is safer for the students and conserves lab time.

Syringes, disposable plastic, 5 or 10 cc: 10

Rulers, plastic with 0.1-cm scale divisions: 10

Thermometers, 150°C, with 1°C-scale divisions: 10

EXPERIMENT 9: SOLUTION FORMATION AND CHARACTERISTICS

Chemicals for Ten Students

Acetic acid, glacial: 1 mL (reagent bottle, beaker, and droppers in hood)

Ammonium chloride (NH_4Cl): 1 g

Citric acid: 1 g

Copper(I) chloride (CuCl): 1 g. Be sure to use copper (I) chloride (cuprous chloride)

Ethylene glycol ($C_2H_4(OH)_2$): 2 mL (dropper bottle). Put 100 mL into a dropper bottle.

Glucose ($C_6H_5(OH)_5CHO$): 1 g

Isopropyl alcohol (C_3H_7OH) 2 mL (dropper bottle)
Monochloroacetic acid ($C_2H_3O_2Cl$): 1 g
Nickel (II) chloride ($NiCl_2$): 1 g. Use solid $NiCl_2 \cdot 6H_2O$.
Octyl alcohol ($C_8H_{17}OH$): 2 mL (dropper bottle)
Phosphoric acid, 85%: 1 mL (reagent bottle, beaker, and droppers in hood)
Potassium nitrate (KNO_3): 15 g
Sodium chloride (NaCl): 15 g
Sodium hydroxide (NaOH): 10 pellets
Stearic acid: 1 g
Sucrose: 2 g
Toluene: 100 mL
1,1,1-Trichloroethane ($C_2H_3Cl_3$): 2 mL (reagent bottle, beaker, and droppers in hood)
Valeric acid: 2 mL (reagent bottle, beaker, and droppers in hood)

Reagents for Ten Students
Silver nitrate ($AgNO_3$), 0.1 *M*: 5 mL (dropper bottle). Dissolve 3.4 g of solid $AgNO_3$ in 200 mL of distilled water.

Miscellaneous Supplies for Ten Students
Cooking oil: 2 mL (dropper bottle)
Paraffin wax (fine shavings): 2 g
pH paper (long-range, 1 to 11): 3-cm pieces: 10. More than one color chart should be available for student use.
pH paper (short-range, 1.4 to 2.8), 3-cm pieces: 10. More than one color chart should be available for student use.
Plastic kitchen wrap: One roll

Disposal Containers
1. 60 mL/ten students: Chem. xxx, Exp. 9, Used Chemicals Part A
2. 100 mL/ten students: Chem. xxx, Exp. 9, Used Chemicals Part B
3. 1.5 mL/ten students: Chem. xxx, Exp. 9, Used 1,1,1-trichloroethane Part B
4. 180 mL/ten students: Chem. xxx, Exp. 9, Used Chemicals Part C
5. 120 mL/ten students: Chem. xxx, Exp. 9, Used Chemicals Part D
6. 200 mL/ten students: Chem. xxx, Exp. 9, Used Chemicals Part E

EXPERIMENT 10:   COLLIGATIVE PROPERTIES OF SOLUTIONS

Chemicals for Ten Students
Ethylene glycol: 150 mL
Sodium chloride (NaCl): 50 g
Sucrose: 220 g

Miscellaneous Supplies for Ten Students
Boiling chips
Dialysis tubing, 8-cm lengths: 20
Ice, crushed: One bucket for the lab
Dried prunes (nonpitted): 20
Strong string, 15-cm lengths: 40

Equipment from Stockroom for Ten Students
Thermometers, 150°C: 10

Disposal Containers
1. 900 mL/ten students: Chem. xxx, Exp. 10, Used Ethyleneglycol Parts B and C

EXPERIMENT 11: REACTION RATES AND EQUILIBRIUM

Chemicals for Ten Students

Iron oxide (Fe_2O_3): 1 g

Manganese dioxide (MnO_2): 1 g

Reagents for Ten Students

Calcium acetate ($Ca(C_2H_3O_2)_2$), saturated: 10 mL (dropper bottle). Add 72 g of solid ($Ca(C_2H_3O_2)_2 \cdot xH_2O$) to 200 mL of distilled water and shake thoroughly. Some solid should remain undissolved.

Hydrochloric acid (HCl), 6 M (dilute): 1 mL (reagent bottle, beaker, and droppers in hood)

Hydrogen peroxide (H_2O_2), 3%: 100 mL. Dilute 20 mL of stabilized 30% H_2O_2 up to a volume of 200 mL with distilled water.

Iron (III) chloride ($FeCl_3$), 0.1 M: 1 mL (dropper bottle). Dissolve 5.4 g of $FeCl_3 \cdot 6H_2O$ in 200 mL of distilled water.

Iron (III) nitrate ($Fe(NO_3)_3$), 0.1 M: 2 mL (dropper bottle). Dissolve 8.0 g of $Fe(NO_3)_3 \cdot 9H_2O$ in 200 mL of distilled water.

Manganese (II) chloride ($MnCl_2$), 0.1 M: 1 mL (dropper bottle). Dissolve 3.9 g of $MnCl_2 \cdot 4H_2O$ in 200 mL of distilled water.

Potassium chloride (KCl), 0.1 M: 1 mL (dropper bottle). Dissolve 1.5 g of KCl in 200 mL of distilled water.

Potassium iodate (KIO_3), 0.03 M: 35 mL (dropper bottle). Dissolve 1.3 g of KIO_3 in 200 mL of distilled water. Make sure sufficient amounts of this reagent are prepared for the entire lab.

Potassium thiocyanate (KSCN), 0.1 M: 2 mL (dropper bottle). Dissolve 2.0 g of KSCN in 200 mL of distilled water.

Silver nitrate, 0.1 M: 1 mL (dropper bottle). Dissolve 3.4 g of solid $AgNO_3$ in 200 mL of distilled water.

Sodium bisulfite ($NaHSO_3$), 0.05 M, containing starch indicator: 25 mL (dropper bottle). Add 10 mL of distilled water to 0.4 g of soluble starch to form a thin paste. Add the paste with stirring to 190 mL of boiling distilled water. Cool the solution and dissolve 1.0 g of $NaHSO_3$ in the mixture. Make sure sufficient amounts of this reagent are prepared for the entire lab.

Sodium chloride (NaCl), 0.1 M: 1 mL (dropper bottle). Dissolve 1.2 g of NaCl in 200 mL of distilled water.

Sodium hydroxide (NaOH), 6 M: 1 mL (reagent bottle, beaker, and droppers in hood)

Miscellaneous Supplies for Ten Students

Ice, crushed: One bucket for the lab

Equipment from Stockroom for Ten Students

Stopwatches: 5

Thermometers, 150°C: 5

Disposal Containers

1. 90 mL/ten students: Chem. xxx, Exp. 11, Used Chemicals Parts A and B
2. 100 mL/ten students: Chem. xxx, Exp. 11, Used Chemicals Part C
3. 80 mL/ten students: Chem. xxx, Exp. 11, Used Chemicals Part D

EXPERIMENT 12: ACIDS, BASES, SALTS, AND BUFFERS

Chemicals for Ten Students

Ammonium chloride (NH_4Cl): 1 g

Iron (Fe), wire or filings: 6 g

Marble chips ($CaCO_3$): 6 g

Sodium acetate ($NaC_2H_3O_2$): 1 g

Sodium carbonate (Na_2CO_3): 1 g

Sodium chloride (NaCl): 1 g

Zinc (Zn): 6 g

Reagents for Ten Students

Acetic acid ($HC_2H_3O_2$), 0.1 M: 10 mL (dropper bottle). Add 3.4 mL of 6 M $HC_2H_3O_2$ to 200 mL of distilled water.

Ammonium chloride (NH_4Cl), 0.1 M: 10 mL (dropper bottle). Dissolve 1.1 g of solid NH_4Cl in 200 mL of distilled water.

Aqueous ammonia (NH_3), 0.1 M: 10 mL (dropper bottle). Add 3.4 mL of 6 M NH_3 to 200 mL of distilled water.

Bromcresol green indicator: 1 mL (dropper bottle). Dissolve 0.20 g of bromcresol green in 200 mL of distilled water.

Hydrochloric acid (HCl), 0.05 M: 60 mL (dropper bottle). Add 1.7 mL of 6 M HCl to 200 mL of distilled water.

Hydrochloric acid (HCl), 1.0 M: 20 mL (dropper bottle). Add 33.4 mL of 6 M HCl to 167 mL of distilled water. Replace dropper with a Teflon-lined screw cap when bottle is put into storage.

Methyl orange indicator: 1 mL (dropper bottle). Dissolve 0.20 g of methyl orange in 200 mL of distilled water.

Methyl red indicator: 1 mL (dropper bottle). Dissolve 0.20 g of methyl red in 200 mL of distilled water.

Methyl violet indicator: 1 mL (dropper bottle). Dissolve 0.20 g of methyl violet in 200 mL of distilled water.

Phenolphthalein indicator: 1 mL (dropper bottle). Dissolve 0.20 g of phenolphthalein in 200 mL of 95% ethanol.

Silver nitrate ($AgNO_3$), 0.1 M: 1 mL (dropper bottle). Dissolve 3.4 g of $AgNO_3$ in 200 mL of distilled water.

Sodium acetate ($NaC_2H_3O_2$), 0.1 M: 30 mL (dropper bottle). Dissolve 2.7 g of solid $NaC_2H_3O_2 \cdot 3H_2O$ in 200 mL of distilled water.

Sodium chloride (NaCl), 0.1 M: 10 mL (dropper bottle). Dissolve 1.2 g of solid NaCl in 200 mL of distilled water.

Sodium hydroxide (NaOH), 0.05 M: 60 mL (dropper bottle). Add 1.7 mL of 6 M NaOH to 200 mL of distilled water.

Sodium hydroxide (NaOH), 1.0 M: 20 mL (dropper bottle). Add 33.4 mL of 6 M NaOH to 167 mL of distilled water. Replace the dropper with a Teflon-lined screw cap when bottle is put into storage.

Miscellaneous Supplies for Ten Students

Aspirin: 1 g. We prefer to leave commercial products in their original containers to add interest.

Baking soda: 1 g (original container)

Buffered aspirin: 1 g (original container)

Cotton (cloth or yarn), 4-cm length: 20

Dishwashing detergent, liquid: 3 mL (dropper bottle or original container)

Household ammonia: 3 mL (dropper bottle or original container)

Laundry detergent, liquid: 3 mL (dropper bottle or original container)

Lemon juice: 3 mL (dropper bottle or original container)

Milk: 35 mL (dropper bottle or original container)

Nylon (cloth or yarn), 4-cm length: 20
Orange juice or oranges: 3 mL (dropper bottle or original container)
pH paper (long-range, 1 to 11), 5-cm pieces: 10. More than one color chart
should be available in the lab.
Wool (cloth or yarn), 4-cm length: 20

Equipment from Stockroom for Ten Students
Test tubes, 15-cm: In some parts of the experiment, each student will need
eight of this size, Some should be available for checkout from the
stockroom.
Test tubes, 7.5-cm: In some parts of the experiment, each student will need
seven of this size. Some should be available for checkout from the
stockroom.

Disposal Containers
1.   160 mL/ten students: Chem. xxx, Exp. 12, Used Chemicals Part B

EXPERIMENT 13:   ANALYSIS OF VINEGAR

Chemicals for Ten Students
Potassium acid phthalate (KHP): 12 g

Reagents for Ten Students
Phenolphthalein indicator, 0.1%: 10 mL (dropper bottle). Dissolve 0.20 g of
solid phenolphthalein in 200 mL of 95% ethanol.
Sodium hydroxide (NaOH), 6 M: 60 mL (reagent bottle, beaker, and droppers in hood)

Miscellaneous Supplies for Ten Students
White vinegar: 400 mL (original container). Alternatively, a solution made
by mixing 20 mL of glacial acetic acid and 380 mL of distilled water
will be satisfactory.

Equipment from Stockroom for Ten Students
Burets, 25 or 50 mL: 10
Pipets, 10 mL: 10
Suction bulbs: 10
Volumetric flasks, 100 mL: 10

EXPERIMENT 14:   DETERMINATION OF K_a FOR WEAK ACIDS

Reagents for Ten Students
Phenolphthalein indicator, 0.1%: 10 mL (dropper bottle). Dissolve 0.20 g of
solid phenolphthalein in 200 mL of 95% ethanol.
Sodium hydroxide (NaOH), 6 M: 35 mL (reagent bottle, beaker, and droppers in hood)
Sodium hydroxide·(NaOH), 0.1 M: Approximately 900 mL. This is an optional solution in the lab. Students may use their solution prepared in
Experiment 13, or they may prepare the solution according to directions given in the experiment. To prepare 1 L, mix together 17.5 mL of
6 M NaOH and 1000 mL of distilled water.

Miscellaneous Supplies for Ten Students
pH paper (long-range, 1 to 11), 3-cm pieces: 10. More than one color chart
should be available in the lab.
pH paper (narrow-range, 1.4 to 2.8, or any narrow range that includes 2),
3-cm pieces: 10. More than one color chart should be available in the lab.

pH paper (narrow-range, 0.0 to 3.0), 3-cm pieces: 10. More than one color chart should be available in the lab.

Unknown acid solution: 500 mL. See Instructor's Manual for preparation directions.

Unknown solid acid: 10 g. See Instructor's Manual for preparation directions.

Equipment from Stockroom for Ten Students
Burets, 25 or 50 mL: 10
Pipets, 10 mL: 10
Suction bulbs: 10

EXPERIMENT 15: THE ACIDIC HYDROGENS OF ACIDS

Reagents for Ten Students
Phenolphthalein indicator, 0.1%: 1 mL (dropper bottle). Dissolve 0.20 g of solid phenolphthalein in 200 mL of 95% ethanol.

Phosphoric acid solution (H_3PO_4), 0.080 M: 350 mL. To make 1 L, use a 10-mL graduated cylinder to carefully measure 5.4 mL of 85% (syrupy) phosphoric acid (specific gravity = 1.7). Empty the cylinder into a 1-L volumetric flask. Rinse the cylinder three times with distilled water and add the rinse water to the contents of the volumetric flask. Add distilled water to fill the volumetric flask to the mark and mix well.

Sodium hydroxide (NaOH), 0.1 M: 1300 mL. This is an optional solution in the lab. Students may use their solution prepared in Experiment 13, or they may prepare the solution according to directions given in the experiment. To prepare 1500 mL, mix 26.2 mL of 6 M NaOH and 1500 mL of distilled water.

Sodium hydroxide (NaOH), 6 M: 25 mL (reagent bottle, beaker, and droppers in hood)

Miscellaneous Supplies for Ten Students
Unknown acid solutions: 350 mL. See Instructor's Manual for preparation directions.

Equipment from Stockroom for Ten Students
Burets, 25 or 50 mL: 10
Pipets, 10 mL: 10
Suction bulbs: 10

EXPERIMENT 16: THE USE OF MELTING POINTS IN THE IDENTIFICATION OF ORGANIC COMPOUNDS

Chemicals for Ten Students
Acetamide: 5 g
Acetoacetanilide: 5 g
Benzophenone: 5 g
Biphenyl: 5 g
Coumarin: 5 g
Ethyl carbamate: 5 g
Indole: 5 g
o-Nitrophenol: 5 g
Stearic acid: 5 g
p-Toluidine: 5 g
Vanillin: 5 g

Miscellaneous Supplies for Ten Students
Boiling chips
Melting-point capillary tubes: 90
Small rubber bands cut from rubber tubing: 20
Split stoppers: 10
Unknowns: 1-g sample (one per student) of one of the following solids:
   acetamide, acetoacetanilide, benzophenone, biphenyl, coumarin, ethyl
   carbamate, indole, stearic acid, or vanillin

Equipment from Stockroom for Ten Students
Thermometers, 150°C: 10

Disposal Containers
1. 90 capillary tubes/ten students: Chem. xxx, Exp. 16, Used Chemi-
    cals. A wide-mouth bottle works best as this container.

EXPERIMENT 17:   ISOLATION AND PURIFICATION OF AN
ORGANIC COMPOUND

Chemicals for Ten Students
Acetylsalicylic acid: 9 g
Ethanol, 95%: 40 mL (dropper bottle)
Glycerol: 700 mL (provide only if students set up their own heating baths).
Impure aspirin: 2.5 g. See Instructor's Manual for preparation directions.
2-Propanol (isopropyl alcohol): 40 mL (dropper bottle)
Sugar (sucrose): 3 g
Toluene: 40 mL (dropper bottle)

Miscellaneous Supplies for Ten Students
Boiling chips
Ice, crushed: One bucket for the lab
Melting-point capillaries (closed at one end): 10
Small rubber bands cut from rubber tubing: 10
Split stoppers: 10 (these might not be needed)

Equipment from Stockroom for Ten Students
Centrifuges, heads should accept 10-cm test tubes: 4 should be available in
    a lab of 20 to 25 students.
Glycerol baths for melting-point determination: 5. We set up five baths in
    hoods for our labs of 24 students. We use 100-mL beakers and fill them
    three-quarters full with glycerol.
Thermometers, 150°C: 10

Disposal Containers
1. 40 mL/ten students: Chem. xxx, Exp. 17, Used Toluene Parts A and B
2. 700 mL/ten students: Chem. xxx, Exp. 17, Used Glycerol. Provide this
    container only if students are required to set up their own glycerol baths.

EXPERIMENT 18:   HYDROCARBONS

Chemicals for Ten Students
Acetone: 15 mL (dropper bottle)
Cyclohexene: 3 mL (dropper bottle)
Glycerol: 700 mL (provide only if students set up their own heating baths).
Hexane: 12 mL (dropper bottle)
Hexene: 10 mL (dropper bottle)
Toluene: 12 mL (dropper bottle)

Reagents for Ten Students

Bromine (Br_2), 1%: 5 mL (brown dropper bottle). Prepare this solution in a hood. Gloves should be worn. Prepare only a slight excess of the amount that will be used during a 1-week period. Replace the dropper cap with a Teflon-lined screw cap when the solution is stored between periods of use. Open the screw cap cautiously after storage because a slight pressure might develop in the container. Dissolve 1 mL of liquid bromine in 100 mL of cyclohexane. Store and use in a hood.

Potassium permanganate ($KMnO_4$), 1%: 3 mL (dropper bottle). Dissolve 1.0 g of solid $KMnO_4$ in 100 mL of distilled water.

Miscellaneous Supplies for Ten Students

Boiling tubes, 6×50 mm: 10 (we used 6×50 mm disposable culture tubes).
Melting-point capillaries (closed at one end): 40
Small rubber bands cut from rubber tubing: 20
Split stoppers: 10 (these might not be needed)
Unknowns: 3-mL sample (one per student) of one of the compounds listed in the table of Part D of the experiment.

Equipment from Stockroom for Ten Students

Gycerol baths for boiling-point determinations: 5. We set up five baths in hoods for our labs of 24 students. We use 100-mL beakers and fill them three-quarters full with glycerol.
Thermometers, 150°C: 10

Disposal Containers

1. 65 mL/ten students: Chem. xxx, Exp. 18, Used Chemicals
2. 700 mL/ten students: Chem: xxx, Exp. 18, Used Glycerol. Provide this container only if students are required to set up their own glycerol baths.

EXPERIMENT 19: REACTIONS OF ALCOHOLS AND PHENOLS

Chemicals for Ten Students

Acetone: 50 mL (dropper bottle)
Antiseptic: 20 mL (dropper bottle) — See instructor's manual for preparation directions.
t-Butanol: 3 mL (dropper bottle)
Cyclohexanol: 3 mL (dropper bottle) — Heat in hot water bath if the alcohol solidifies.
Ethanol: 8 mL (dropper bottle)
Ethylene glycol: 3 mL (dropper bottle)
1-Propanol: 6 mL (dropper bottle)
2-Propanol: 6 mL (dropper bottle)

Reagents for Ten Students

Ceric nitrate solution: 25 mL (dropper bottle). Dissolve 80 g of solid ceric ammonium nitrate (($NH_4)_2Ce(NO_3)_6$) in 200 mL of 2 M nitric acid. Prepare the 2 M nitric acid by adding 67 mL of 6 M (dilute) HNO_3 to 133 mL of distilled water.
Chromic acid (CrO_3), 10%: 3 mL (dropper bottle). Dissolve 22 g of solid CrO_3 in 200 mL of 3 M (dilute) H_2SO_4. Replace dropper cap with a Teflon-lined screw cap when the bottle is put into storage.
Ferric chloride ($FeCl_3$), 1%: 5 mL (dropper bottle). Dissolve 3.4 g of $FeCl_3 \cdot 6H_2O$ in 200 mL of distilled water.
Iodine–potassium iodide (I_2–KI): 50 mL (dropper bottle). Dissolve 8.3 g of potassium iodide (KI) and 12.7 g of iodine (I_2) in 200 mL of distilled water. Grind the solid KI and I_2 together in a mortar before adding water.

Lucus reagent: 25 mL (dropper bottle). Dissolve 195 g of solid $ZnCl_2$ in 206 mL of cold, concentrated hydrochloric acid (HCl). This gives about 200 mL of reagent. Replace the dropper cap with a Teflon-lined screw cap when the bottle is put into storage.

Liquified phenol, 88%: 2 mL (dropper bottle). Liquified phenol from chemical suppliers may be used, or make the solution by mixing together 176 g of solid phenol and 24 mL of water in a beaker. Gently heat the mixture with stirring until a liquid is produced. Replace the dropper cap with a Teflon-lined screw cap when the bottle is stored.

Resorcinol, 35%: 1 mL (dropper bottle). Dissolve 70 g of solid resorcinol in 130 mL of distilled water.

Sodium hydroxide (NaOH), 10%: 100 mL (dropper bottle). Dilute 85 mL of 6 M NaOH to 200 mL using distilled water.

Miscellaneous Supplies for Ten Students

pH paper (wide-range, 1 to 11), 5-cm pieces: 10. More than one color chart should be available in the lab.

Unknowns: 3-mL sample (one per student). See the Instructor's Manual for the list of compounds.

Disposal Containers
1. 75 mL/ten students: Chem. xxx, Exp. 19, Used Chemicals Part A
2. 40 mL/ten students: Chem. xxx, Exp. 19, Used Chemicals Part B
3. 20 mL/ten students: Chem. xxx, Exp. 19, Used Chemicals Part C
4. 60 mL/ten students: Chem. xxx, Exp. 19, Used Chromic Acid.
5. 90 mL/ten students: Chem. xxx, Exp. 19, Used Chemicals Part E
6. 40 mL/ten students: Chem. xxx, Exp. 19, Used Chemicals Part F
7. 80 mL/ten students: Chem. xxx, Exp. 19, Used Chemicals Part G

EXPERIMENT 20:   REACTIONS OF ALDEHYDES AND KETONES

Chemicals for Ten Students
Acetone: 65 mL (dropper bottle)
2-Butanone (methylethyl Retone): 5 mL (dropper bottle)
Butyraldehyde: 10 mL (dropper bottle)
Flavoring agent: 10 mL (dropper bottle)
p-Tolualdehyde: 1 mL (dropper bottle)

Reagents for Ten Students
Ammonia (NH_3 or NH_4OH), 6 M: 30 mL (reagent bottle, beaker, and droppers in hood).

Chromic acid (CrO_3), 10%: 3 mL (dropper bottle). Dissolve 22 g of solid CrO_3 in 200 mL of 3 M (dilute) H_2SO_4.

2,4-Dinitrophenylhydrazine: 20 mL (dropper bottle). Add 27 mL of concentrated H_2SO_4 to 5.3 g of solid 2,4-dinitrophenylhydrazine contained in a small beaker. Stir the mixture with a glass stirring rod, then pour the mixture cautiously into 40 mL of distilled water contained in a 500-mL beaker. Add 135 mL of 95% ethanol to the mixture in the 500-mL beaker and stir. Rinse the original small beaker several times with solution from the 500-mL beaker. Pour the rinse solutions back into the 500-mL beaker. After thorough mixing, the solution in the 500-mL beaker should be filtered before it is put into a dropper bottle.

Iodine–potassium iodide (I_2–KI): 50 mL (dropper bottle). Dissolve 8.3 g of potassium iodide (KI) and 12.7 g of iodine (I_2) in 200 mL of distilled water. Grind the solid KI and I_2 together in a mortar before adding water.

Silver nitrate ($AgNO_3$), 0.1 M: 30 mL (dropper bottle). Dissolve 3.4 g of $AgNO_3$ in 200 mL of distilled water.

Sodium hydroxide (NaOH), 10%: 8 mL (dropper bottle). Dilute 42 mL of 6 M NaOH to 200 mL using distilled water.

Sodium hydroxide (NaOH), 6 M: 30 mL (reagent bottle, beaker, and droppers in hood).

Miscellaneous Supplies for Ten Students

Unknowns: 2-mL sample (one per student). See the Instructor's Manual for the list of compounds.

Disposal Containers

1. 25 mL/ten students: Chem. xxx, Exp. 20, Used Chemicals Part A
2. 60 mL/ten students: Chem. xxx, Exp. 20, Used Chemicals Part C
3. 95 mL/ten students: Chem. xxx, Exp. 20, Used Chemicals Part D

EXPERIMENT 21: REACTIONS OF CARBOXYLIC ACIDS, AMINES, AND AMIDES

Chemicals for Ten Students

Acetamide: 1 g

Acetic acid, glacial: 3 mL (reagent bottle, beaker, and droppers in hood)

Alka-Seltzer®: 8 g (3 tablets)

Benzamide: 1 g

Benzoic acid: 1 g

Butyric acid: 3 mL (dropper bottle in hood)

Caproic acid: 2 mL (dropper bottle in hood)

N,N-dimethylaniline: 2 mL (dropper bottle in hood)

Triethylamine: 2 mL (dropper bottle in hood)

Reagents for Ten Students

Hydrochloric acid (HCl), 6 M (dilute): 50 mL (reagent bottle, beaker, and droppers in hood).

Sodium bicarbonate ($NaHCO_3$), saturated: 10 mL (dropper bottle). Mix together 24 g of solid $NaHCO_3$ and 200 mL of distilled water. Shake to dissolve. Some solid should remain undissolved.

Sodium hydroxide (NaOH), 3 M: 15 mL (dropper bottle). Mix together 100 mL of 6 M NaOH solution and 100 mL of distilled water.

Sodium hydroxide (NaOH), 6 M: 25 mL (reagent bottle, beaker, and droppers in hood).

Miscellaneous Supplies for Ten Students

Centrifuges, heads should accept 10-cm test tubes: We put four in our labs of 24 students.

Unknowns: 2-mL sample (one per student). See the Instructor's Manual for the list of compounds.

pH paper (wide-range, 1 to 11), 5-cm pieces: 10. More than one color chart should be available in the lab.

Disposal Containers

1. 30 mL/ten students: Chem. xxx, Exp. 21, Used Chemicals Part A
2. 65 mL/ten students: Chem. xxx, Exp. 21, Used Chemicals Part B
3. 40 mL/ten students: Chem. xxx, Exp. 21, Used Chemicals Part D
4. 15 mL/ten students: Chem. xxx, Exp. 21, Used Chemicals Part E
5. 35 mL/ten students: Chem. xxx, Exp. 21, Used Chemicals Part F

EXPERIMENT 22: THE SYNTHESIS OF ASPIRIN AND OTHER ESTERS

Chemicals for Ten Students

Acetic acid, glacial: 2 mL (reagent bottle, beaker, and droppers in hood)

Acetic anhydride: 8 mL (dropper bottle in hood). Replace the dropper cap with a Teflon-lined screw cap when bottle is stored.

n-Butyl alcohol: 3 mL (dropper bottle)

Butyric acid: 2 mL (dropper bottle in hood). Replace the dropper cap with a Teflon-lined screw cap when bottle is stored.

Ethyl alcohol: 15 mL (dropper bottle)

Isoamyl alcohol: 3 mL (dropper bottle)

Methyl alcohol: 3 mL (dropper bottle)

n-Propyl alcohol: 3 mL (dropper bottle)

Oil of wintergreen (methyl salicylate): 1 mL (dropper bottle)

Salicylic acid: 4 g

Reagents for Ten Students

Iron (III) chloride ($FeCl_3$), 1%: 1 mL (dropper bottle). Dissolve 3.4 g of $FeCl_3 \cdot 6H_2O$ in 200 mL of distilled water.

Sulfuric acid (H_2SO_4), 18 M (conc.): 4 mL (reagent bottle, beaker, and droppers in hood)

Miscellaneous Supplies for Ten Students

Aspirin tablets, crushed: 10 (in original container)

Centrifuges, heads should accept 10-cm test tubes: We put four in our labs of 24 students.

Ice, crushed: one bucket for the lab

Thermometers, 150°C: 10

Disposal Containers

1. 65 mL/ten students: Chem. xxx, Exp. 22, Used Chemicals

EXPERIMENT 23: IDENTIFYING FUNCTIONAL GROUPS IN UNKNOWNS

Chemicals for Ten Students

Acetic acid, glacial: 2 mL (reagent bottle, beaker, and droppers in hood)

Acetone: 40 mL (dropper bottle)

Benzoic acid: 1 g

Butyraldehyde: 3 mL (dropper bottle)

N,N-dimethylaniline: 2 mL (dropper bottle in hood). Replace the dropper cap with a Teflon-lined screw cap when bottle is stored.

Ethanol: 2 mL (dropper bottle)

Hexene: 5 mL (dropper bottle)

Triethylamine: 2 mL (dropper bottle in hood). Replace the dropper cap with a Teflon-lined screw cap when the bottle is stored.

Reagents for Ten Students

Ammonia (NH_3 or NH_4OH), 6 M: 23 mL (reagent bottle, beaker, and droppers in hood).

Bromine (Br_2), 1%: 2 mL (brown dropper bottle). Prepare this solution in a hood. Gloves should be worn. Prepare only a slight excess of the amount that will be used during a 1-week period. Replace the dropper cap with a Teflon-lined screw cap when the solution is stored between periods of use. Open the screw cap cautiously after storage because slight pressure might develop in the container. Dissolve 1 mL of liquid bromine in 100 mL of cyclohexane.

Ceric nitrate solution: 12 mL (dropper bottle). Dissolve 80 g of solid ceric ammonium nitrate ((NH_4)$_2$Ce(NO_3)$_6$) in 200 mL of 2 M nitric acid. Prepare the 2 M nitric acid by adding 67 mL of 6 M (dilute) HNO_3 to 133 mL of distilled water.

Chromic acid (CrO_3), 10%: 2 mL (dropper bottle). Dissolve 22 g of solid CrO_3 in 200 mL of 3 M (dilute) H_2SO_4. Replace the dropper cap with a Teflon-lined screw cap when the bottle is stored.

2,4-Dinitrophenylhydrazine: 5 mL (dropper bottle). Add 27 mL of concentrated H_2SO_4 to 5.3 g of solid 2,4-dinitrophenylhydrazine contained in a small beaker. Stir the mixture with a glass stirring rod, then pour the mixture cautiously into 40 mL of distilled water contained in a 500-mL beaker. Add 135 mL of 95% ethanol to the mixture in the 500-mL beaker and stir. Rinse the original small beaker several times with solution from the 500-mL beaker. Pour the rinse solutions back into the 500-mL beaker. After thorough mixing, the solution in the 500-mL beaker should be filtered before it is put into a dropper bottle.

Ferric chloride ($FeCl_3$), 1%: 2 mL (dropper bottle). Dissolve 3.4 g of solid $FeCl_3 \cdot 6H_2O$ in 200 mL of distilled water.

Hydrochloric acid (HCl), 6 M (dilute): 10 mL (reagent bottle, beaker, and droppers in hood).

Liquified phenol, 88%: 2 mL (dropper bottle). Liquified phenol from chemical suppliers may be used, or make the solution by mixing together 176 g of solid phenol and 24 mL of water in a beaker. Gently heat the mixture with stirring until a liquid is produced. Replace the dropper cap with a Teflon-lined screw cap when the bottle is stored.

Silver nitrate ($AgNO_3$), 0.1 M: 20 mL (dropper bottle). Dissolve 3.4 g of solid $AgNO_3$ in 200 mL of distilled water.

Sodium hydroxide (NaOH), 10%: 5 mL (dropper bottle). Dilute 42 mL of 6 M NaOH solution to 200 mL using distilled water.

Sodium hydroxide (NaOH), 6 M: 20 mL (reagent bottle, beaker, and droppers in hood)

Miscellaneous Supplies for Ten Students

pH paper (wide-range, 1 to 11), 5-cm pieces: 10. More than one color chart should be available in the lab.

Unknowns, water-insoluble: 2 mL (if liquid) or 0.4 g (if solid); one per student. See Instructor's Manual for preparation directions.

Unknowns, water-soluble: 2 mL (if liquid) or 0.4 g (if solid); one per student. See Instructor's Manual for preparation directions.

Disposal Containers

1. 50 mL/ten students: Chem. xxx, Exp. 23, Used Chemicals Part A
2. 20 mL/ten students: Chem. xxx, Exp. 23, Used Chromic Acid Part A
3. 70 mL/ten students: Chem. xxx, Exp. 23, Used Chemicals Part B

EXPERIMENT 24: SYNTHETIC POLYMERS

Chemicals for Ten Students

Benzoyl peroxide: 0.2 g. *This compound is explosive and only the amount needed should be available.*

Glycerol: 40 mL

Glyoxal, 40%: 16 mL (dropper bottle). This is obtained as a 40% solution from the supplier (Sigma Chemical Co.)

Methyl methacrylate: 50 mL (dropper bottle). Store in refrigerator between uses.

Phthalic anhydride: 100 g

Poly (sodium acrylate): 2 g. We obtain this material from Bartlett Services, 2708 Steeple Lane, Muscatine, Iowa, 52791. They sell it under the name Supersorb.

Resorcinol: 10 g

Sodium acetate: 5 g

Reagents for Ten Students

Adipoyl chloride, 5%: 100 mL. Dissolve 4 mL of adipoyl chloride in 145 mL of hexane.

Borax, 4%: 20 mL (dropper bottle). Dissolve 8 g of solid sodium borate ($Na_2B_4O_7 \cdot 10H_2O$) in 190 mL of distilled water. The solid dissolves slowly and should be shaken several times during a few days or put on a magnetic stirrer for several hours.

Hexamethylenediamine (also named 1,6-hexanediamine), 5%: 100 mL. Dissolve 5 g of solid hexamethylenediamine and 2 g of solid $Na_2CO_3 \cdot 10H_2O$ in 95 mL of distilled water.

Hydrochloric acid (HCl), 3 M: 1 mL (dropper bottle). Pour 100 mL of 6 M (dilute) HCl into 100 mL of distilled water. Replace the dropper cap with a Teflon-lined screw cap when the bottle is stored.

Poly (vinyl alcohol), 4%: 100 mL. The use of a hot plate equipped with a magnetic stirrer greatly facilitates this preparation. Dissolve 1 g of solid sodium benzoate (a preservative) in 1 L of distilled water that is being stirred by a magnetic stirrer. Add, with continued stirring, 40 g of solid poly (vinyl alcohol) (use only a nearly 100% hydrolyzed form). Heat the resulting suspension, with continued stirring, until the temperature reaches 80°–90°C. The solid should have dissolved to form a nearly clear solution. Food coloring may be added to the solution if desired (we use 3 drops of green and 8 drops of blue per liter to give a blue-green color).

Sodium chloride (NaCl), saturated: 50 mL (dropper bottle). Mix together 75 g of solid NaCl and 200 mL of distilled water. Shake to dissolve. Some solid should remain undissolved.

Miscellaneous Supplies for Ten Students

Aluminum foil, 10-cm squares for mold: 10

Paper clips: 10

Disposal Containers

1.   150 mL/ten students: Chem. xxx, Exp. 24, Used Chemicals Part C

EXPERIMENT 25:   DYES, INKS, AND FOOD COLORINGS

Chemicals for Ten Students

Acetic acid, glacial: 2 mL (reagent bottle, beaker, and droppers in hood)

Acetone: 24 mL (dropper bottle)

Ethanol, 95%: 60 mL (dropper bottle)

β-Naphthol: 1 g

p-Nitroaniline: 1 g

Phenol: 1 g

Resorcinol: 1 g

Reagents for Ten Students

Hydrochloric acid (HCl), 12 M (conc.): 70 mL (reagent bottle, beaker, and droppers in hood).

Malachite green: 100 mL. Dissolve 0.25 g of malachite green in 100 mL of distilled water. If the water insoluble base form of indicator is used, dissolve the solid in 3-4 mL of dilute (6 M) HCl before adding water.

Primuline: 50 mL. Dissolve 0.2 g of primuline in 50 mL of distilled water.

Sodium hydroxide (NaOH), 6 M: 30 mL (reagent bottle, beaker, and droppers in hood).

Sodium nitrite ($NaNO_2$), 1.0 M: 45 mL. Dissolve 3.4 g of $NaNO_2$ in 50 mL of distilled water. This solution should be prepared immediately before lab.

Miscellaneous Supplies for Ten Students

Boiling chips

Cotton cheesecloth, 5-cm squares: 40. Strips of multifiber test fabric give good results. We use No. 10 available from Testfabrics, Inc., P.O. Drawer O, Blackford Ave., Middlesex, NJ 08846.

Chromatography paper, Whatman #1, 9 cm × 7 cm: 10

Chromatography paper, strips, 1 cm × 15 cm: 10

Filter paper, 5.5 cm: 10

Food coloring, three different colors; 5 drops (small vials or bottles). We use food colorings that are available in grocery stores and mix the dyes according to their directions to get three different colors. We use tape to attach a small test tube or culture tube to each vial of coloring material and put 2 or 3 dropping capillaries in each tube to be used with that color. The dropping capillaries are made according to the directions given in Step 1 of Part C of the experiment.

Ice, crushed: One bucket for the lab

Ink: 1 to 3 black felt-tip pens. Only certain inks work well; Paper Mate's nylon fiber point with black ink (863-11) separates beautifully into reds and blues.

Kool-Aid®, four different flavors: 10 drops (small dropper bottles or vials with droppers provided). We use flavors of Kool-Aid that contain two or more food colorings such as grape, orange, great bluedini and lemon-lime. Check labels to be certain two or more food colorings are present. Add 10 mL of distilled water to the contents of each package, mix well, then pour off liquid from undissolved solid. Use the liquid.

Melting-point capillaries (open at both ends): 40

Equipment from Stockroom for Ten Students

Pencils: 5. To keep students from using pens on their chromatography paper, it is a good idea to have some pencils available.

Rulers, 15 cm with mm markings: 10

Test tubes, 10 cm: 40 (each student will need a total of 16).

Thermometers, 150°C: 10

Disposal Containers

1. 45 mL/ten students: Chem. xxx, Exp. 25, Used Chromatography Solvent Part B
2. 112 mL/ten students: Chem. xxx, Exp. 25, Used Chemicals Part D
3. 125 mL/ten students: Chem. xxx, Exp. 25, Used Chemicals Part E
4. 575 mL/ten students: Chem. xxx, Exp. 25, Used Chemicals Part F

EXPERIMENT 26: A STUDY OF CARBOHYDRATES

Reagents for Ten Students

Aniline acetate: 3 mL (dropper bottle). To 17 mL of distilled water, add with stirring 7 mL of glacial acetic acid and 10 mL of aniline. This solution should be prepared the day of the lab period in which it is used.

Barfoed's reagent: 80 mL (dropper bottle). Dissolve 13.2 g of copper acetate ($Cu(C_2H_3O_2)_2 \cdot H_2O$) in 200 mL of distilled water. Filter, if necessary, and then add 1.9 mL of glacial acetic acid.

Benedict's reagent: 80 mL (dropper bottle). With the aid of heat, dissolve 35 g of sodium citrate and 20 g of anhydrous sodium carbonate in about 160 mL of distilled water. Filter the solution and, with stirring, slowly add 5.0 g of copper sulfate ($CuSO_4 \cdot 5H_2O$) dissolved in 20 mL of distilled water. Dilute the resulting solution to 200 mL with distilled water.

Fructose, 4%: 50 mL (dropper bottle). Dissolve 6 g of fructose in 150 mL of distilled water.

Gelatin dessert, 10%: 50 mL (dropper bottle). Dissolve 15 g of lemon-flavored powder in 150 mL of distilled water.

Glucose, 4%: 50 mL (dropper bottle). Dissolve 6 g of glucose in 150 mL of distilled water.

Honey, 10%: 50 mL (dropper bottle). Dissolve 15 g of honey in 150 mL of distilled water.

Hydrochloric acid (HCl), 12 M (conc.): 100 mL (reagent bottle, beaker, and droppers in hood).

Lactose, 4%: 50 mL (dropper bottle). Dissolve 6 g of lactose in 150 mL of distilled water.

Seliwanoff's reagent: 160 mL. Dissolve 0.2 g of resorcinol in 400 mL of 6 M HCl.

Sucrose, 4%: 50 mL (dropper bottle). Dissolve 6 g of sucrose in 150 mL of distilled water.

Xylose, 4%: 70 mL (dropper bottle). Dissolve 8 g of xylose in 200 mL of distilled water.

Yeast suspension, 3%: 240 mL. Add 12 g of dry yeast to 400 mL of distilled water and stir the suspension. We have found that foil packets of active dry yeast are very convenient to store and use. We mix one packet (¼ oz) with every 240 mL of distilled water.

Miscellaneous Supplies for Ten Students
Boiling chips
Unknowns: 8 mL sample (one per student) of one of the following 4% sugar solutions: fructose, glucose, lactose, sucrose, or xylose

Disposal Containers
1. 160 mL/ten students: Chem. xxx, Exp. 26, Used Chemicals Part B
2. 160 mL/ten students: Chem. xxx, Exp. 26, Used Chemicals Part C
3. 160 mL/ten students: Chem. xxx, Exp. 26, Used Chemicals Part D
4. 60 mL/ten students: Chem. xxx, Exp. 26, Used Chemicals Part E

EXPERIMENT 27:   PREPARATION OF SOAP BY LIPID SAPONIFICATION

Chemicals for Ten Students
Dodecanol: 20 mL (dropper bottle)
Mineral oil: 20 mL (dropper bottle)
Sodium chloride (NaCl): 200 g

Reagents for Ten Students
Calcium chloride ($CaCl_2$), 1%: 40 mL (dropper bottle). Dissolve 2.6 g of $CaCl_2 \cdot 2H_2O$ in 200 mL of distilled water.

Hydrochloric acid (HCl), 6 M (dilute): 5 mL (reagent bottle, beaker, and droppers in hood).

Iron (III) chloride ($FeCl_3$), 1%: 40 mL (dropper bottle). Dissolve 3.4 g of $FeCl_3 \cdot 6H_2O$ in 200 mL of distilled water.

Magnesium chloride ($MgCl_2$), 1%: 40 mL (dropper bottle). Dissolve 4.3 g of $MgCl_2 \cdot 6H_2O$ in 200 mL of distilled water.

Phenolphthalein indicator: 5 mL (dropper bottle). Dissolve 0.20 g of phenolphthalein in 200 mL of 95% ethanol.
Sodium hydroxide (NaOH), 6 M: 50 mL (reagent bottle, beaker, and droppers in hood).
Sulfuric acid (H_2SO_4), 18 M (conc.): 10 mL (reagent bottle, beaker, and droppers in hood).

Miscellaneous Supplies for Ten Students
Cheesecloth, 15-cm squares: 60
Ice, crushed: One bucket for the lab
Vegetable oil: 20 mL (dropper bottle)

EXPERIMENT 28: ISOLATION OF NATURAL PRODUCTS: TRIMYRISTIN AND CHOLESTEROL

Chemicals for Ten Students
Acetic acid, glacial: 10 mL (reagent bottle, beaker, and droppers in hood)
Acetic anhydride: 2 mL (reagent bottle, beaker, and droppers in hood)
Acetone: 25 mL (dropper bottle)
Hexane: 60 mL (dropper bottle)
2-Propanol: 50 mL (dropper bottle)
Sulfuric acid (H_2SO_4), (conc.): 10 drops (reagent bottle, beaker, and droppers in hood)

Reagents for Ten Students
Cholesterol, 1%: 2 mL (dropper bottle in hood). Dissolve 1 g of solid cholesterol in 100 mL of glacial acetic acid. Replace the dropper cap with a Teflon-lined screw cap when the bottle is stored.

Miscellaneous Supplies for Ten Students
Cheesecloth, 8-cm squares: 40
Egg yolk, cooked: 20 g (about two egg yolks from hard-boiled eggs)
Melting-point capillaries (closed at one end): 10
Nutmeg, powdered: 10 g

Disposal Containers
1. 10 mL/ten students: Chem. xxx, Exp. 28, Used Cholesterol Test Materials Part C
2. 1 to 2 g/ten students: Chem. xxx, Exp. 28, Prepared Cholesterol Part C.

EXPERIMENT 29: AMINO ACIDS AND PROTEINS

Chemicals for Ten Students
Acetic acid, glacial: 3 mL (reagent bottle, beaker, and droppers in hood)
Ethanol, 95%: 100 mL
Isopropyl alcohol (2-propanol): 15 mL (dropper bottle)
Ninhydrin, in spray can: These are available from chemical suppliers, or prepare 0.2% ninhydrin in n-butanol and couple a container to a propellant can.

Reagents for Ten Students
Alanine, 4%: 45 mL (dropper bottle). Dissolve 4 g of solid alanine in 100 mL of distilled water.
Arginine, 4%: 5 mL (dropper bottle). Dissolve 4 g of solid arginine in 100 mL of distilled water.
Asparagine, 2%: 0.5 mL (dropper bottle). Dissolve 2 g of solid asparagine in 100 mL of distilled water.

Aspartic acid, 4%: 5 mL (dropper bottle). Dissolve 4 g of solid aspartic acid in 100 mL of distilled water.

Copper sulfate ($CuSO_4$), 0.5%: 1 mL (dropper bottle). Dissolve 1.6 g of $CuSO_4 \cdot 5H_2O$ in 200 mL of distilled water.

Egg albumin, 2%: 100 mL. Prepare 1% sodium nitrate, $NaNO_3$, by adding 2.0 g of $NaNO_3$ to 200 mL of distilled water. To the resulting solution, add 4 g of fresh egg whites and gently stir the mixture. A large egg yields about 30 g of egg white. Store the resulting solution in a refrigerator between uses.

Ether–ethanol, 1:1 ratio: 60 mL. Add 100 mL of ethyl ether to 100 mL of absolute alcohol.

Gelatin, 2%: 30 mL (dropper bottle). Dissolve 2 g of solid gelatin in 100 mL of warm distilled water. It is best to make this the day of lab.

Glycine, 2%: 0.5 mL (dropper bottle). Dissolve 2.0 g of glycine in 100 mL of distilled water. Be certain you do not use the hydrochloride salt.

Hydrochloric acid (HCl), 0.1%: 10 mL (dropper bottle). Dilute 1.0 mL of 6 M HCl to 200 mL using distilled water.

Lead acetate ($Pb(C_2H_3O_2)_2$), 5%: 1 mL (dropper bottle). Dissolve 12.4 g of $Pb(C_2H_3O_2)_2 \cdot 3H_2O$ in 200 mL of distilled water.

Mercury (II) chloride ($HgCl_2$), 5%: 1 mL (dropper bottle). Dissolve 5.0 g of $HgCl_2$ in 95 mL of distilled water. Be certain to use the correct chloride of mercury.

Methanol, 80%: 30 mL (dropper bottle). Add 40 mL of distilled water to 160 mL of methanol.

Millon's reagent: 5 mL (dropper bottle). Dissolve 33 g of mercury in 50 mL of concentrated nitric acid by gently heating the mixture under a hood. Dilute the resulting solution with 165 mL of distilled water.

Proline, 4%: 0.5 mL (dropper bottle). Dissolve 4 g of proline in 100 mL of distilled water.

Silver nitrate ($AgNO_3$), 5%: 1 mL (dropper bottle). Dissolve 2.5 g of $AgNO_3$ in 48 mL of distilled water.

Sodium hydroxide (NaOH), 0.1%: 10 mL (dropper bottle). Dilute 1.0 mL of 6 M NaOH to 200 mL using distilled water.

Sodium hydroxide (NaOH), 6 M: 5 mL (reagent bottle, beaker, and droppers in hood)

Miscellaneous Supplies for Ten Students

Cheesecloth, 15-cm squares: 40

Chromatography paper, 9 cm × 7 cm: 10. Cut from Whatman #1 chromatography paper, catalog no. 3001917.

Melting-point capillaries, open at both ends: 20

Milk: 200 mL

pH paper (narrow-range, any series that covers the entire range—for example, 1 to 3, 3 to 5, 5 to 8, 8 to 11—will work): 5 strips for each narrow range, per student.

Unknowns: 3-drop samples (small vial), one per student. The unknown solutions should contain one, two, or three of the following amino acids at the concentration indicated: asparagine (2%), glycine (2%), or proline (4%). Be sure the concentration of each amino acid is at the value indicated. For example, an unknown containing both asparagine and proline should contain 2 g of asparagine and 4 g of proline dissolved in 100 mL of distilled water.

pH paper (wide-range, 1 to 11), 5-cm pieces: 10. More than one color chart should be available in the lab.

Equipment from Stockroom for Ten Students

Oven: Students will need access to one heated to 100°C.

Pencils: It is a good idea to have some available for those students who only brought a pen to lab.

Rulers, 15 cm with mm markings: 10

Scissors: Several pairs should be available from the stockroom or in the lab.

Disposal Containers

1. 30 mL/ten students: Chem. xxx, Exp. 29, Used Chromatography Solvent
2. 200 mL/ten students: Chem. xxx, Exp. 29, Used Ether–Ethanol Solution Part B
3. 30 mL/ten students: Chem. xxx, Exp. 29, Used Biuret Test Materials Part E
4. 45 mL/ten students: Chem. xxx, Exp. 29, Used Millon Test Materials Part E
5. 60 mL/ten students: Chem. xxx, Exp. 29, Used Heavy Metal Test Materials Part F

EXPERIMENT 30:   ENZYMES: NATURE'S CATALYSTS

Reagents for Ten Students

Benedict's reagent: 40 mL (dropper bottle). With the aid of heat, dissolve 10.4 g of sodium citrate and 6.0 g of anhydrous sodium carbonate in 50 mL of distilled water. Filter the solution and, while stirring slowly, add 1.0 g of copper sulfate ($CuSO_4 \cdot 5H_2O$) dissolved in 6 mL of distilled water.

Gelatin, double strength: 40 mL (dropper bottle). Follow the directions on a package of flavored gelatin dessert or unflavored gelatin, except use half of the amount of water. Heat the mixture to liquify it before each lab session.

Hydrogen peroxide, 3%: 10 mL (dropper bottle). Dilute 20 mL of 30% H_2O_2 to 200 mL using distilled water.

Litmus milk: 60 mL (dropper bottle). Add 0.5 g of powdered litmus to 100 mL of milk. Shake the solution thoroughly and then filter through cotton.

Meat tenderizer, 4%: 10 mL (dropper bottle). We use Schilling's Unseasoned Meat Tenderizer. Add 4 g of meat tenderizer to 100 mL of distilled water and mix well.

Pancreatin, 5%: 10 mL (dropper bottle). Add 5 g of pancreatin to 100 mL of distilled water and stir the mixture.

Pepsin (buffered), 1%: 10 mL (dropper bottle). Add 1 g of pepsin, 0.14 g of sodium phosphate ($Na_2HPO_4 \cdot 12H_2O$, disodium salt), and 2.06 g of citric acid ($C_6H_8O_7 \cdot H_2O$) to distilled water and dilute to 100 mL.

Phenol red indicator: 1 mL (dropper bottle). Dissolve 1 pellet (0.1 g) of sodium hydroxide (NaOH) in 100 mL of distilled water. Mix well, then put 10 mL of the solution into a clean 250 mL beaker. Add 0.1 g of solid phenol red to the 250 mL beaker and stir with a grinding action to dissolve most of the solid. Dilute to 200 mL with distilled water. A little solid might remain undissolved.

Rennin, 5%: 4 mL (dropper bottle). Add 2.5 g of rennin to 48 mL of distilled water and stir the mixture.

Sucrose, 5%: 20 mL (dropper bottle). Dissolve 5 g of sucrose in 100 mL of distilled water.

Trypsin (buffered), 1%: 10 mL (dropper bottle). Add 0.5 g of trypsin, 3.42 g of sodium phosphate ($Na_2HPO_4 \cdot 12H_2O$, disodium salt), and 0.04 g of citric acid ($C_6H_8O_7 \cdot H_2O$ to distilled water and dilute to 50 mL.

Urea, 5%: 2 mL (dropper bottle). Dissolve 10.5 g of urea in 200 mL of distilled water.

Urease, 1%: 2 mL (dropper bottle). Dissolve 1 g of solid urease (activity of 100 units or greater per gram) in 100 mL of distilled water.

Yeast suspension, 3%: 20 mL. Add 3 g of dry yeast to 100 mL of distilled water and stir the suspension. We have found that foil packets of active dry yeast are very convenient to store and use. We mix one packet (¼ oz) with every 200 mL of distilled water. It is adviseable to make this fresh each lab day.

Miscellaneous Supplies for Ten Students

Cheesecloth, 10-cm squares: 40

Ice, crushed: One bucket for the lab

Milk, treated with buttermilk: 200 mL. To each quart of whole pasteurized milk, add 20 mL of buttermilk and allow the milk to stand 2 hours before using.

Potato, peeled, raw, 2-cm cubes: 10

Sand: 20 g

Equipment from Stockroom for Ten Students

Thermometers, 150°C: 10

Disposal Containers

1. 50 mL/ten students: Chem. xxx, Exp. 30, Used Benedict's Test Materials Part D

EXPERIMENT 31: FACTORS THAT INFLUENCE ENZYME ACTIVITY

Reagents for Ten Students

Buffer, pH 4: 10 mL (dropper bottle). Dissolve 5.58 g of sodium phosphate ($Na_2HPO_4 \cdot 12H_2O$, disodium salt) and 2.76 g of citric acid ($C_6H_8O_7 \cdot H_2O$) in distilled water and dilute to 200 mL.

Buffer, pH 7: 10 mL (dropper bottle). Dissolve 11.80 g of sodium phosphate ($Na_2HPO_4 \cdot 12H_2O$, disodium salt) and 0.74 g of citric acid ($C_6H_8O_7 \cdot H_2O$) in distilled water and dilute to 200 mL.

Catechol, 0.01 M: 100 mL (dropper bottle). Dissolve 0.22 g of catechol in 200 mL of distilled water.

1,4-Cyclohexanediol, 0.01 M: 10 mL (dropper bottle). Dissolve 0.23 g of 1,4-cyclohexanediol in 200 mL of distilled water.

Hydrochloric acid (HCl), 0.1 M: 10 mL (dropper bottle). Dilute 3.4 mL of 6 M HCl to 200 mL using distilled water.

Lead nitrate ($Pb(NO_3)_2$), 5%: 5 mL (dropper bottle). Dissolve 10.0 g $Pb(NO_3)_2$ in 190 mL of distilled water.

Phenol, 0.01 M: 10 mL (dropper bottle). Dissolve 0.20 g of phenol in 200 mL of distilled water.

Phenylthiourea, saturated solution: 5 mL (dropper bottle). Add 0.8 g of phenylthiourea to 200 mL of distilled water and shake thoroughly. Some solid should remain on the bottom.

Sodium carbonate (Na_2CO_3), 0.1 M: 10 mL (dropper bottle). Dissolve 2.12 g of anhydrous Na_2CO_3 in 200 mL of distilled water.

Sodium fluoride (NaF), 2%: 300 mL. Dissolve 10 g of NaF in 500 mL of distilled water.

Trypsin, 5%: 5 mL (dropper bottle). Add 2.5 g of trypsin to 48 mL of distilled water.

Miscellaneous Supplies for Ten Students
Cheesecloth, 15-cm squares: 30
Ice, crushed: One bucket for the lab
Potato, raw, peeled, in small pieces: 100 g
Sand: 50 g

Equipment from Stockroom for Ten Students
Thermometers, 150°C: 10

Disposal Containers
1. 160 mL/ten students: Chem. xxx, Exp. 31, Used Chemicals Part B
2. 90 mL/ten students: Chem. xxx, Exp. 31, Used Chemicals Part C
3. 45 mL/ten students: Chem. xxx, Exp. 31, Used Chemicals Part D
4. 140 mL/ten students: Chem. xxx, Exp. 31, Used Chemicals Part E
5. 60 mL/ten students: Chem. xxx, Exp. 31, Used Chemicals Part F
6. 80 mL/ten students: Chem. xxx, Exp. 31, Used Chemicals Part G

EXPERIMENT 32: VITAMIN C CONTENT OF FOODS, PART I

Reagents for Ten Students
Acetic acid, 10%: 120 mL (dropper bottle). Dilute 57 mL of 6 M acetic acid to 200 mL using distilled water.

N-bromosuccinimide (NBS), 0.0011 M: 1.5 L. Dissolve 0.33 g of NBS in 8 mL of acetone and dilute to 1.5 L using distilled water.

Oxalic acid, 1%: 1.2 L. Dissolve 21 g of oxalic acid dihydrate in 1500 mL of distilled water.

Potassium iodide (KI), 4%: 300 mL. Dissolve 12.6 g of KI in 300 mL of distilled water.

Starch indicator, 4%: 15 mL (dropper bottle). Add a small amount of distilled water to 8 g of soluble starch to form a thin paste. Add the paste, with stirring, to 200 mL of boiling distilled water.

Vitamin C, standard, 0.50 mg/mL: 150 mL. Dissolve 0.075 g of vitamin C and 2.1 g of oxalic acid in distilled water and dilute the resulting solution to 150 mL using distilled water. This solution can be stored in a refrigerator for up to 4 or 5 days. If no refrigerator is available, the solution must be made fresh daily.

Miscellaneous Supplies for Ten Students
Cabbage, raw: 100 g
Cheesecloth, 15-cm squares: 90
Orange juice, fresh (or oranges): 150 mL
Reconstituted orange juice: 150 mL
Vitamin C unknowns: 15 mL per student. See the Instructor's Manual for the concentrations.

Equipment from Stockroom for Ten Students
Burets, 25 mL or 50 mL: 10
Pipets, 10 mL: 10
Suction bulbs: 10

EXPERIMENT 33: VITAMIN C CONTENT OF FOODS, PART II: SAMPLES FROM HOME

Reagents for Ten Students

Acetic acid, 10%: 140 mL (dropper bottle). Dilute 57 mL of 6 M acetic acid to 200 mL using distilled water.

N-Bromosuccinimide (NBS), 0.0011 M: 1.8 L. Dissolve 0.44 g of NBS in 11 mL of acetone and dilute to 2 L using distilled water.

Oxalic acid, 1%: 1.8 L. Dissolve 28 g of oxalic acid dihydrate in 2000 mL of distilled water.

Potassium iodide (KI), 4%: 350 mL. Dissolve 14.7 g of KI in 350 mL of distilled water.

Starch indicator, 4%: 18 mL (dropper bottle). Add a small amount of distilled water to 8 g of soluble starch to form a thin paste. Add the paste, with stirring, to 200 mL of boiling distilled water.

Vitamin C, standard, 0.50 mg/mL: 150 mL. Dissolve 0.075 g of vitamin C and 2.1 g of oxalic acid in distilled water and dilute the resulting solution to 150 mL using distilled water. This solution can be stored in a refrigerator for up to 4 or 5 days. If no refrigerator is available, the solution must be made fresh daily.

Miscellaneous Supplies for Ten Students

Samples of fruit, vegetables, or juices. We ask students to bring enough samples from home so they can do three comparison studies involving a total of six titrations.

Equipment from the Stockroom for Ten Students

Burets, 25 mL or 50 mL: 10
Pipets, 10 mL: 10
Suction bulbs: 10

Appendix C Table of Atomic Weights and Numbers

Name	Symbol	Atomic Number	Atomic Weight	Name	Symbol	Atomic Number	Atomic Weight
Actinium	Ac	89	(227)	Mercury	Hg	80	200.6
Aluminum	Al	13	26.98	Molybdenum	Mo	42	95.94
Americium	Am	95	(243)	Neodymium	Nd	60	144.2
Antimony	Sb	51	121.8	Neon	Ne	10	20.18
Argon	Ar	18	39.95	Neptunium	Np	93	237.0
Arsenic	As	33	74.92	Nickel	Ni	28	58.69
Astatine	At	85	(210)	Nielsbohrium	Ns	107	(262)
Barium	Ba	56	137.3	Niobium	Nb	41	92.91
Berkelium	Bk	97	(247)	Nitrogen	N	7	14.01
Beryllium	Be	4	9.012	Nobelium	No	102	(259)
Bismuth	Bi	83	209.0	Osmium	Os	76	190.2
Boron	B	5	10.81	Oxygen	O	8	16.00
Bromine	Br	35	79.90	Palladium	Pd	46	106.4
Cadmium	Cd	48	112.4	Phosphorus	P	15	30.97
Calcium	Ca	20	40.08	Platinum	Pt	78	195.1
Californium	Cf	98	(251)	Plutonium	Pu	94	(244)
Carbon	C	6	12.01	Polonium	Po	84	(210)
Cerium	Ce	58	140.1	Potassium	K	19	39.10
Cesium	Cs	55	132.9	Praseodymium	Pr	59	140.9
Chlorine	Cl	17	35.45	Promethium	Pm	61	(145)
Chromium	Cr	24	52.00	Protactinium	Pa	91	231.0
Cobalt	Co	27	58.93	Radium	Ra	88	226.0
Copper	Cu	29	63.55	Radon	Rn	86	(222)
Curium	Cm	96	(247)	Rhenium	Re	75	186.2
Dysprosium	Dy	66	162.5	Rhodium	Rh	45	102.9
Einsteinium	Es	99	(252)	Rubidium	Rb	37	85.47
Erbium	Er	68	167.3	Ruthenium	Ru	44	101.1
Europium	Eu	63	152.0	Rutherfordium	Rf	104	(261)
Fermium	Fm	100	(257)	Samarium	Sm	62	150.4
Fluorine	F	9	19.00	Scandium	Sc	21	44.96
Francium	Fr	87	(223)	Seaborgium	Sg	106	(263)
Gadolinium	Gd	64	157.3	Selenium	Se	34	78.96
Gallium	Ga	31	69.72	Silicon	Si	14	28.07
Germanium	Ge	32	72.59	Silver	Ag	47	107.9
Gold	Au	79	197.0	Sodium	Na	11	22.99
Hafnium	Hf	72	178.5	Strontium	Sr	38	87.62
Hahnium	Ha	105	(262)	Sulfur	S	16	32.06
Hassium	Hs	108	(269)	Tantalum	Ta	73	180.9
Helium	He	2	4.003	Technetium	Tc	43	98.91
Holmium	Ho	67	164.9	Tellurium	Te	52	127.6
Hydrogen	H	1	1.008	Terbium	Tb	65	158.9
Indium	In	49	114.8	Thallium	Tl	81	204.4
Iodine	I	53	126.9	Thorium	Th	90	232.0
Iridium	Ir	77	192.2	Thulium	Tm	69	168.9
Iron	Fe	26	55.85	Tin	Sn	50	118.7
Krypton	Kr	36	83.80	Titanium	Ti	22	47.90
Lanthanum	La	57	138.9	Tungsten	W	74	183.9
Lawrencium	Lr	103	(262)	Uranium	U	92	238.0
Lead	Pb	82	207.2	Vanadium	V	23	50.94
Lithium	Li	3	6.941	Xenon	Xe	54	131.3
Lutetium	Lu	71	175.0	Ytterbium	Yb	70	173.0
Magnesium	Mg	12	24.31	Yttrium	Y	39	88.91
Manganese	Mn	25	54.94	Zinc	Zn	30	65.37
Meitnerium	Mt	109	(266)	Zirconium	Zr	40	91.22
Mendelevium	Md	101	(258)				

A value in parentheses is the mass number of the isotope of longest half-life.